Quantum Communication and Cryptography

Quantum Communication and Cryptography

Walter O. Krawec
School of Computing
University of Connecticut
Storrs, CT, United States

ELSEVIER SYNGRESS

Syngress is an imprint of Elsevier
125 London Wall, London EC2Y 5AS, United Kingdom
50 Hampshire Street, 5th Floor, Cambridge, MA 02139, United States

ISBN: 978-0-443-21532-2

For information on all Syngress publications
visit our website at https://www.elsevier.com/books-and-journals

Publisher: Mara Conner
Acquisitions Editor: Chris Katsaropoulos
Editorial Project Manager: Soumya Yadav
Production Project Manager: Jenela Priscy
Cover Designer: Miles Hitchen

For Abby

Contents

Preface xi

Acknowledgements xv

1. Introduction 1
   - 1.1. Quantum communication 3
   - 1.2. Quantum key distribution 7
   - 1.3. Alternative security models 9
   - 1.4. Closing comments 11
   - References 12

2. Background: quantum communication and information theory 13
   - 2.1. Basic linear algebra 13
   - 2.2. Quantum states 21
   - 2.3. Measuring a quantum state 23
   - 2.4. State evolution 32
   - 2.5. Combining systems: tensor product 35
   - 2.6. Quantum information theory 43
   - References 60

3. Modern cryptography 61
   - 3.1. Encryption 62
   - 3.2. Public key encryption 76
   - 3.3. Authentication and integrity 79
   - References 81

4. Quantum key distribution (QKD) protocols 83
   - 4.1. Goal of QKD and the BB84 protocol 83
   - 4.2. Stages of a QKD protocol 87

4.3. Basic QKD protocols 88
4.4. Basic post processing techniques 97
References 106

5. Security of QKD 109
5.1. Security goal and composable security 109
5.2. Adversarial models and noise 112
5.3. Bounding the secret key length 118
5.4. Finite key security 129
References 139

6. Alternative quantum key distribution protocols 141
6.1. Two-way QKD 141
6.2. Semi-quantum cryptography 152
6.3. Twin field QKD 163
References 165

7. High-dimensional quantum cryptography 169
7.1. High-dimensional BB84 169
7.2. Round Robin protocol 173
7.3. High dimensional semi-quantum key distribution 175
7.4. Cryptographic applications of quantum random walks 182
7.5. Group key protocols 189
7.6. Closing remarks 193
References 193

8. Alternative security models and cryptographic primitives 195
8.1. Quantum random number generation 195
8.2. Certified deletion 205
8.3. Bounded storage and noisy storage 208
8.4. Quantum pseudo-random states 211
8.5. Game theoretic security 214
References 217

9. Practical quantum cryptography 221
9.1. Device imperfections: key-rate derivation 221
9.2. Decoy state BB84 226
9.3. Practical device considerations 232
9.4. Quantum networks 235
References 239

10. New and future directions: quantum public key cryptography 243
10.1. New security assumptions and constructions 243
10.2. Quantum public key encryption 249
10.3. Closing remarks 258
References 258

Index 259

Preface

This book is meant to serve as an introduction to quantum cryptography. While a large portion is devoted to quantum key distribution (QKD), other areas within the field are also covered, including new research directions such as quantum public key systems. Much of the material in this book I have used at some point or another, for various lectures and classes I've given in this subject. While this book would, I think, serve as a useful text for an advanced undergraduate course, or early graduate course, on quantum cryptography, I have tried to also make it accessible to the independent reader and student. The text is largely self-contained, and does not assume any previous knowledge of quantum computing or cryptography, though some background in basic linear algebra and probability theory would be helpful, along with some knowledge of complex numbers.

This book does not discuss any of the seminal quantum algorithms, except in passing. For instance Shor's algorithm and Grover's search algorithm, though vital pillars in quantum computing, are not discussed. Instead, this book focuses entirely on cryptography and using quantum information and quantum resources to secure information. I also do not cover post-quantum cryptography, though I do have a chapter on modern cryptography to introduce the reader to general concepts from classical cryptography.

Quantum cryptography is a vast field, and growing rapidly each year. As such, I cannot possibly include everything and had to be subjective in my choice of material. However, I think that what is covered here can provide a foundation of knowledge for the reader to pursue new research, and understand current research, in the field. I have tried to give sufficient details on security proofs to allow the reader to understand their subtleties, and to be able to take this knowledge and apply it elsewhere. I have also tried to give high-level introductions to the various topics, but also delve into technical details. A reader who is interested in learning some of the landscape, without necessarily being interested in going deeper, should find something here, as should the reader wanting to further understand the technical theory behind protocol design and security proofs.

Book organization

This book consists of ten chapters, the first three of which provide general introductory and background information. Chapter 2 is a self-contained tutorial on basic quantum communication and quantum information theory. Basic terminology is introduced, qubits, Dirac notation, the mathematical model of quantum states, and so on. Quantum information theory is also introduced, including von Neumann entropy, and quantum min entropy (the latter of which is very useful in quantum cryptography). This chapter is meant to contain all the basic quantum theory you will need to understand the rest of the book. If the reader is already familiar with this material, this chapter can probably be skipped, though the information theory section may contain topics not covered in a general quantum computing course or text, and so should be at least glanced at.

Chapter 3 serves as a brief introduction to the field of modern cryptography, purely from a classical stand point. Quantum cryptography builds off of this field, and so it is, I think, vital to have some appreciation for the definitions and concepts used in modern cryptography. Topics ranging from basic encryption (both symmetric key and public key), authentication, and also pseudo-randomness are covered here. Of particular importance are the definitions of security for these cryptographic primitives.

In Chapters 4 and 5, we take a look at quantum key distribution (QKD). For a long while, QKD was synonymous with quantum cryptography, though by now QKD is only one part of the field. Nonetheless, it is a vital part. Even if one is not interested in this particular application, the security proof techniques, and much of the protocol construction ideas, find themselves in other topics. Thus, it is vital to understand basic QKD and also how security proofs operate, to really study quantum cryptography. Chapter 4 discusses basic protocol design, while Chapter 5 discusses security proof techniques. For the latter, we'll look at both asymptotic and finite key security proofs.

Chapters 6 and 7 look into alternative QKD protocol designs, including two-way QKD, semi-quantum cryptography, and also twin field QKD. Each of these provides unique properties and insights. Some are purely theoretical, and some are designed to have a significant practical advantage. Protocols, and security proofs will be discussed here, giving the reader a larger tool-set to tackle other problems in quantum cryptography.

Chapter 8 moves away from QKD entirely and discusses newer quantum cryptographic primitives. This includes random number generation (including so-called device independent protocols), certified deletion, and alternative security models. Quantum pseudorandom state generators are also discussed, moving us away from "information theoretic" security, and towards "computational security."

Most of the text is theoretical in nature, however Chapter 9 discusses some real-world issues that arise when implementing quantum cryptographic systems. Of course, the focus, here, is on QKD since that is the primary applied quantum technology today, in the cryptographic setting. Thus, decoy state BB84 is discussed, and how it can be used to mitigate certain practical device constraints. Quantum networking will also be discussed, including repeater and trusted node technology.

Chapter 10 ends the book with a look at a very new, and exciting, area of research, namely quantum public key systems. Here, I'll discuss some computational based assumptions that are used, and also discuss protocol design for quantum public key encryption systems. Many of the definitions of security, here, are based on their classical counterparts, which are discussed in Chapter 3. However, the constructions rely on provably weaker assumptions, than is required classically, making quantum cryptography, beyond QKD, a vital and exciting research area. This chapter only covers some of the many fascinating new research directions taking place, with the hope of inspiring a future researcher into investigating this, or other areas, of quantum cryptography.

How to use this book

In general, I wrote this book to serve as a reference for an advanced undergraduate, or early graduate, course on quantum cryptography. It is meant to be largely self-contained for such a purpose, and many of the chapters are based on notes I've prepared over the years teaching similar courses. If the course is meant to serve students without any background in quantum computing or quantum information, Chapter 2 should be covered in some detail. If this course is meant as a follow-up course to a more general quantum computing class, then most of this chapter could be skipped, though Section 2.6 should probably be looked at, as it covers the quantum information theory required to prove security, topics which are not always present in a general quantum computing class.

I would generally move on to topics in Chapter 4, and provide general proofs of security as discussed in Chapter 5. I would also discuss practical issues, and quantum networking techniques, discussed in Chapter 9. For more advanced courses, I've covered material from Chapters 8 and 10. Material from other chapters can be covered as time permits, based on student and instructor's interest.

If this book is used as a supplement to a modern cryptography course, where the instructor wishes to introduce some topics from quantum cryptography without going into full detail, I would advise Chapter 4, 8, and 10. Especially the latter, as the proof techniques, there, should be very familiar to a student of modern (classical) cryptography.

For the individual reader who is looking to gain a working knowledge of quantum cryptography, again, this book is meant to be mostly self-contained, and does not assume any background in quantum computing. I would start with the first Chapter which sets the stage with some of the basic ideas; Chapter 2 should be read in detail, to provide the technical background required to understand the security proofs later. I would then follow the above outline, reading in detail Chapters 4 and 5, and then proceeding to read subsequent chapters, based on the readers interest.

Finally, for the reader who wishes to just gain a general understanding of quantum cryptography, I would focus on reading high-level details from Chapters 1, 4, and then subsequent chapters, referring back to Chapters 2 and 3, only as needed. Though each chapter contains technical details, they can also be read at a high level to gain some understanding of the landscape. Many of the technical results in those chapters may be skipped in this case.

Acknowledgements

There are many people whom I must thank. I have to thank my entire family, first and foremost. My wife, Jennifer, and my children, for their continued love and support. My parents for all they have done for me, and continue to do for me. My siblings, and especially my sister for all the weekend computer games, and for being a ready ear to listen to both my gripes, and my bad jokes. And my in-laws for all the help and support they give my wife, children, and I.

I'm very grateful to Dr. Bing Wang, for proof-reading earlier drafts of this book, and her comments and suggestions which greatly improved the text. She provided very valuable feedback, and spent a considerable amount of time reading through this book. She has also been a great friend and colleague over these past years. Thank you!

I'd also like to thank my PhD advisor, Dr. Antonio Nicolosi, who encouraged me to study quantum cryptography as a grad student. Also Dr. Lee Spector, for writing a book on quantum computing which caught my eye, initially, and first got me interested in quantum computing in general.

Finally, I'd like to thank Chris Davis, for initially reaching out to me with the idea of writing a book like this one. The editorial team at Elsevier, for their help in preparing this manuscript. Especially Soumya Yadav, who was incredibly supportive, and very patient with me, despite my continued request for deadline extensions, and my very slow responses to emails.

I'm sure I missed some people, and for that I apologize.

Thank you all!

1 Introduction

Cryptography is used everywhere today by just about everyone. If you're reading this online no doubt many cryptographic protocols were used to bring the file to your screen. If you're like me and prefer a physical book, even then cryptography was used for the sale, inventory control, and/or bookkeeping of the original vendor. Almost every website today, supports an "https" connection which makes use of public key and private key encryption and authentication protocols. What started as simply an "art" of hiding messages (e.g., encryption), is now a mathematical science, involving rigorous proofs of security, carefully constructed definitions, and the use, if necessary, of minimal, and well-defined, computational assumptions. These are the core pillars of modern day cryptography.

Today, we are generally using classical cryptographic systems, involving the transmission and manipulation of "zeros" and "ones." Messages are classical bit strings and the algorithms used are all running on classical systems, using classical computation. With a few notable exceptions most such classical systems require at least some computational assumptions for their proof of security. Some of the assumptions are fairly "minimal." For instance, the existence of a one-way function (OWF), discussed more later, allows for the secure construction of several private-key cryptographic systems (e.g., encryption, and message authentication, just to name some). Other systems, such as public key cryptography, generally require stronger computational assumptions. While many of the assumptions are believed by many to be true (especially the minimal ones, such as the existence of OWFs), we have no solid rigorous proof and any such proof would require a significant breakthrough in computational complexity theory.

While there are some important classical cryptographic protocols that can be proven secure without assumptions, this is usually the exception and often lead to systems that are impractical today. For example, One Time Pad (OTP) encryption can encrypt any message in a way that is "perfectly secret." That is, if you encrypt a message with this system, no adversary, no matter how much time it has to run, and no matter what computational resources it has available (including any possible future technological breakthrough) can learn your message. However, for this system to work, your secret key must be as large, or larger, than the message you're encrypting. Furthermore, the key can only be used once and then it must be discarded. This leads to issues with any practical implementation: if you want to use OTP encryption, you need to be able to safely distribute, or establish, a very large shared secret key with the party you're communicating with. Today, our secret keys for symmetric key encryption are typically 128 bits or 256-bits long. Such a "small" key can encrypt any size message (e.g., it can be used to encrypt a 1 TB message). On the other hand, if you want to encrypt and send a 1 TB message using OTP, you need to also safely and secretly send or distribute a 1 TB long secret key!

Quantum Communication and Cryptography. https://doi.org/10.1016/B978-0-44-321532-2.00008-5

What is this "secret key"? A cryptographic algorithm performs some task, but in general there is something that must be secret. Historically, it was often the case that the *algorithm* that performs the security task was kept secret. However today, we generally make the algorithm public knowledge and instead rely on a small secret number or *key*, which is the secret information required to, for example, send and read messages securely. This notion of making the algorithm public, but relying on a "short" secret token or key is generally known as Kerckhoffs's principle. By following this principle we can design a secure system and use it for multiple applications and parties and then we only need to worry about distributing, and keeping safe, this one random key. Sending this key to a recipient party is called the *key distribution* problem and it is one of the areas that quantum cryptography can greatly help with.

As mentioned above, we have an encryption protocol that is secure without any computational assumptions (i.e., perfectly secret): OTP. However it requires a large secret key. We do *not* have a classical protocol that can distribute a secret key securely without making some form of computational assumption on an adversary's capabilities (i.e., without assuming certain mathematical problems are hard to solve, algorithmically). As it turns out, this is not due to a lack of imagination, but instead it is a mathematical fact that one cannot create a secure secret key distribution protocol that uses only classical information and communication, unless some kind of computational assumption is made on the capabilities of the adversary. In the quantum setting this is not the case, and key distribution is possible without computational assumptions.

Beyond key distribution, there are also other tasks that are either impossible to perform classically, or can only be done by making strong assumptions on an adversary's computational abilities (assumptions which may be invalidated at some later point in time due to some future technological or algorithmic breakthrough). The landscape, however, changes once one augments classical cryptographic systems with quantum information and quantum resources. Suddenly key distribution is possible without the need for computational assumptions. Other tasks such as public key encryption are possible assuming only OWFs (which is impossible in classical cryptography where stronger assumptions are required). Finally, novel cryptographic primitives are suddenly allowed, such as *certified deletion* [1] where you can actually have proof that a server deleted your file on request (which is impossible in the classical setting since classical data can always be copied without you knowing).

This book is meant as an introduction to the fascinating field of quantum cryptography. No prior knowledge of quantum computation or communication is required as these topics will be introduced in sufficient depth in Chapter 2, with additional topics introduced as needed. Of course if the reader is already familiar with quantum computation, Chapter 2 may be skipped (though Section 2.6 may be useful as that covers quantum information theory). Of note there is no mention or discussion of quantum algorithms (e.g., Shor's algorithm); there are, by now, several useful textbooks on such topics. Instead, this book's focus is on the cryptographic side of quantum science.

A large portion of the book is devoted to so-called *quantum key distribution* (QKD), both protocol design and also security proof methodologies. This is due to the fact that QKD is perhaps the seminal quantum cryptographic application; it is also commercially viable today. Finally, and importantly, many of the methods for analyzing the security of QKD leak into other fields of quantum cryptography, so a solid understanding of QKD (especially their security proofs) is vital to understanding the larger field of quantum cryptography. However, this book will also delve into topics beyond QKD, such as random number generation, bit commitment, alternative security models, and public key encryption, just to list a few topics. This book also primarily focuses on the theoretical study of quantum cryptography; however practical implementation issues are also discussed. It is hoped that the material in this book will serve as a useful foundation to the current student and future researcher, while also serving as inspiration to encourage the further development of this fascinating field.

1.1 Quantum communication

Quantum communication involves the sending, and receiving, of *quantum bits* or *qubits.* A classical bit can be in one of two possible states, namely zero "0" or one "1." A quantum bit can also be in a "zero" or "one" state, which we denote $|0\rangle$ and $|1\rangle$ respectively (that is read "ket zero" and "ket one"). However, a qubit can also be in a *superposition* of all possible states of the form:

$$|\psi\rangle = \alpha\,|0\rangle + \beta\,|1\rangle\,, \tag{1.1}$$

where $\alpha, \beta \in \mathbb{C}$ such that $|\alpha|^2 + |\beta|^2 = 1$. Thus, unlike a classical bit, there are actually infinitely many states a qubit can be in.

Another difference between qubits and classical bits is in how data is read from them. On receipt of a classical bit, one may read that bit to learn exactly its value; furthermore, reading a classical bit does not disturb it. The fact that you saw the classical bit was in the state "0" does not alter it in any way. Qubits, however, are different. The act of reading a qubit (called *measuring* a qubit), is a potentially destructive process that produces random output. In particular, given a qubit of the form shown in Eq. (1.1), reading this qubit in the so-called *computational basis,* yields an outcome of zero with probability $|\alpha|^2$; otherwise, it yields an outcome of one, with probability $|\beta|^2$ (which, due to the normalization constraint, mentioned earlier, is $1 - |\alpha|^2$).

Not only is the measurement outcome random, it actually causes *state collapse.* That is, if you measure Eq. (1.1) and observe "zero," the state actually becomes $|0\rangle$. The original superposition state of $\alpha\,|0\rangle + \beta\,|1\rangle$ is destroyed. If you had actually observed "one" the state collapses to (becomes) $|1\rangle$. Measurements are destructive and irreversible. As it turns out, there are also many ways to measure a qubit (infinitely many, in fact), and one can even make "partial" measurements. It is important to note, though perhaps it is not yet obvious, that a superposition of this form is *not* a "classical mixture." Namely, the state is not "$|0\rangle$" with probability $|\alpha|^2$ or "$|1\rangle$" with probability $|\beta|^2$ before measurement. Superposition is something different; it is not a classical mixture. Hopefully this will become more clear later in this book.

Note that we said, above, we were measuring in the "computational basis." As we will see in Chapter 2, qubits are modeled as two-dimensional complex vectors. Thus, one can talk about a basis $\{|v_0\rangle, |v_1\rangle\}$ for this space (we technically require orthonormal bases, but this is not relevant for our high-level discussion, here). The state $|0\rangle$ and $|1\rangle$ are typically called the computational basis, but there are infinitely many other bases. One other important basis is the *Hadamard basis* spanned by states denoted $|+\rangle$ and $|-\rangle$, which are defined, with respect to the computational basis, as:

$$|+\rangle = \frac{1}{\sqrt{2}}|0\rangle + \frac{1}{\sqrt{2}}|1\rangle \tag{1.2}$$
$$|-\rangle = \frac{1}{\sqrt{2}}|0\rangle - \frac{1}{\sqrt{2}}|1\rangle .$$

In general, if one is measuring in some basis $|v_0\rangle$ and $|v_1\rangle$, the measurement device will return (or observe) one of the two basis states with some probability. State collapse, then, will cause the entire state to collapse to the observed basis state. To compute the probability, one first writes the qubit with respect to that basis (i.e., $|\psi\rangle = x|v_0\rangle + y|v_1\rangle$), and then the probability of observing $|v_i\rangle$ is simply the norm of the scalar coefficient squared. For instance, the probability of observing $|v_0\rangle$ in the above state, would be $|x|^2$; if that is the actual observation, the state collapses to $|v_0\rangle$.

Returning to the computational and Hadamard bases, let's assume $|\psi\rangle = |+\rangle = \frac{1}{\sqrt{2}}|0\rangle + \frac{1}{\sqrt{2}}|1\rangle$. Measuring this state in the computational basis will yield an observation of $|0\rangle$ or $|1\rangle$, each with probability $1/2$. If one actually observes, say, $|1\rangle$, the state collapses (becomes) $|1\rangle$ and any subsequent measurement will always be $|1\rangle$. On the other hand, if you were to take $|\psi\rangle$ and, instead of measuring in the computational basis, you measure in the Hadamard basis, you will always see $|+\rangle$.

Now, it can be shown, and this will make sense in Chapter 2, that $|0\rangle$ and $|1\rangle$ can actually be written as superpositions (linear combinations, really), of Hadamard states. Namely:

$$|0\rangle = \frac{1}{\sqrt{2}}|+\rangle + \frac{1}{\sqrt{2}}|-\rangle \tag{1.3}$$

$$|1\rangle = \frac{1}{\sqrt{2}}|+\rangle - \frac{1}{\sqrt{2}}|-\rangle \tag{1.4}$$

Now, let's return to our previous example with $|\psi\rangle = |+\rangle$. Measuring this in the computational basis yields $|0\rangle$ or $|1\rangle$, each with equal probability. Let's say we actually observed $|1\rangle$, and so the state, now, actually is $|1\rangle$. However, since $|1\rangle = \frac{1}{\sqrt{2}}|+\rangle - \frac{1}{\sqrt{2}}|-\rangle$, if we now measure this system in the Hadamard basis, the outcome will be either $|+\rangle$ or $|-\rangle$, each with probability $1/2$. Notice that measurements are destructive – if you measure in the "wrong" basis you will get a random result and you cannot "rewind" the measurement to try again. This will be vital, later, when discussing the security of quantum communication. In general, classical data will be hidden in the quantum state by choosing different bases. An adversary, who doesn't know the right basis, will not be able to measure correctly. Any attempt to measure will likely cause state collapse – a disturbance that can be detected.

Additional properties unique to qubits can be proven. For instance, the so-called *no cloning theorem* states that, essentially, one cannot make a copy of an arbitrary, unknown, qubit. Again, this is very different from classical information, where it is very easy to make copies (just hit CTRL-C, CTRL-V for instance!). Given any arbitrary, unknown, quantum state $|\psi\rangle$, you cannot make a perfect copy of it. This may be very annoying as a computer-user, but it is very useful in cryptography: if we send data encoded as random quantum states (i.e., we "hide" our classical data in one or more qubits), an adversary cannot simply make a copy of our data to attack at a later time. Instead, all attacks against quantum cryptographic systems have to be *active attacks*. There is no concept of "copy now, and attack later." Furthermore, due in part to the destructive nature of measurements, any such attack will leave a fingerprint, or trace, behind, which can be detected.

1.1.1 Entanglement

Besides superposition, *entanglement* is another quantum phenomenon that is highly useful in quantum cryptography. Consider, now, the case where we have two qubits (two physical particles) and they are in the joint state:

$$\frac{1}{\sqrt{2}}|0,0\rangle_{A,B} + \frac{1}{\sqrt{2}}|1,1\rangle_{A,B}\,. \tag{1.5}$$

Intuitively, from our earlier discussion, if we were to measure *both* particles together, we should see $|0,0\rangle$ with probability $|1/\sqrt{2}|^2 = 1/2$; otherwise we will see $|1,1\rangle$. Furthermore, state collapse causes both particles to collapse to the observed state (either $|0,0\rangle$ or $|1,1\rangle$). However we must keep in mind that these are two separate, physical, systems, and so we can actually separate them (in theory by great distances). Let's send the "A" system to Pluto and keep the "B" system here on Earth. Let's also send someone to Pluto to perform a measurement. If the person on Pluto measures the A system, this is just a single qubit and so can only have a "zero" or "one" outcome. Let's say that this person observed "zero" (which happens half the time). State collapse still occurs; however, it occurs on the entire joint system, even though the two particles are separated. Thus, if Alice, on Pluto, sees "zero," her qubit collapses to $|0\rangle$, but also the entire state (including Bob's particle B here on Earth) also collapses to $|0\rangle$ in this case! Thus, if Bob were to measure his system, he will see exactly the same outcome that Alice sees, no matter how far apart they are.

This is interesting, but if this were the end of the story, entanglement would not be very useful. Indeed, one can simulate the above experiment classically. First, someone, instead of creating two entangled particles as in Eq. (1.5), will simply flip a coin, and if it's heads, will write the number "0" on two pieces of paper; otherwise the number "1" will be written. Each piece of paper is put in a separate envelope and sealed. One envelope is kept on Earth (the B envelope), while the A envelope is mailed to Pluto (the postage required for such a shipment would be tremendous no doubt, but it's fine to think about these things nonetheless). Now, Alice on Pluto later opens her envelope and she will see a zero or one, that is, as far as she is aware, random. Bob also sees a random zero or one. And, of course, both Alice and Bob "see" the same value, exactly as in the previous quantum experiment. Thus, in this sense, entanglement doesn't get us anywhere.

However, this is not the entire power of quantum entanglement! In fact, quantum entanglement can produce experimental outcomes that cannot be simulated classically as we did above. As an example, let's consider the CHSH game [2] (named after Clauser, Horne, Shimony, and Holt, and discussed, in more detail, in Section 8.1.2). The game consists of two cooperative players, Alice and Bob, and a Referee. The game is such that Alice and Bob will either both win or both lose. First, Alice and Bob are allowed to communicate with one another as much as they like. For instance, they can strategize how to win, they can write down some random information, they can have access to some "randomness in an envelope," and so on. If Alice and Bob are allowed quantum information, they can also pass along some qubits to one another, including some entangled bits, as in Eq. (1.5).

Following this communication stage, Alice and Bob are separated by a distance large enough so that they cannot communicate. That is, Alice and Bob are not allowed to talk any more, but they can of course use any communication they saved from earlier. At this point, the Referee will choose two random bits: x and y; the value of x will be sent to Alice and the value of y will be sent to Bob. Alice will not know y and, likewise, Bob will not know x. Since they are no longer allowed to communicate, they cannot send their values to each other either.

Given their random values, now, Alice must choose a bit a and Bob must choose a bit b. How they choose these bits can be based on their earlier communication. But, importantly, they are not allowed any additional communication, once they receive their "challenge" x and y. These bits are sent to the Referee.

Alice and Bob both *win* the CHSH game if the following equation is true:

$$a \oplus b = xy, \tag{1.6}$$

where "$\oplus$" is the XOR operation and by xy we mean regular multiplication. Since x and y are both bits, $xy = 1$ only when both x and y are one (otherwise, for all other cases, $xy = 0$). Thus, Alice and Bob win if they choose the same bit (i.e., $a = b$) on all cases where $xy = 0$. When $xy = 1$, they must choose different bits to win. Said differently, they must output the same bit in all cases, except for the one case where $x = y = 1$ (since that is the only time $xy = 1$).

It can be proven that, no matter how much classical communication and randomness they are allowed to share at the beginning of the game (before x and y are chosen), the best chance of winning for Alice and Bob is only 75%. That is, if they play the game multiple times, they can win 3/4 of them on average. However, if Alice and Bob are allowed to share just one entangled pair (Eq. (1.5)), and this is done before the challenge bits x

and y are chosen, suddenly they are able to win with approximately 85% probability. This shows that entanglement is doing something that we cannot simulate with purely classical information and randomness – there is something interesting, and new, here!

We will use entanglement in various quantum cryptographic protocols. QKD is a prime example; however entanglement will also be useful in other cryptographic systems, and also in quantum network settings as entanglement allows for *quantum teleportation.* Quantum teleportation allows an arbitrary qubit state $|\psi\rangle$ to be "teleported" or transferred to another location without actually physically moving it. This is useful in quantum networks, and distributed quantum computation. It can also be used by an adversary, as we discuss later, allowing her to perform rather sophisticated attacks against quantum systems (however, these attacks can still be detected and must still be active!).

1.2 Quantum key distribution

For a long while, the term quantum cryptography was actually synonymous with quantum key distribution (QKD). Only recently has quantum cryptography developed to such a large extent that QKD is just one of many possible protocols. However, it is still a very important one to understand, and many of the methods and proof techniques used for QKD, extend throughout to other quantum cryptographic protocols.

QKD is a solution to the key distribution problem, discussed earlier. Let's say we have two parties, Alice and Bob who would like to perform some cryptographic task (e.g., encryption of messages). To do so, they require a *secret key.* This secret key must be random, and must be safely distributed, so that an adversary has no information on it. Today, we use classical systems (such as public key cryptography) to solve this task; however, as discussed earlier, public key systems require computational assumptions in order to be proven secure. Should these assumptions fail, the underlying cryptographic primitive is broken. Furthermore, since classical data can be copied, adversaries, now, are copying data all the time, to attack at a later point. If ever the underlying assumption is found incorrect, any data sent using that protocol would be potentially at risk.

For instance, several classical systems are known to be insecure against large-scale quantum computers. While there are alternative candidate constructions (so-called "post quantum" cryptographic systems), these systems still require computational assumptions (e.g., that it is hard for a computer, either classical or quantum, to solve a particular mathematical puzzle). As discussed, computational assumptions are a requirement for secure key distribution, if one is restricted to classical communication.

QKD solves the key distribution problem, not by relying on computational assumptions, but on physical ones. For example, one requires that certain laws of quantum mechanics are correct, and that user devices are secure and trustworthy. QKD is achieved by encoding classical data onto qubits, observing the error rates in the transmission (which gives a hint as to the attack strategy of an adversary), and then using classical methods to extract a final secret key, even if there is an adversary "snooping" on the data.

To see, at a high level, how this can be done, let's assume Alice wants to send (or distribute) a secret key $k \in \{0, 1\}^n$ to Bob. We'll use k_i to mean the i'th bit of k. To do so, for each $i = 1, \cdots, n$, she will choose a random basis, either the computational, or the Hadamard (see Section 1.1), and send k_i encoded in that basis. Essentially, she "hides" the bit information by choosing different qubit encoding strategies.

For example, if her i'th key bit k_i is a one, and her basis choice for that bit is the Hadamard basis, Alice will send to Bob the qubit state $|-\rangle$ (which is a "one" encoded in the Hadamard basis; a zero in that basis would be $|+\rangle$). Alternatively, if her key bit were zero, and she chose the computational basis, she would send a $|0\rangle$.

The qubit travels to Bob – however, before it arrives, the adversary captures it. Now, due to the no cloning theorem, the adversary, commonly called Eve, cannot simply make a copy to attack later. Instead, she must attack right now. But what can she do? Her goal is to learn k_i and to do so, she must interact with the system in some way. Let's say she tries to measure – but how should she measure? She can measure in the computational basis or the Hadamard basis (or maybe some other basis, but let's restrict our attention, for now, to this particular attack strategy). Based on our earlier discussion, if Eve measures in the "wrong" basis (the opposite basis Alice used), then she will get a random outcome. She will also cause a disturbance, and the original state that Alice sent will collapse to a random state in the wrong basis!

Bob, when he finally gets the qubit, will ask Alice for the basis information and Alice can now send this information to Bob. Importantly, this information is not sent in a secret/encrypted manner! Instead, it is sent publicly (one needs authentication, but that detail will be dealt with later in this book). However, thanks to the no cloning theorem, it is "safe" to do so, since Eve already attacked. Bob, when he learns the correct basis, will perform the correct measurement (which Eve could not do with certainty), and record the outcome. This should match k_i. Of course, if Eve is attacking, Bob's measurement will produce random data (or *noise*). Thus, Alice and Bob will also, occasionally, send "decoys" to check for this kind of attack (i.e., send decoys to check the noise level in the quantum channel, thus determining what an adversary may be up to).

The above is a somewhat high-level description of QKD and there are many subtleties that must be addressed, and will be addressed in this book. However, the main takeaway for now, is, Alice and Bob may "hide" classical data in a qubit state. The qubit cannot be copied to attack later and so an adversary has to attack actively. Any such attack, however, must necessarily create at least some disturbance in the qubit which may be later detected. Importantly, even if an attacker creates disturbance, there are classical methods so as to allow the transmitted information to be distilled into a secret key. If there is too much of a disturbance, then parties simply abort. Thus, in a sense, one either gets a secret key from QKD, or one knows to abort, in which case no sensitive information is lost, because the key has not been used for anything; Alice discards her choices of k_i and chooses again for a later session.

Importantly, just as with quantum computing, quantum cryptography requires both quantum *and* classical information processing. QKD requires quantum operations to distill a "raw key." But it is classical computing that will distill that raw key into a final secret key. The same is true for any other quantum cryptographic protocol: there is a quantum element, but it must be combined with classical computation, too.

1.3 Alternative security models

QKD does not require computational assumptions for security. Sometimes, however, it is interesting to see what is possible if one were to allow for such assumptions. Indeed, even in the quantum setting, there are several tasks that are impossible to do securely, without making assumptions. One of the difficulties of working in the field of quantum cryptography is that, first, you must design a suitable protocol and prove it's secure (just as in the classical case), but secondly, you must also show that there is some advantage, whether practical or theoretical, over its classical counterpart. For QKD the advantage is clear. However, when using computational assumptions to build systems, one must be sure that either the application is unique, or that the assumptions required are weaker than what's required classically (otherwise, what's the point of using quantum?).

One fascinating area of recent study is in the development of quantum public key systems. For this, consider an one-way function (OWF) f. Such a function is "easy" to compute, but "hard" to invert. Namely, given x, it is easy to compute $f(x)$. But given y such that $y = f(x)$ for some random x, it is very hard (i.e., computationally inefficient), to find some z such that $f(z) = y$ (z may be the same, or different, from x depending on the function). Whether or not OWFs actually exist is unknown. However, their existence is necessary and sufficient for many classical symmetric key cryptographic systems. However, for public key encryption, stronger assumptions are required, at least in the classical setting.

What is a public key encryption protocol? Section 3.2 covers this in more detail, but, essentially, such a protocol consists of a secret key sk and public key pk. These two keys are related, somehow, mathematically. The public key is given to everyone (including an adversary), while the secret key is kept secret. Given a copy of the public key, one may encrypt a message. However, only the secret key can be used to decrypt (unlock) the message. This allows many parties to send encrypted messages to one party, and is a vital building block for much of our modern-day communication infrastructure.

That OWFs cannot be used, in a generic fashion, to build a classical public key system has been known for several decades [4,3]. However, rather interestingly, a OWF is sufficient to build a public key encryption protocol, if one also includes some quantum data! In particular, a quantum public key system (discussed more, in Chapter 10), utilizes a classical secret key, and a *quantum* public key. Many copies of the quantum public key are produced by the secret key holder (this does not violate no cloning, since the secret key holder knows the state exactly and should have an efficient quantum algorithm to build a public key state). Anyone with one of the quantum public keys may utilize it to encrypt a message. Only the secret key holder, as in the classical case, can decrypt it.

Computational assumptions are not the only assumptions one may make to an adversary to see what interesting cryptography can be produced. Instead, additional physical assumptions may be made, instead of computational ones. For instance, one may assume that the adversary can do anything within the laws of quantum physics (including having a long run-time to attack), but the adversary has a "noisy" quantum memory. That is, if the adversary stores data as quantum states for too long, it eventually decoheres, and becomes, essentially, random classical bits. Such a physical assumption, which is certainly not too unrealistic today, allows for the development of many additional quantum cryptographic protocols, that are impossible to prove secure in the standard model (even with quantum resources). Oblivious transfer (OT) is one such primitive.

OT is a cryptographic protocol that consists of two parties: Alice and Bob. Unlike key distribution, where Alice and Bob trust one another, here in OT, it is assumed that one of Alice or Bob is the adversary and is trying to cheat the other. In this protocol, Alice has two messages m_0 and m_1 while Bob has a bit b. The goal is for Bob to learn m_b (the message he wants), and no more (i.e., he should not learn m_{1-b}). Furthermore, Alice should never learn Bob's choice of message b. This task cannot be done, even with quantum bits, without some additional assumptions; if we assume the adversary has "practical" quantum memory, then suddenly it is possible. While it may seem to some that OT is just a "game," it is actually a vital building block for more advanced and useful, cryptographic protocols (classical and quantum), and so its study is of high importance. We'll discuss this more, in Section 8.3.

The above discussion involved security models where additional assumptions were placed on the adversary (either computational assumptions, or additional physical assumptions). However, one may go in the other direction, and actually give the adversary more capabilities. For instance, the *device independent* model of quantum cryptography assumes that, not only can the adversary do whatever she wants to the transmitted quantum data (within the laws of physics), but she may also have complete control over the source and measurement devices themselves! That is, Alice and Bob do not need to trust their devices – they can be manufactured by an adversary and still you will be able to do some secure cryptography from them.

Take, for instance, random number generation. Now, imagine you buy a box with a button on it, and a single-digit screen. The box is advertised so that, every time you push the button, a random "zero" or "one" is shown on the screen. You are told, quite confidently, that the box really does what it's supposed to: namely choose a random bit, totally at random, and totally secret known only to whoever sees the screen. Do you trust it? How can you? An adversarial manufacturer could, for example, pre-program the device with some randomness and the box will simply output the next bit on this "tape." The manufacturer also has a copy of the pre-programmed randomness, so will know exactly what "random" bits you have. This could be dangerous if these random bits are used for, say, a secret key!

However, if the box is allowed to utilize quantum resources, you can actually test it, yourself, to see that it is indeed giving you true randomness, that no adversary could have pre-programmed or become aware of. Even though you have no idea what's inside the box, you can perform a statistical test to ensure that the box is somehow using quantum resources the way it's supposed to. Technically, now, you buy two boxes from the adversary and they must be using entanglement between themselves. Essentially, you have these two boxes "play" the CHSH game, described earlier, to ensure that quantum resources, specifically an entangled state of the form shown in Eq. (1.5), are being used by the boxes, to produce the random outcomes. Then, due to certain properties of entangled states, one can be sure that the random bits you're getting are actually safe, even though you never actually look inside the boxes. This will be discussed, more, in Section 8.1.2.

1.4 Closing comments

I have tried to make this book accessible to anyone with some basic linear algebra and probability theory. The topics in quantum mechanics needed to at least understand the fundamentals are all covered, as are also those topics needed in classical cryptography. Having an appreciation, and understanding, of both is, I think, required to fully understand quantum cryptography. The field, however, is quite vast at this point, and so I had to be selective in which topics I covered. Not everything will be here. However, the knowledge gained should provide the reader with enough material to be able to understand, at least at a high level, other topics in quantum cryptography that were not covered here.

There is sometimes a divide in some corners of the world, between those who study "quantum" cryptography (such as QKD), and those who study "post-quantum" cryptography (which is classical cryptography, with stronger mathematical assumptions). It is my belief that both fields have strong merits, and I will not try to unduly promote one system as more advantageous than another. QKD, and quantum cryptography in general, has several advantages over classical cryptography, including post-quantum cryptography; it also has several disadvantages, which will be discussed throughout this book. The same is true for post-quantum systems. In reality, it will likely be that both systems should be utilized side-by-side and such research is being investigated by several groups. This book will only cover quantum cryptography and will not touch on any topics in post-quantum. The methods and protocols are very different, that books in both fields are required, just as both fields are required to secure our future communication infrastructure.

It is my hope that this book will serve as an inspiration for some to become more interested in quantum cryptographic research. I started my own career in quantum cryptography purely by accident: as a graduate student, I would wander the library often, looking for something interesting to read. One day, a book with an interesting title caught my eye, "Automatic Quantum Computer Programming: A Genetic Programming Approach" by Lee Spector. I took out the book, and read it over my summer break and was immediately hooked. I hope this book will also hook a few into further studying and developing this field.

References

[1] Anne Broadbent, Rabib Islam, Quantum encryption with certified deletion, in: Theory of Cryptography: 18th International Conference, TCC 2020, Durham, NC, USA, November 16–19, 2020, Proceedings, Part III 18, Springer, 2020, pp. 92–122.
[2] John F. Clauser, Michael A. Horne, Abner Shimony, Richard A. Holt, Proposed experiment to test local hidden-variable theories, Physical Review Letters 23 (15) (1969) 880.
[3] Russell Impagliazzo, A personal view of average-case complexity, in: Proceedings of Structure in Complexity Theory. Tenth Annual IEEE Conference, IEEE, 1995, pp. 134–147.
[4] Russell Impagliazzo, Steven Rudich, Limits on the provable consequences of one-way permutations, in: Proceedings of the Twenty-First Annual ACM Symposium on Theory of Computing, 1989, pp. 44–61.

2

Background: quantum communication and information theory

In this chapter, we will introduce the basic concepts and notation from quantum computing and information theory, required to understand the rest of this book. Many other valuable resources exist to supplement the reading here. Of particular note is the excellent textbook [9].

Road map: This chapter begins with a review of some basic linear algebra, in Section 2.1. While this material may be familiar, it is introduced using Dirac notation, which we will use throughout this book, and so it may be helpful to review this section even if you have a working knowledge of the underlying material.

In Section 2.2, we will start introducing basic quantum systems, quantum states, and the concept of superposition. Measurements will be discussed in Section 2.3, including basis measurements, and general measurement operators. Section 2.4 will discuss state evolution via unitary operators, while Section 2.5 will discuss tensor product and how we combine multiple quantum states. Entanglement will also be discussed in that section, along with partial measurements.

Finally, Section 2.6 discusses quantum information theory. We begin with some basic Shannon information theory. Density operator formalism is discussed in Section 2.6.1, including measurements, state evolution, partial trace, and trace distance. Finally, quantum entropy is discussed in Section 2.6.2, including von Neumann entropy and quantum min entropy.

2.1 Basic linear algebra

As it turns out, all the mathematics behind quantum computing and cryptography, from modeling quantum states to arguing about security, has a foundation in linear algebra. So, in this section we'll review some of the basics that you'll need to understand the rest of the text. Readers familiar with linear algebra may consider glossing over this section. We'll be introducing the basic concepts, however, using *Dirac notation* which is the main notation system used in quantum computing research. *Thus, even if you are familiar with basic linear algebra, you may want to at least skim this section to become familiar with the notation we'll be using throughout.* Alternatively, you may want to skip ahead to Section 2.2 and refer back to this section when something doesn't make sense in the notation. This

Quantum Communication and Cryptography. https://doi.org/10.1016/B978-0-44-321532-2.00009-7

section is not meant to be an exhaustive survey of all linear algebra topics needed, but many of the important issues are covered here.

The most basic element in linear algebra is the *vector*. Abstractly, vectors are objects that can be added together and also multiplied with a scalar. For our purposes, the vectors we care about are n-dimensional complex vectors, that is n-tuples with each element a complex number. We denote the space of such vectors as $\mathbb{C}^n$. A vector will be written using Dirac notation as a *ket* denoted $|\cdot\rangle$, where the $\cdot$ can be any label you like. Let's consider the "simple" case of $n = 2$. A vector, then, can be written as:

$$|\texttt{my-vector}\rangle = \begin{pmatrix} \alpha \\ \beta \end{pmatrix}, \tag{2.1}$$

where α and β are two complex numbers (numbers of the form $x + yi$ where $i^2 = -1$). Above, the vector is called "my-vector." Moving forward, we'll be using Greek letters to denote vectors typically, for example $|\psi\rangle$ which is traditional in quantum research – however, feel free to call your vectors whatever you like.

Given two vectors $|\psi\rangle$ and $|\phi\rangle$ one may add them together to form a new vector $|\psi\rangle + |\phi\rangle$. If one writes them as tuples (as in Eq. (2.1)), then the resulting vector is simply the addition of each coordinate together, namely:

$$\underbrace{\begin{pmatrix} a \\ b \end{pmatrix}}_{|\psi\rangle} + \underbrace{\begin{pmatrix} x \\ y \end{pmatrix}}_{|\phi\rangle} = \underbrace{\begin{pmatrix} a+x \\ b+y \end{pmatrix}}_{|\psi\rangle+|\phi\rangle}. \tag{2.2}$$

The above naturally generalizes to any dimension n. Besides addition, one also has *scalar multiplication*: namely, given a complex number $z \in \mathbb{C}$ and a vector $|\psi\rangle$, we can write $z\,|\psi\rangle$ to mean:

$$z\underbrace{\begin{pmatrix} a \\ b \end{pmatrix}}_{|\psi\rangle} = \underbrace{\begin{pmatrix} za \\ zb \end{pmatrix}}_{z|\psi\rangle} \tag{2.3}$$

Above, we used both Dirac notation, and also Cartesian-style vectors (e.g., arrays of numbers). From an educational point of view, it is a delicate matter as to how one should exactly view a vector $|\psi\rangle$. In general, we recommend thinking of them as Cartesian vectors only when first learning, but to try very hard to never write out, explicitly, what the vector is as a tuple – instead it is much more elegant, and generally easier, to work directly with Dirac notation and to not use tuples. However, there are times when you will want to work with coordinates/tuples and indeed it may be easier to do so. There is no hard and fast rule – in general working with Dirac notation (i.e., vectors abstractly) is usually easier in most situations; however when first starting, and even for some advanced problems, working with Cartesian coordinate vectors will be easier. Moving forward we'll work in both systems for educational purposes, slowly switching away from Cartesian coordinates as we move on.

Vector arithmetic follows a lot of the properties we're used to working with. In particular, the distributive property holds, namely for any scalar $z \in \mathbb{C}$ and vectors $|u\rangle$, $|v\rangle$ (of the same dimension), then:

$$z(|u\rangle + |v\rangle) = z\,|u\rangle + z\,|v\rangle\,. \tag{2.4}$$

Other important properties will become apparent as we move forward.

2.1.1 Vector spaces and bases

One of the most important concepts in linear algebra, in particular when working with vectors, is the *vector space*. A vector space V over $\mathbb{C}$ is a set of *vectors* such that certain properties are satisfied. Of importance is that the vector space is closed under addition and scalar multiplication. This means for any two vectors $|u\rangle$ and $|v\rangle$ from the set V, if we add them together to form a new vector $|w\rangle = |u\rangle + |v\rangle$, it holds that $|w\rangle \in V$. Furthermore, if we take any $\alpha \in \mathbb{C}$, then the product $\alpha\,|u\rangle$ also lives in V.

One example of a vector space is simply $\mathbb{C}^n$. Indeed, take any two vectors from this space and add them together, you will end up with an n-dimensional vector in $\mathbb{C}^n$; similarly if you were to multiply with any complex number α. A slightly more complicated example (though only slightly so) would be the vector space $V = \{(\alpha, 0)^T \;:\; \alpha \in \mathbb{C}\}$. Note that, here, we use v^T to mean the transpose of v, while later we will use v^* to mean the conjugate transpose.

Clearly, the above is a *vector subspace* of $\mathbb{C}^2$. That it is closed under addition and scalar multiplication is straightforward to see. An example of a set that is *not* a vector space would be $\{(\alpha, 0) \;:\; \alpha \text{ is an even integer}\}$. While this is closed under addition, it is not closed under scalar multiplication (just multiply by $\frac{1}{2}$). As we progress, we will encounter varying vector spaces and subspaces, though in this text they will all ultimately be subspaces of $\mathbb{C}^n$ for some n.

Now, consider a vector space V (which we may think of as a subspace of $\mathbb{C}^n$). A *spanning set* is a set $\mathcal{S} \subset V$ such that for any $|v\rangle \in V$, we can write $|v\rangle$ as a linear combination of vectors in $\mathcal{S}$. Namely, we can find scalars such that:

$$|v\rangle = \sum_{|s\rangle \in \mathcal{S}} \alpha_s\,|s\rangle\,. \tag{2.5}$$

Given such a set, we say $\mathcal{S}$ *spans* V.

A *basis* is a set $\mathcal{B} = \{|b_1\rangle\,, |b_2\rangle\,, \cdots\} \subset V$ that spans V, and such that no vector in $\mathcal{B}$ may be written as a linear combination of other vectors in $\mathcal{B}$. That is, for every $|b_i\rangle \in \mathcal{B}$, we cannot find scalars α_j such that:

$$|b_i\rangle = \sum_{j \neq i} \alpha_j\,|b_j\rangle \tag{2.6}$$

In a way, a basis is a minimal spanning set. Namely, start with any spanning set. If there exists a vector in the set that can be written as a linear combination of the others, then one may discard it (it is, essentially, redundant, as it can already be written as other vectors in the set).

As an example, the so-called *standard basis* of $\mathbb{C}^n$ is $\{(1,0,0,\cdots,0)^T, (0,1,0,\cdots,0)^T, \cdots, (0,0,0,\cdots,1)^T\}$. That any vector $|v\rangle \in \mathbb{C}^n$ can be written as a linear combination of vectors from this set, is trivial. It is also clear that no vector in this basis can be written as a linear combination of the others. Note that there are n vectors in this set. This is no coincidence. In fact, we define the *dimension* of a vector space V to be the number of vectors in a basis of that space. While there may be infinitely many bases for any one vector space, they will always contain the same number of vectors!

Exercise 2.1. Consider the following problems:

1. Show that $\{(1,1)^T, (1,-1)^T\}$ is a basis for $\mathbb{C}^2$.
2. Show that $\{(2,0)^T\}$ is a basis for the vector subspace $V = \{(\alpha,0)^T \ : \ \alpha \in \mathbb{C}\}$ (which was discussed earlier in this subsection).

Bases are important building blocks of vector spaces; in a way they are the "atoms" of a vector space. They will be very important later, when we discuss quantum states and measurements.

2.1.2 Inner products

Let V be a vector space. We say V is an *inner product space* if there exists a function $f : V \times V \to \mathbb{C}$ (i.e., a function that takes two vectors as inputs and outputs a complex number) which satisfies the following properties:

1. The function must be linear in its second argument, namely: $f(|v\rangle, \sum_i \alpha_i |w_i\rangle) = \sum_i \alpha_i f(|v\rangle, |w_i\rangle)$.
2. If you reverse the order of the inputs, the outputs are complex conjugates of one another; namely $f(|v\rangle, |w\rangle) = (f(|w\rangle, |v\rangle))^*$.
3. Finally, the inner product of a vector with itself is always a real, non-negative number, namely $f(|v\rangle, |v\rangle) \geq 0$, for any vector $|v\rangle$; equality holds, only if $|v\rangle$ is the zero vector.

We are dealing with vectors involving complex numbers, namely $\mathbb{C}^n$ (or some subspace of $\mathbb{C}^n$). In this case, we may use the standard *dot-product* as our inner product function. Given $|v\rangle = (\alpha_1, \cdots, \alpha_n)^T$ and $|w\rangle = (\beta_1, \cdots, \beta_n)^T$, then the inner product is $f(|v\rangle, |w\rangle) = \sum_i \alpha_i^* \beta_i$. The reader may readily verify that this definition satisfies the above three necessary properties.

The notation $f(\cdot,\cdot)$ is non-standard and, in fact, rather "clunky." Dirac notation admits a beautiful simplification for handling inner product. Given a vector $|v\rangle = (\alpha_1, \cdots, \alpha_n)^T$, we define by $\langle v|$ (read "bra v"), to be the conjugate transpose of $|v\rangle$. Namely:

$$|v\rangle = \begin{pmatrix} \alpha_1 \\ \vdots \\ \alpha_n \end{pmatrix} \Longrightarrow \langle v| = (\alpha_1^*, \cdots, \alpha_n^*). \tag{2.7}$$

Note that, if using Cartesian vectors, then $\langle v|$ is a row vector, while $|v\rangle$ is a column vector. Consider, now, the matrix product $\langle v| \times |w\rangle$ (for some other n-dimensional vector $|w\rangle = (\beta_1, \cdots, \beta_n)^T$). This works out to be a 1×1 matrix which, of course, we may take to be simply a scalar value. The product is exactly the inner product:

$$\langle v| \times |w\rangle = (\alpha_1^*, \cdots, \alpha_n^*) \times \begin{pmatrix} \beta_1 \\ \vdots \\ \beta_n \end{pmatrix} = \sum_i \alpha_i^* \beta_i. \tag{2.8}$$

Inner products are such a common operation, we will compress the notation and write, simply $\langle v|w\rangle$. This is a "braket" (a play on the word braket, hence the name for this Dirac "BraKet" notation).

For computing inner products, one may mentally switch back and forth between thinking of a braket as an inner product and thinking of it as the multiplication of two vectors (one of which you must take the conjugate transpose first). For example, let's define:

$$|+\rangle = \frac{1}{\sqrt{2}}|0\rangle + \frac{1}{\sqrt{2}}|1\rangle, \tag{2.9}$$

(the above is an important vector that will show up often); and now, let's compute $\langle 0|+\rangle$:

$$\langle 0|+\rangle = \langle 0| \times \left(\frac{1}{\sqrt{2}}|0\rangle + \frac{1}{\sqrt{2}}|1\rangle\right) = \frac{1}{\sqrt{2}}\langle 0|0\rangle + \frac{1}{\sqrt{2}}\langle 0|1\rangle. \tag{2.10}$$

If you recall that we define $|0\rangle$ and $|1\rangle$, typically, as the standard basis, one may readily verify that $\langle 0|0\rangle = 1$ and $\langle 0|1\rangle = 0$. Thus:

$$\langle 0|+\rangle = \frac{1}{\sqrt{2}}. \tag{2.11}$$

Above, we took advantage of the linearity of matrix multiplication; however we could also have taken advantage of linearity of inner product for the above computation. In general, when one sees a braket, you should think inner product; but if it's easier to compute something by thinking of it as a matrix multiplication, go ahead and do so. The very curious reader may want to read more of the theory behind this notation and the reasoning behind it, for that we recommend [9].

Now, above, we said that the inner product of $|0\rangle$ and $|1\rangle$ were zero, but this assumed we used the standard basis $|0\rangle = (1,0)^T$ and $|1\rangle = (0,1)^T$. What if we were to use other vector representations for these states? Well, as it turns out, we will from now on simply define the states $|0\rangle$ and $|1\rangle$ to be *orthonormal.* Namely, they will always be *orthogonal* and they will always be *normalized:*

Definition 2.1.1. Two vectors $|x\rangle$ and $|y\rangle$ are *orthogonal* if their inner product is zero, namely if:

$$\langle x|y\rangle = \langle y|x\rangle = 0. \tag{2.12}$$

A vector $|w\rangle$ is *normalized* if its inner product is one, namely:

$$\langle w|w\rangle = 1. \tag{2.13}$$

A vector is *sub-normalized* if its inner product is smaller than one.

Given any vector $|v\rangle$ (which may not be normalized), one may always *normalize it* by dividing by the square root of the inner product of itself. Namely, the normalized form of $|v\rangle$ is:

$$\frac{|v\rangle}{\sqrt{\langle v|v\rangle}}. \tag{2.14}$$

Note that if $|v\rangle$ is already normalized, then the normalized form is identical to $|v\rangle$. Otherwise, the new vector will always have inner product one with itself.

Returning to our $|0\rangle$ and $|1\rangle$ states, if we define them as orthonormal, then it won't matter what tuple vector representation we use, we will always know that each one is normalized and orthogonal. If we then define other vectors with respect to these, we will always be able to compute inner products of those new vectors also.

Exercise 2.2. Let $|+\rangle = \frac{1}{\sqrt{2}}(|0\rangle + |1\rangle)$ and $|-\rangle = \frac{1}{\sqrt{2}}(|0\rangle - |1\rangle)$. Compute $\langle +|+\rangle$, $\langle +|-\rangle$, and $\langle -|-\rangle$. What can you say about these two vectors? You may use the fact that $|0\rangle$ and $|1\rangle$ are orthonormal.

While we discussed general bases above, one very important type of basis used in quantum computing and cryptography will be *orthonormal bases*. Let V be an inner product space (i.e., a vector space with an inner product) and let $\mathcal{B} = \{|b_1\rangle, |b_2\rangle, \cdots, |b_n\rangle\}$ be a subset of vectors. Then we say $\mathcal{B}$ is an *orthonormal basis* of V if:

1. $\mathcal{B}$ is a basis of V.
2. Each vector is pair-wise orthogonal. Namely, for every $i \neq j$, it holds that $\langle b_i|b_j\rangle = 0$.
3. Each vector is normalized. Namely, for each i, it holds that $\langle b_i|b_i\rangle = 1$.

Note that, conditions two and three, above, can be combined by writing:

$$\langle b_i|b_j\rangle = \delta_{i,j}, \tag{2.15}$$

where $\delta_{i,j}$ is the Kronecker delta function; namely $\delta_{i,j} = 1$, only if $i = j$ and is zero otherwise.

Before leaving, we'll mention one important inequality involving inner product, namely the Cauchy-Schwarz inequality. Let $|v\rangle$ and $|w\rangle$ be two vectors, then it holds that:

$$|\langle v|w\rangle| \leq \sqrt{\langle v|v\rangle \langle w|w\rangle}. \tag{2.16}$$

This will be useful in many situations in this book.

2.1.3 Linear operators

Let V and W be two vector spaces. A *linear operator* is a function A mapping vectors from V to vectors in W, namely $A : V \to W$, that is also linear in the sense that:

$$A\left(\sum_i \alpha_i |v_i\rangle\right) = \sum_i \alpha_i A(|v_i\rangle). \tag{2.17}$$

To avoid overly cumbersome notation, we will write $A|v\rangle$ instead of $A(|v\rangle)$. Thus, the above, becomes simply:

$$A \sum_i \alpha_i |v_i\rangle = \sum_i \alpha_i A |v_i\rangle. \tag{2.18}$$

One commonly understood case of a linear operator is a matrix and the use of matrix multiplication. Namely, if A is an $m \times n$ matrix, then A will map (column) vectors from $\mathbb{C}^n$ to vectors in $\mathbb{C}^m$. However, representing A as a matrix, is only one way to represent a linear operator. And, while in our case, matrix representations are not missing anything (i.e., all linear operators from $\mathbb{C}^n$ to $\mathbb{C}^m$ can be represented as a matrix), such representations become cumbersome as the dimension of the systems involved increase.

Instead of matrix representations, we will often represent linear operators by their action on basis states. Note that, due to linearity, if we explicitly define the action of A on basis vectors, then we will know exactly how A behaves on *any* vector. Thus, if $\{|b_1\rangle, \cdots, |b_n\rangle\}$ is a basis, and if we know what $A|b_i\rangle$ is, for every i, then we will know exactly how A behaves on $|\psi\rangle$ since we can always write $|\psi\rangle = \sum_i \alpha_i |b_i\rangle$, for some scalars α_i. The action will simply be:

$$A|\psi\rangle = \sum_i \alpha_i A |b_i\rangle. \tag{2.19}$$

Exercise 2.3. Let Z be a linear operator defined by:

$$Z|0\rangle = |0\rangle$$
$$Z|1\rangle = -|1\rangle.$$

Then, what is $Z|+\rangle$ and $Z|-\rangle$ (where $|\pm\rangle$ were defined in Exercise 2.2). Write out an equivalent matrix representation of Z:

$$Z = \begin{pmatrix} a & b \\ c & d \end{pmatrix} \tag{2.20}$$

(i.e., fill in numbers for a, b, c, and d)

Another way to represent an operator is through *outer product representation.* Let $|v\rangle$ be a vector in some vector space V, and $|w\rangle$ be a vector in vector space W; we define a linear operator $|w\rangle\langle v|$ which maps vectors from V to W via the following action:

$$|w\rangle\langle v||\psi\rangle = \langle v|\psi\rangle |w\rangle. \tag{2.21}$$

That is, on input some vector $|\psi\rangle \in V$, the output will be $|w\rangle \in V$, with a scalar coefficient of the inner product $\langle v|\psi\rangle$. Note that, again, Dirac notation shines – if you forget that the meaning of $|w\rangle\langle v|$ is a linear operator with the action discussed above, then, if you are ever faced with computing $|w\rangle\langle v|\,|\psi\rangle$, just think of it as vector multiplication (where, recall $\langle v|$ is the conjugate transpose of $|v\rangle$)! In this case: $|w\rangle \times \langle v| \times |\psi\rangle = |w\rangle\,(\langle v|\psi\rangle) = \langle v|\psi\rangle\,|w\rangle$, where, for the last equality, we took advantage of the fact that $\langle v|\psi\rangle$ is a scalar. In general, then, a linear operator may be defined by summing over multiple operators of the form $\alpha_{v,w}\,|w\rangle\langle v|$ (for different vectors $|v\rangle$ and $|w\rangle$ and a scalar $\alpha_{v,w}$).

Given a vector space V, one very important, and at the same time trivial, operator is the *identity operator* denoted I_V (or, simply, I if the context is clear). This operator simply maps any vector to itself, namely $I\,|\psi\rangle = |\psi\rangle$ for all $|\psi\rangle \in V$. Let $\{|b_1\rangle, \cdots, |b_n\rangle\}$ be an orthonormal basis and define the operator $U = \sum_i |b_i\rangle\langle b_i|$. Then, it turns out, $U = I$ (i.e., U is the identity operator). This is known as the *completeness relation* and will be an important tool in analyzing and proving results later. To see why U is the identity operator, first note that $U\,|b_i\rangle = |b_i\rangle$. This can be seen using Eq. (2.21):

$$U\,|b_i\rangle = \left(\sum_j |b_j\rangle\langle b_j|\right)|b_i\rangle = \sum_j |b_j\rangle\langle b_j|\,|b_i\rangle = \sum_j \langle b_i|b_j\rangle\,|b_j\rangle = |b_i\rangle\,. \tag{2.22}$$

The last equality follows from the fact that, since the basis vectors are orthonormal, $\langle b_i|b_j\rangle = 0$ whenever $i \neq j$ (and is one otherwise). Now, let $|\psi\rangle$ be any state. Since $\{|b_i\rangle\}$ form a basis, we can write $|\psi\rangle = \sum_i \alpha_i\,|b_i\rangle$. Then:

$$U\,|\psi\rangle = \sum_i \alpha_i U\,|b_i\rangle = \sum_i \alpha_i\,|b_i\rangle = |\psi\rangle\,. \tag{2.23}$$

Since $|\psi\rangle$ was arbitrary, the result is proven.

In summary, we have three ways to represent a linear operator: as a matrix, by its action on basis states, and by its outer-product form. Any of the three may be used; depending on the application, usually one method is easier, or more natural, than another. The reader will encounter each of the methods throughout this book.

We'll close this section by defining certain types of special operators that will be important later. One important type of linear operator is known as a *unitary operator*:

Definition 2.1.2. A *unitary operator* U acting on some complex vector space V (which we will assume is finite dimensional here), is an operator such that:

$$UU^* = U^*U = I, \tag{2.24}$$

where I is the identity operator and U^* is the conjugate transpose of U.

Another important type of linear operator is known as a *Hermitian operator*:

Definition 2.1.3. An operator H is *Hermitian* if $H^* = H$, where H^* is the conjugate transpose of H (the adjoint operator).

Definition 2.1.4. A *positive operator, A*, is one where, for any vector $|\zeta\rangle$, it holds that:

$$\langle\zeta|A|\zeta\rangle > 0. \tag{2.25}$$

A *positive semi-definite* operator is the same, but where $\langle\zeta|A|\zeta\rangle \geq 0$.

2.1.4 Matrix functions

Let A be a linear operator. Then, an *eigenvector* is a non-zero vector $|\phi\rangle$ such that $A|\phi\rangle = \lambda|\phi\rangle$, where $\lambda \in \mathbb{C}$ is the corresponding *eigenvalue*. An operator A is said to be *diagonalizable* if it can be written as:

$$A = \sum_i \lambda_i |\phi_i\rangle\langle\phi_i|, \tag{2.26}$$

where $|\phi_i\rangle$ are orthonormal eigenvectors with corresponding eigenvalues λ_i. Note that any *normal operator* (i.e., one where $AA^* = A^*A$) is diagonalizable.

Now, given some function $f : \mathbb{C} \to \mathbb{C}$, and some diagonalizable operator A, we can define a matrix function $f(A)$ by extending the function f to work on operators. The definition will simply be:

$$f(A) = \sum_i f(\lambda_i)|\phi_i\rangle\langle\phi_i|, \tag{2.27}$$

where, again, λ_i and $|\phi_i\rangle$ are the eigenvalues and orthonormal eigenvectors of operator A. For example, consider:

$$Z = 1 \cdot |0\rangle\langle 0| + (-1) \cdot |1\rangle\langle 1| = \begin{pmatrix} 1 & 0 \\ 0 & -1 \end{pmatrix} \tag{2.28}$$

Then, consider the function $f(x) = e^{\pi x}$ and, so, $f(Z) = e^{\pi Z}$ is defined to be:

$$e^{\pi Z} = e^{\pi}|0\rangle\langle 0| + e^{-\pi}|1\rangle\langle 1| = \begin{pmatrix} e^{\pi} & 0 \\ 0 & e^{-\pi} \end{pmatrix}. \tag{2.29}$$

2.2 Quantum states

Quantum states in a closed system are modeled as normalized vectors in some complex vector space, and are denoted $|\cdot\rangle$ using Dirac notation, where the "$\cdot$" can be any label. To be more precise, they are elements of a complex Hilbert space – however, since all the states we deal with in this text are finite dimensional, we can restrict our definitions to vectors in $\mathbb{C}^n$ for some n. The space our state lives in is generally called the *state space* of the system, while the actual vector is called the *state*, or *state vector*. Normalized, recall, implies that $\langle\psi|\psi\rangle = 1$. This is often known as the first postulate of quantum computing [9]. By "closed system" we mean the state is completely self-contained and not interacting with some outside system. Later, we will explore open quantum systems – vital for modeling quantum communication and cryptographic protocols.

Quantum States: A closed quantum system can be modeled as a normalized vector (called the *state*) living in some Hilbert space (called the *state space*). In this text, all systems are finite dimensional, so the state space is $\mathbb{C}^n$ for some n (the dimension of the state).

A *qubit* or *quantum bit* is a two-dimensional quantum state, i.e., some state $|\psi\rangle \in \mathbb{C}^2$. Qubits are really just mathematical abstractions of some physical particle. This is not dissimilar from a classical bit (a 0 or 1) which is also a mathematical abstraction; indeed, when we discuss algorithms and protocols in classical computer science, we often don't care too much about *how* bits are implemented only that they can be. Of course, this is not always the case – when one starts investigating hardware issues of computers, then one is very interested in the physical implementation. However, bits, by themselves, are abstract place holders for some physical thing: a thing that can be in a "high" or "low" state. There are many ways to implement a bit: for example a high or low voltage signal; a charged or discharged capacitor; or a full or empty glass of water. When designing an algorithm or cryptographic protocol, mostly one does not actually care about the exact physical implementation and, so, one usually just abstractly works with bits.

The thinking is similar in quantum computing: a qubit is just our abstraction of some physical object; however just like there are several ways to physically implement a classical bit, there are several ways to physically implement a qubit. Normally, when designing and analyzing algorithms and cryptographic protocols at a high-level, we don't care too much how they're implemented. This, of course, is not a general rule and there are often times where you need to consider the physical implementation. Nonetheless, for the sake of our discussion in this text, qubits will remain abstract, and we will only refer to specific physical implementations when needed or if it helps with the understanding.

Let's build a better understanding, now, of qubits and their many differences to classical bits (c-bits). One of the very first differences comes in the number of possible states a c-bit versus a qubit may exist in. A c-bit can be in only one of two possible states at a single time: it is either zero or it's one. A single qubit, on the other hand can exist in infinitely many states.

Consider a basic qubit in the abstract: $|\psi\rangle$. Since this is a two dimensional system, we can write it as a vector $|\psi\rangle = \alpha|0\rangle + \beta|1\rangle$, where the $|0\rangle$ and $|1\rangle$ states form the Z orthonormal basis (also called the computational basis) as defined in Section 2.1. This state exists in a *superposition* of the state $|0\rangle$ and the state $|1\rangle$. Normalization requires that $\langle\psi|\psi\rangle = |\alpha|^2 + |\beta|^2 = 1$. If $\alpha = 1$, then $|\psi\rangle = |0\rangle$; otherwise, if $\beta = 1$, then $|\psi\rangle = |1\rangle$. The α and β coefficients (which are complex numbers) are called the *probability amplitudes.* Thus, a qubit, like a c-bit, can exist in a "zero" or "one" state. However, α and β are any complex numbers subject to the before mentioned normalization constraint. There are actually infinitely many possibilities for the qubit to exist in. This property of superposition is the first major difference between a c-bit and a quantum bit (or general, quantum state).

2.3 Measuring a quantum state

The second major difference in qubits versus c-bits is how we extract information from them. With a c-bit, we simply read the bit, extract its state, and leave it undisturbed. Qubits, however, are different – reading a qubit, called *measuring* it, is a process that produces a potential random output and potentially destroys the original state.

Information may be extracted from a quantum state through measurements. These measurements lead to a discrete classical output, the result of which is potentially random and also potentially destructive to the state.

Consider the following state:

$$|+\rangle = \frac{1}{\sqrt{2}}(|0\rangle + |1\rangle). \tag{2.30}$$

The above is actually an important state which we'll refer to often – it's often called the "Hadamard plus state" or the "X basis plus state" (we'll use both interchangeably, or, simply $|+\rangle$). This state is in a superposition of both the $|0\rangle$ state and the $|1\rangle$ state. In a way, it is both the zero and one state simultaneously, though it is not quite as simple as that, especially when it comes time to *extract* information from a qubit, a process known as *measuring.*

A quantum state $|\psi\rangle$ remains in a "closed box" perhaps evolving its state according to some process we describe later in Section 2.4, perhaps in a superposition of multiple states, but, ultimately, remaining sealed away from the environment. Eventually, with any quantum algorithm or quantum cryptographic protocol, one must "look" at the state to extract information from it. This process, called measuring, is the quantum analogue of reading a c-bit. However, while a quantum state may exist in a multitude of superposition states (e.g., the $|+\rangle$ state above, existing in a way as both zero and one), measurements cannot necessarily extract all information from a quantum state (e.g., cannot extract exact information on α and β for an arbitrary state). Instead, a quantum state, through measurement, is converted into classical data with discrete values (again, everything is finite dimensional in this text). This process is random and also leads to *state collapse.*

Consider a c-bit $x \in \{0, 1\}$. If this is stored in some classical memory system, one may read the bit at any time to get a deterministic output; furthermore the act of looking at the bit should not affect its state – it remains x. Consider, now, a quantum state $|\psi\rangle = \alpha|0\rangle + \beta|1\rangle$. "Looking" at this state (i.e., measuring it) in the $|0\rangle$, $|1\rangle$ basis (we will see later, that we can look at a state in different bases – but for now, let's stick to the standard computational basis) causes the measurement device to output *not* "$\alpha|0\rangle + \beta|1\rangle$" as a whole (i.e., it will not output exact information on the complex numbers α and β); instead the measurement will output either "$|0\rangle$" or "$|1\rangle$" – a discrete, essentially classical outcome. This is a probabilistic process with the measurement device outputting (or *observing*) $|0\rangle$ with probability $|\alpha|^2$; otherwise, with probability $|\beta|^2$, the measurement device outputs $|1\rangle$.

This is the reason behind the name *probability amplitude*: the coefficient of a basis vector is related to the probability of observing that particular basis element. The argument extends to multiple dimensions. For instance, if we have a state $\alpha|0\rangle + \beta|1\rangle +$

$\gamma\,|2\rangle$, then the probability of observing a "2" is $|\gamma|^2$. Note that the normalization constraint ensures this forms a valid probability distribution.

Exercise 2.4. Consider the following, three questions:

1. Consider the $|+\rangle$ state. What is the probability of observing $|1\rangle$ after a measurement?
2. Next, consider the following state:

$$|-\rangle = \frac{1}{\sqrt{2}}(|0\rangle - |1\rangle). \tag{2.31}$$

   What is the probability of observing $|1\rangle$ after measuring that state?
3. Finally, consider the state $|1\rangle$. What is the probability of observing $|1\rangle$ after measurement?

Beyond the fact that measurements are now random (unless the state happens to be one of the basis states, though we'll comment more on this later), they also are destructive. The act of looking at a quantum state potentially disturbs it. In particular, if after making an observation, if your measurement observes outcome $|i\rangle$, then the state actually becomes $|i\rangle$. That is, the state *collapses* to the observed state vector; this is known as *state collapse.* Measurements, therefore are: (1) random, (2) destructive and (3) non reversible (once measured, you cannot go back, reverse the measurement, and return to the state prior to measurement).

For example, let's say we start with the state $|+\rangle$. If we measure it, we will observe $|0\rangle$ or $|1\rangle$, both with equal probability. However, measurement is a "one-shot" deal. Once the qubit (originally in state $|+\rangle$) is measured, it collapses to the observed state. Thus, if our measurement produced an observation of "1," the state now really is $|1\rangle$, the original $|+\rangle$ is "destroyed" and the state becomes $|1\rangle$. Any subsequent measurements will produce outcome "1" with probability one.

Exercise 2.5. Imagine you are given a qubit $|\psi\rangle$ but not told anything about it (i.e., you do not know anything about its probability amplitudes). If you perform a measurement on the state and the outcome is "0" what does this tell you about the state? Can you learn the original probability amplitudes? What if you are given as many copies of $|\psi\rangle$ as you want – what can be said then? Note that you cannot copy an unknown qubit yourself – we'll discuss this later though, in Section 2.5.4.

2.3.1 Basis measurements

So far, quantum states and measurements, seem to provide a highly frustrating scenario for quantum information scientists. On the one hand, there are infinitely many states a single qubit can be in. On the other, to actually get information from a state, a measurement must be performed which creates an inherently classical outcome and, furthermore, causes the state to collapse to an, essentially, classical system (zero or one). Is there any real advantage to quantum states and superposition? The answer is, actually, yes though it will take a little more groundwork before it can be truly seen and appreciated. However, harnessing the power of quantum information, within the rules provided to us (e.g., within the realm of infinite state possibilities, but highly restricted measurement rules) is the ultimate challenge in any quantum related problem.

One important fact to point out is that measurements in the "zero/one" basis (the computational basis) are not the only form of measurement available to you. Indeed, one may measure in *any* basis of your state space! This opens up interesting possibilities as, now, there are infinitely many types of measurements one could perform on a qubit.

Consider a qubit, and let $\mathcal{B} = \{|b_0\rangle, |b_1\rangle\}$ be an orthonormal basis for the two dimensional qubit state space (namely, $\mathbb{C}^2$). It turns out, one can take a qubit state $|\psi\rangle$, and measure in the $\mathcal{B}$ basis.

How does this measurement work? As an example, let $|\psi\rangle$ be a qubit; since $\mathcal{B}$ is a basis, we may write the state as a linear combination of vectors in this basis, namely:

$$|\psi\rangle = x\,|b_0\rangle + y\,|b_1\rangle\,, \tag{2.32}$$

for some $x, y \in \mathbb{C}$ with $|x|^2 + |y|^2 = 1$. Then, measuring in the $\mathcal{B}$ basis produces a discrete outcome: the measurement device observes either "$|b_0\rangle$" or "$|b_1\rangle$." Furthermore, the result is random, with the measurement observing $|b_0\rangle$ with probability $|x|^2$; otherwise the measurement device will observe $|b_1\rangle$, and that will occur with probability $|y|^2$. We still have state collapse: on observing "b_i" (for either $i = 0, 1$), the state, which was originally a superposition of these two basis vectors, will collapse to (i.e., become) the observed vector $|b_i\rangle$. Any subsequent measurement in the same basis will result in the same outcome after this collapse.

Of course, the above measurement rule of squaring the corresponding coefficient, only works when the input state is written with respect to the measurement basis. However, of course, this can always be done. Since $\mathcal{B}$ is a basis, given any state $|\psi\rangle$, one can always write that state vector as a linear combination (superposition) of basis states in $\mathcal{B}$. As an example, let $|\psi\rangle = \alpha\,|0\rangle + \beta\,|1\rangle$, and consider the *Hadamard basis*:

Definition 2.3.1. The *Hadamard basis* also known as the X basis, consists of two vectors $|+\rangle$ and $|-\rangle$ defined to be:

$$|+\rangle = \frac{1}{\sqrt{2}}\,|0\rangle + \frac{1}{\sqrt{2}}\,|1\rangle$$

$$|-\rangle = \frac{1}{\sqrt{2}}\,|0\rangle - \frac{1}{\sqrt{2}}\,|1\rangle$$

Now, let's take our state $|\psi\rangle$ and measure in the Hadamard basis. To do so, we must change basis and write $|\psi\rangle = x\,|+\rangle + y\,|-\rangle$. To do this, we will use the following identities:

$$|0\rangle = \frac{1}{\sqrt{2}}\,|+\rangle + \frac{1}{\sqrt{2}}\,|-\rangle \tag{2.33}$$

$$|1\rangle = \frac{1}{\sqrt{2}}\,|+\rangle - \frac{1}{\sqrt{2}}\,|-\rangle$$

One can derive the above, by taking the definition of $|+\rangle$ and $|-\rangle$ and "solving for" $|0\rangle$ and $|1\rangle$.

Using the above identities, we find:

$$|\psi\rangle = \alpha\,|0\rangle + \beta\,|1\rangle = \alpha\left(\frac{1}{\sqrt{2}}|+\rangle + \frac{1}{\sqrt{2}}|-\rangle\right) + \beta\left(\frac{1}{\sqrt{2}}|+\rangle - \frac{1}{\sqrt{2}}|-\rangle\right)$$
$$= \frac{\alpha+\beta}{\sqrt{2}}|+\rangle + \frac{\alpha-\beta}{\sqrt{2}}|-\rangle .$$

From this, we see that our "x" is $(\alpha+\beta)/\sqrt{2}$ and our y is $(\alpha-\beta)/\sqrt{2}$. And, so, the probability of observing $|+\rangle$, given $|\psi\rangle = \alpha\,|0\rangle + \beta\,|1\rangle$ is:

$$Pr(\text{"}+\text{"}) = \left|\frac{\alpha+\beta}{\sqrt{2}}\right|^2 = \frac{|\alpha|^2 + |\beta|^2 + \alpha\beta^* + \alpha^*\beta}{2} = \frac{1}{2} + Re(\alpha\beta), \tag{2.34}$$

where, above, we used the fact that $|\alpha|^2 + |\beta|^2 = 1$ and that, for any complex number z, it holds that $z + z^* = 2Re(z)$. We also used the fact that, for any two complex numbers z, w it holds that $(zw^*)^* = z^*w$. We also used the fact that $|z|^2 = zz^* = z^*z$.

Exercise 2.6. Let $p \in [0, 1]$ be a real number and consider the following state:

$$|\psi\rangle = \sqrt{p}\,|0\rangle + \sqrt{1-p}\,|1\rangle \tag{2.35}$$

Answer the following questions:

1. What is the probability of observing $|0\rangle$ for a given p?
2. What is the probability of observing $|-\rangle$ for a given p?
3. Graph the probability distribution of observing $|0\rangle$ with the x-axes of your plot being the value of p.
4. Same as above, but now graph the probability distribution of observing $|-\rangle$.

Thus, to measure $|\psi\rangle$ in basis $\mathcal{B}$, one must write out $|\psi\rangle$ as a superposition (linear combination) of basis vectors in $\mathcal{B}$. Once done, the probability of observing any particular basis element in $\mathcal{B}$ involves plucking out the corresponding coefficient of that vector (the probability amplitude) and take its norm squared.

Changing basis, though always possible, is not the only way to determine the probability distribution. Instead, one may look at the inner product. Let $\mathcal{B} = \{|b_0\rangle, \cdots, |b_{d-1}\rangle\}$ be the orthonormal basis we want to measure in and let $|\psi\rangle$ be the d-dimensional state we want to measure. Then, if we denote by $p(b_i)$ to be the probability of observing "b_i" given this input state, it holds that:

$$p(b_i) = |\langle b_i|\psi\rangle|^2 = |\langle\psi|b_i\rangle|^2. \tag{2.36}$$

Note the distribution depends, of course, on the input state $|\psi\rangle$. The above equation is important to remember as it provides a much more natural way to compute the probabilities of measurement outcomes. One need only compute the inner product of the basis vector and the state you're measuring. The "larger" the inner product, the more likely you are of observing that particular basis state. This gives a nice intuitive geometric interpretation of measurements.

But why does Eq. (2.36) agree with our earlier interpretation of measurements – namely that the probability of making an observation is based on the probability amplitude, or the scalar coefficient "attached" to the basis vector in its linear combination. Well, let's take $|\psi\rangle$ and since $\mathcal{B}$ is a basis, we may write it as:

$$|\psi\rangle = \sum_{j=0}^{d-1} \alpha_j |b_j\rangle. \tag{2.37}$$

From our earlier discussion, we know the probability of observing "b_i" is $|\alpha_i|^2$. So, to prove that Eq. (2.36) matches this interpretation, we need to show that $\alpha_i = \langle b_i|\psi\rangle$. However, note that:

$$\langle b_i|\psi\rangle = \langle b_i| \times \left(\sum_{j=0}^{d-1} \alpha_j |b_j\rangle\right) = \sum_j \alpha_j \langle b_i|b_j\rangle = \sum_j \alpha_j \delta_{i,j} = \alpha_i, \tag{2.38}$$

where, above, $\delta_{i,j} = 1$ if $i = j$ and is zero otherwise. The above follows since the $\{|b_i\rangle\}$ form an orthonormal basis. Thus, the two interpretations of measurement calculations – one using inner product and one using the scalar coefficients – give exactly the same values and may be used interchangeably. In general, the inner product form will be easier in the long run to use as you may start to easily run across similar calculations.

As a concrete example, let's compute the probability of observing a $|-\rangle$ given that an X basis measurement is performed on the state $|1\rangle$. For this, we compute:

$$\langle 1|-\rangle = \frac{1}{\sqrt{2}}(\langle 1|0\rangle - \langle 1|1\rangle) = -\frac{1}{\sqrt{2}}. \tag{2.39}$$

Thus, the probability of making this observation is $|\langle 1|-\rangle|^2 = 1/2$. (*Don't forget to take the norm squared at the end!*) As a second example, let's make an X basis measurement of the state $|\psi\rangle = \sqrt{\frac{3}{4}}|0\rangle + \sqrt{\frac{1}{4}}|1\rangle$. What's the probability of observing "+"? We could make a change of basis as we did before and determine the correct probability amplitude. Instead, let's just use the inner product rule:

$$\langle +|\psi\rangle = \sqrt{\frac{3}{4}}\langle +|0\rangle + \sqrt{\frac{1}{4}}\langle +|1\rangle = \sqrt{\frac{3}{4}} \times \frac{1}{\sqrt{2}} + \sqrt{\frac{1}{4}} \times \frac{1}{\sqrt{2}}. \tag{2.40}$$

Note that, above, we made use of the fact that $\langle +|0\rangle = 1/\sqrt{2}$ and $\langle +|1\rangle = 1/\sqrt{2}$. This can be memorized (as it's useful), or easily derived. Continuing:

$$\langle +|\psi\rangle = \sqrt{\frac{3}{8}} + \sqrt{\frac{1}{8}} \Longrightarrow p(+) = |\langle +|\psi\rangle|^2 = \frac{1}{2} + \frac{\sqrt{3}}{4}. \tag{2.41}$$

Exercise 2.7. Let $|\phi\rangle = \frac{1}{2}|+\rangle + \sqrt{\frac{3}{4}}|-\rangle$. Compute using both the inner product method and the change of basis method, the following:

1. $p(0) =$
2. $p(-) =$

Exercise 2.8. Let $|\psi\rangle$ be some d-dimensional state and let $\mathcal{B} = \{|b_0\rangle, \cdots, |b_{d-1}\rangle\}$ be any orthonormal basis. Show that $|\psi\rangle$ may be written in the following form:

$$|\psi\rangle = \sum_{j=0}^{d-1} \langle b_j|\psi\rangle |b_j\rangle. \tag{2.42}$$

Note that, as before, anytime a measurement in any basis is made, the state collapses to the observed basis vector.

2.3.2 Measurement operators

As it turns out, even more complicated measurements are possible, along with additional methods to compute the probabilities without having to directly make a change of basis. Let $|\psi\rangle$ be a quantum state living in some state space $\mathcal{H}$. Let $\{M_x\}_{x\in\chi}$ be operators acting on the same state space which are indexed by some set χ (which, for our purposes, is finite). We will soon see that the set χ consists of observable outcomes that the measurement device might produce. Then, we say these operators are *measurement operators* if:

$$\sum_{x\in\chi} M_x^* M_x = I, \tag{2.43}$$

where I is the identity operator on $\mathcal{H}$. Then, we may measure $|\psi\rangle$ using these measurement operators. We will observe some outcome $x \in \chi$ with probability:

$$p(x) = \langle\psi|M_x^* M_x|\psi\rangle. \tag{2.44}$$

The post measured state collapses to:

$$\frac{M_x|\psi\rangle}{\sqrt{p(x)}}. \tag{2.45}$$

Measurement operators, those satisfying Eq. (2.43), generalize the basis measurements we've been considering so far. For example, a Z basis measurement is defined by measurement operators $M_0 = |0\rangle\langle 0|$ and $M_1 = |1\rangle\langle 1|$. Note that these are valid measurement operators since:

$$M_0^* M_0 + M_1^* M_1 = |0\rangle\langle 0||0\rangle\langle 0| + |1\rangle\langle 1||1\rangle\langle 1| = |0\rangle\langle 0| + |1\rangle\langle 1| = I, \tag{2.46}$$

where for the last equality we took advantage of the completeness relation discussed in Section 2.1.3. Now, let $|\psi\rangle = \alpha|0\rangle + \beta|1\rangle$, our typical, friendly, qubit state. From our earlier discussion, we know that the probability of observing $|0\rangle$ will be $|\alpha|^2$. As it turns out, this agrees with the measurement operator formalism. In particular:

$$p(0) = \underbrace{\left(\alpha^* \langle 0| + \beta^* \langle 1|\right)}_{\langle\psi|} \underbrace{|0\rangle \langle 0| |0\rangle \langle 0|}_{M_0^* M_0} \underbrace{(\alpha |0\rangle + \beta |1\rangle)}_{|\psi\rangle}$$
$$= \alpha^* \langle 0|0\rangle \alpha \langle 0|0\rangle = \alpha^* \alpha = |\alpha|^2. \tag{2.47}$$

Note that the beauty of Dirac notation allowed us to readily compute the above in a very natural way by thinking of the above as a product of row and column vectors, leading to a 1×1 scalar. To be truly formal, one must recall the definition of an operator of the form $|v\rangle \langle w|$, from Section 2.1.3.

In general, measurement operators allow us to perform more ambitious measurements, beyond projective basis measurements. Measurement operators may also have more outcome possibilities than the dimension of the system. For example, one may design measurement operators which attempt to distinguish two non orthogonal states, or output an "inconclusive" outcome. We'll return to this shortly after discussing POVMs.

POVMs are *Positive Operator-Valued Measures* and they follow from the measurement operator formalism, however they are only useful if one is interested in the probability of an outcome occurring and not the actual post measured state. (Perhaps *only useful* is a bit strong as one can construct measurement operators from a POVM and thus compute the post measured state.) A POVM is a collection of operators $\{E_x\}_{x\in\chi}$, indexed over some outcome set χ, such that:

1. Each E_x acts on the state space of the system being measured.
2. Each E_x is a Hermitian, positive semi-definite operator (i.e., $E_x \geq 0$; see Eq. (2.25)).
3. Finally, $\sum_{x\in\chi} E_x = I$.

Given a POVM $\{E_x\}$, and a quantum state $|\psi\rangle$, the probability of observing some outcome x is found to be:

$$p(x) = \langle \psi | E_x | \psi \rangle. \tag{2.48}$$

That the POVM formalism follows from measurement operators is easily seen. Let $\{M_x\}$ be a collection of measurement operators. Then, define $E_x = M_x^* M_x$. This is a POVM since each $E_x \geq 0$ (see Exercise 2.9). Furthermore, since $\{M_x\}$ are valid measurement operators, it follows that $I = \sum_x M_x^* M_X = \sum_x E_x$. Finally, if one were to measure using $\{M_x\}$, by Eq. (2.44), the probability of observing outcome x is easily seen to match Eq. (2.48).

Exercise 2.9. Show that each E_x is positive semi-definite, where $E_x = M_x^* M_x$, as discussed in the preceding paragraph.

Note that POVMs only have a rule to compute the probability of an outcome, but they say nothing of the post measured state. Often, though not always, this is sufficient, especially when dealing with information theoretic problems as we only care about the classical outcome of a measurement and, once a measurement is performed, we no longer care about the quantum state itself. (This is, of course, not always the case.) However, given a POVM, one may construct a collection of measurement operators using the rule $M_x = \sqrt{E_x}$, where $\sqrt{E_x}$ is the matrix function discussed in Section 2.1.4. This construction, of course, is not unique and there may be other measurement operators that "match" the given POVM.

2.3.3 Example: state discrimination

POVMs are very useful in many cases. For instance, let's consider the following experiment: say a source prepares one of two quantum states, either it prepares a $|0\rangle$ or it prepares a $|+\rangle$. When given a qubit (one of the two), your goal is to determine which state was sent to you. Now, as it turns out, this task is impossible on its own, which we can easily prove using POVM formalism. Let's assume we could design some kind of measurement which can always distinguish between $|0\rangle$ and $|+\rangle$. Well, given the above discussion, we might as well only consider POVMs; thus, let's assume there exist a POVM $\{E_0, E_+\}$ satisfying the following:

1. $\langle 0|E_0|0\rangle = 1$. That is, if the state really is $|0\rangle$, then the measurement should produce an outcome "0" with probability one.
2. $\langle +|E_+|+\rangle = 1$. Same as above, but for the case when the input state is $|+\rangle$.
3. $\langle +|E_0|+\rangle = \langle 0|E_+|0\rangle = 0$. That is, the measurement should never report the wrong answer.

However, this leads to contradiction. Indeed, consider $0 = \langle 0|E_+|0\rangle$. By Exercise 2.10, it must hold that $\langle 1|E_+|0\rangle = \langle 0|E_+|1\rangle = 0$. But, then, since we also need $1 = \langle +|E_+|+\rangle$, and since $|+\rangle = \frac{1}{\sqrt{2}}(|0\rangle + |1\rangle)$, we have:

$$1 = \langle +|E_+|+\rangle = \frac{1}{2}(\langle 0|E_+|0\rangle + \langle 1|E_+|1\rangle + \langle 0|E_+|1\rangle + \langle 1|E_+|0\rangle$$
$$= \frac{1}{2}\langle 1|E_+|1\rangle \leq \frac{1}{2},$$

where, for the last inequality, we used the fact that $\langle 1|E_+|1\rangle$ represents the probability that the measurement would produce outcome "+" given an input state of $|1\rangle$ which cannot be larger than one. This leads to the contradiction that $1 \leq \frac{1}{2}$ and, so, such a POVM cannot exist.

Exercise 2.10. Let E_+ be a Hermitian positive semi-definite operator. Show that if $\langle 0|E_+|0\rangle = 0$, then it must be that $\langle 1|E_+|0\rangle = 0$. Hint: Define the operator $\sqrt{E_+}$ and consider $\sqrt{E_+}\,|0\rangle$. Use the fact that $\langle v|E_+|w\rangle = \langle v|\sqrt{E_+}\sqrt{E_+}|w\rangle$, along with the fact that, for any vector $|x\rangle$, if $\langle x|x\rangle = 0$, then $|x\rangle$ is the zero vector.

Thus, this POVM cannot exist. *However* if we change the rules of the game to include a third possible measurement outcome, namely an "I don't know" outcome, then suddenly the task becomes possible. Now, we have a POVM denoted by operators $\{E_0, E_+, E_?\}$, with the following constraints needed:

1. $\langle 0|E_0|0\rangle > 0$
2. $\langle +|E_+|+\rangle > 0$
3. $\langle +|E_0|+\rangle = \langle 0|E_+|0\rangle = 0.$

Now, the measurement is never wrong in that it will never report an incorrect state; however it no longer has to always report a conclusive outcome. Instead, it is allowed an additional inconclusive, or "I don't know" result, namely $E_?$. We can define this POVM in the following way:

$$E_0 = p\,|-\rangle\langle -| \tag{2.49}$$

$$E_+ = p\,|1\rangle\langle 1| \tag{2.50}$$

$$E_? = I - E_0 - E_+. \tag{2.51}$$

Note that we define E_0 by having it "look" for a state orthogonal to $|+\rangle$ (which will never be observed if the state actually is $|+\rangle$). Similar for E_+. These operators "look" for states orthogonal to the other one. It is clear that this satisfies our requirements so long as $p > 0$. Defining $E_?$ as we did is a nice "trick" to make them sum to identity (a requirement for POVMs). The only thing we need to prove, then, is that they are all positive operators; namely that for some p, the operators above are positive. Clearly if $p = 0$ this holds trivially, but then our measurement always outputs "I don't know." Can we do better?

Let's consider E_0 first. Let $|\psi\rangle$ be any vector (not even necessarily normalized). Then: $\langle\psi|E_0|\psi\rangle = p\,\langle\psi|-\rangle\langle -|\psi\rangle = p|\langle\psi|-\rangle|^2 \geq 0$. Similarly for E_+. Thus, for any $p > 0$, E_0 and E_+ are both positive operators; the trouble case is the $E_?$ operators since, if p is too large, this becomes a negative operator. However:

$$\begin{aligned}\langle\psi|E_?|\psi\rangle = \langle\psi|(I - E_0 - E_+)|\psi\rangle &= \langle\psi|\psi\rangle - \langle\psi|E_0|\psi\rangle - \langle\psi|E_+|\psi\rangle \\ &= \langle\psi|\psi\rangle - p|\langle\psi|-\rangle|^2 - p|\langle\psi|1\rangle|^2.\end{aligned}$$

By the Cauchy-Schwarz inequality (see Eq. (2.16)), it holds that $|\langle\psi|-\rangle|^2 \leq \langle\psi|\psi\rangle\langle -|-\rangle = \langle\psi|\psi\rangle$. Thus:

$$\langle\psi|E_?|\psi\rangle \geq \langle\psi|\psi\rangle - 2p\,\langle\psi|\psi\rangle = (1-2p)\,\langle\psi|\psi\rangle\,. \tag{2.52}$$

The right hand side of the above expression is non-negative whenever $1 - 2p \geq 0$ or whenever $p \leq \frac{1}{2}$. Thus, we can conclude a POVM exists which satisfies our distinguishing requirements, so long as we set p between zero and $1/2$. The higher the value of p, the "better" this POVM is at distinguishing.

POVMs are powerful constructs and very useful for proving (1) measurements do not exist which satisfy some requirements or (2) constructing a "convoluted" measurement.

Exercise 2.11. Repeat the above state discrimination example for two arbitrary states $|\psi\rangle$ and $|\phi\rangle$ where $0 < |\langle\psi|\phi\rangle| < 1$.

2.4 State evolution

The next topic we cover is *state evolution* which specifies how a quantum state may change with time, without a (destructive) measurement being performed. This will dictate what processes we can apply to a quantum state, besides directly observing it through a measurement. State evolution is a non-destructive and, in fact, reversible process.

Given a quantum state $|\psi\rangle$, then, the state may change in time (or *evolve*) to a state $|\phi\rangle$ where $|\phi\rangle = U|\psi\rangle$. Here, U is a unitary operator (see Definition 2.1.2) and, recall, the notation $U|\psi\rangle$ means applying linear operator U to the state vector $|\psi\rangle$ (or multiplying matrix U with vector $|\psi\rangle$ if U is represented in matrix form). Said differently, one may change a quantum state $|\psi\rangle$ to a new state $|\phi\rangle$ only if there exists a unitary operator U such that $|\phi\rangle = U|\psi\rangle$. Of course, the transformation may be difficult to implement practically – from a quantum cryptographic perspective, this may matter for Alice and Bob, but may not matter for the adversary, Eve, depending on the security model.

Before describing some basic unitary operators we will use throughout this book, let's first discuss how to describe a unitary operator. There are three main methods and to demonstrate each, and to show the difference with each, we'll consider a particular unitary operator known as the *Hadamard Operator* denoted H. The first method is the matrix representation method. Here we will assume the standard basis for the Z basis, namely $|0\rangle = (1, 0)^T$ and $|1\rangle = (0, 1)^T$. Then, the matrix representation of a unitary operator is simply, well, a matrix – in the qubit case it is a 2×2 matrix. The Hadamard is defined in this form as:

$$H = \frac{1}{\sqrt{2}}\begin{pmatrix} 1 & 1 \\ 1 & -1 \end{pmatrix} \tag{2.53}$$

The *action* of the operator, then, on a state $|\psi\rangle$ is simply the matrix multiplication of the operator with the vector, namely the matrix product of $U|\psi\rangle$. For the Hadamard operator, we see the following identities:

$$H|0\rangle = \frac{1}{\sqrt{2}}\begin{pmatrix} 1 \\ 1 \end{pmatrix} = \frac{1}{\sqrt{2}}(|0\rangle + |1\rangle) = |+\rangle$$

$$H|1\rangle = \frac{1}{\sqrt{2}}\begin{pmatrix} 1 \\ -1 \end{pmatrix} = \frac{1}{\sqrt{2}}(|0\rangle - |1\rangle) = |-\rangle$$

That is, H maps $|0\rangle$ to the $|+\rangle$ state while mapping $|1\rangle$ to the $|-\rangle$ state.

Exercise 2.12. Show that $H|+\rangle = |0\rangle$ and $H|-\rangle = |1\rangle$. Note that not all unitary operators are their own inverse, however every unitary operator is invertible.

This is one way to represent a unitary operator. A second way involves simply describing the operator's action on basis states. Note that linear operators are completely defined based on their action on basis states. Thus, this gives a clean way to describe an operator, using only Dirac notation, without the need for vector notation and matrices (which, as the systems increase in size, can lead to unbearably large matrices!). For the Hadamard, we may equivalently define it in the following way:

$$H|0\rangle = |+\rangle \tag{2.54}$$
$$H|1\rangle = |-\rangle$$

And that's it! In general, we will refrain from using matrix and vector notation and use only Dirac notation when possible and, so, will prefer this method of describing operators over the matrix representation method.

Exercise 2.13. Show that $H|-\rangle = |1\rangle$ using Eq. (2.54), and not the matrix representation. To illustrate, we will show how $H|+\rangle = |0\rangle$ can be shown:

$$H|+\rangle = H\left(\frac{1}{\sqrt{2}}|0\rangle + \frac{1}{\sqrt{2}}|1\rangle\right) \tag{2.55}$$

$$= \frac{1}{\sqrt{2}}H|0\rangle + \frac{1}{\sqrt{2}}H|1\rangle \tag{2.56}$$

$$= \frac{1}{\sqrt{2}}|+\rangle + \frac{1}{\sqrt{2}}|-\rangle = |0\rangle, \tag{2.57}$$

where the last equality follows from Eq. (2.33).

Finally, the third representation method is the outer-product representation method. Here, we write U as a sum of simple operators of the form $|v\rangle\langle w|$, for different vectors $|v\rangle$ and $|w\rangle$. Recalling the action of an operator of the form $|v\rangle\langle w|$ on some vector $|x\rangle$ from Eq. (2.21), this allows us to write U as sums of such terms. To illustrate, consider, again, the Hadamard, which we now write as:

$$H = |+\rangle\langle 0| + |-\rangle\langle 1|. \tag{2.58}$$

Note, in particular, we wrote it as a sum of $|\text{“output”}\rangle\langle\text{“input”}|$ where we want a particular "input" (e.g., $|0\rangle$) to map to a particular "output" (e.g., $|+\rangle$). This way of thinking will always work if the "input" and "output" are orthonormal vectors.

What are some example qubit (i.e., 2×2) unitary operators that are commonly used? The following are a few that we'll use throughout (besides the Hadamard operator which we've discussed in depth so far):

- *Pauli Z Operator*:

$$Z = \sigma_Z = |0\rangle\langle 0| - |1\rangle\langle 1| \tag{2.59}$$

- *Pauli X Operator*:

$$X = \sigma_X = |1\rangle\langle 0| + |0\rangle\langle 1| \tag{2.60}$$

- *Pauli Y Operator*:

$$Y = \sigma_Y = i\,|1\rangle\langle 0| - i\,|0\rangle\langle 1| \tag{2.61}$$

There are several others that we'll use and we'll introduce them as needed. Of course, we also need to consider unitary operators on two (or more) qubits. However, before describing that, we will need to show how to formally model multiple qubits. This is done using the tensor product, which we'll define and discuss in Section 2.5.

2.4.1 Unitary change in basis

Let's say we're working in an n-dimensional space and let $\{|v_1\rangle, \cdots, |v_n\rangle\}$ be an orthonormal basis. Let U be some unitary operator acting on this space. Then, abstractly, we may write the output of $U|v_i\rangle$ as simply being some vector we'll call $|w_i\rangle$. Namely, $|w_i\rangle = U|v_i\rangle$. What can we say about $|w_i\rangle$? As it turns out, the vectors $\{|w_1\rangle, \cdots, |w_n\rangle\}$ also form an orthonormal basis.

Indeed, unitary operators *must preserve inner norm*. For any vectors $|x\rangle$ and $|y\rangle$, let $|v_x\rangle = U|x\rangle$ and $|v_y\rangle = U|y\rangle$. Then, it holds that:

$$\langle x|y\rangle = \langle x|I|y\rangle = \langle x|U^*U|y\rangle = \langle v_x|v_y\rangle . \tag{2.62}$$

If those vectors happen to be orthonormal, then:

$$\delta_{i,j} = \langle v_i|v_j\rangle = \langle v_i|I|v_j\rangle = \langle v_i|U^*U|v_j\rangle = \langle w_i|w_j\rangle , \tag{2.63}$$

where $\delta_{i,j}$ is the Kronecker delta function. The above follows since $U^*U = I$ by definition. From the above, it is not difficult to show that all the $|w_i\rangle$ are linearly independent also; thus they form an orthonormal basis.

Exercise 2.14. Show that if $\{|w_1\rangle, \cdots, |w_n\rangle\}$ are orthonormal, then they are linearly independent.

Thus, unitary operators are simply changes in basis from one orthonormal basis to another. Indeed, given any two orthonormal bases, one may always find a unitary operator that maps basis vectors from the first to basis vectors from the second (hint: consider the outer-product formulation as in Eq. (2.21)).

This gives an alternative view to measurements. Consider a state $|\psi\rangle$ and we want to measure in some orthonormal basis $\{|w_1\rangle, \cdots, |w_n\rangle\}$. Then, instead of looking at $|\langle w_i|\psi\rangle|^2$, we can instead apply a unitary operator U which takes $|w_i\rangle$ to the computational state $|i\rangle$. Namely apply $U = \sum_i |i\rangle\langle w_i|$ to $|\psi\rangle$ and then measure in the Z basis.

Why does this work? Well, since $\{|w_i\rangle\}$ is an orthonormal basis, we can write $|\psi\rangle = \sum_i \alpha_i |w_i\rangle$. Then, we know that the probability of observing basis vector $|w_i\rangle$ is simply $|\langle w_i|\psi\rangle|^2 = |\alpha_i|^2$. However, by applying the above operator U, we have:

$$U|\psi\rangle = \sum_i \alpha_i U|w_i\rangle = \sum_i \alpha_i |i\rangle . \tag{2.64}$$

Thus, now, the probability of observing $|i\rangle$ is, again, simply $|\alpha_i|^2$. This gives an alternative method for a basis measurement: instead of measuring in some orthonormal basis, one may first apply a unitary operator and then measure in the standard computational basis. The probability distribution will be identical. Note that the post-measured state will be a computational state in the latter case; however applying U^* will fix that.

Why is this important? From a computational standpoint, we want to argue about "how complicated" a quantum algorithm is. Normally a quantum algorithm consists of applying several "small" unitary operators to a quantum state and then measuring it. However, this measurement may be complicated to describe! So, instead, if we force all algorithms to measure in the computational basis, we waive our hand over that issue – instead the complexity of the measurement will be absorbed into the initial circuitry by attempting to construct a change of basis unitary. This allows us to gauge the complexity of a quantum process, in terms only of the unitary circuitry and not the final measurement.

2.5 Combining systems: tensor product

Let's say we have a quantum state $|\psi\rangle$ and another quantum state $|\phi\rangle$. How do we model the joint system of two states together? This is done, mathematically, using tensor product and is denoted $|\psi\rangle \otimes |\phi\rangle$. The tensor space is a very abstract thing – for our sake we will not be too concerned with all the properties and the rigorous definition of what it means to tensor two vectors or two vector spaces together. Instead, we will rely on the basic terminology needed to analyze basic quantum algorithms and cryptographic protocols, introducing more advanced concepts as needed.

What does $|\psi\rangle \otimes |\phi\rangle$ "look like" mathematically? As it turns out, tensor product obeys many of the natural multiplicative rules you're already used to. If $|\psi\rangle = \alpha|0\rangle + \beta|1\rangle$ and $|\phi\rangle = x|0\rangle + y|1\rangle$, then we may think of $|\psi\rangle \otimes |\phi\rangle$ as the product of the two in the following way:

$$|\psi\rangle \otimes |\phi\rangle = (\alpha|0\rangle + \beta|1\rangle) \otimes (\gamma|0\rangle + \delta|1\rangle) = \alpha\gamma|0\rangle \otimes |0\rangle + \alpha\delta|0\rangle \otimes |1\rangle + \beta\gamma|1\rangle \otimes |0\rangle + \beta\delta|1\rangle \otimes |1\rangle . \tag{2.65}$$

Notice the similarity to, say, polynomial multiplication: if you had a polynomial $f(x, y) = \alpha x + \beta y$ and $g(x, y) = \gamma x + \delta y$, then the polynomial $f(x, y) \times g(x, y)$ is easily seen to be $\alpha\gamma x^2 + \alpha\delta xy + \beta\gamma yx + \beta\delta y^2$. Thinking of "$x$" as $|0\rangle$ and x^2 as $|0\rangle \otimes |0\rangle$, you can begin to see the relation. This may be a handy trick to remember the rules of tensor: for the most part, you can think about multiplying polynomials together. Though, keep in mind, we do not have commutativity, for instance $|0\rangle \otimes |1\rangle \neq |1\rangle \otimes |0\rangle$.

Now, notationally, Eq. (2.65) is incredibly clunky. Tensor product is going to be a fairly common operation and so we don't want to have to write $\otimes$ continually (any more than we want to write $\times$ when doing standard multiplication). So, we often drop the $\otimes$ and simply write $|\psi\rangle\,|\phi\rangle$ to mean $|\psi\rangle \otimes |\phi\rangle$. We may even go so far (and often do) as to write $|\psi, \phi\rangle$ to mean the same thing. Using this notation, we can write the above more succinctly as:

$$|\psi, \phi\rangle = \alpha\gamma\,|0, 0\rangle + \alpha\delta\,|0, 1\rangle + \beta\gamma\,|1, 0\rangle + \beta\delta\,|1, 1\rangle\,, \tag{2.66}$$

which is clearly preferred to the original form. Note that when dealing with multiple 0's and 1's inside a single ket (indicating multiple qubits together), we often may even forgo writing the comma; namely the above may be: $|\psi, \phi\rangle = \alpha\gamma\,|00\rangle + \alpha\delta\,|01\rangle + \beta\gamma\,|10\rangle + \beta\delta\,|11\rangle$. Note this only makes sense when we specifically have 0's and 1's; dropping the comma does not work well for, say, $|\psi\phi\rangle$ which we will never, deliberately, do.

Let's think, now, more abstractly, about what it would mean to have two qubits. A single qubit can be in a state $|0\rangle$, $|1\rangle$, or a superposition of both as discussed earlier. Then, it only makes sense that if we had *two* qubits, those two qubits could be in a state $|0, 0\rangle$, $|0, 1\rangle$, $|1, 0\rangle$, or $|1, 1\rangle$. However, as it turns out, these two qubits can also be in a superposition of all four states:

$$\alpha_{0,0}\,|0, 0\rangle + \alpha_{0,1}\,|0, 1\rangle + \alpha_{1,0}\,|1, 0\rangle + \alpha_{1,1}\,|1, 1\rangle\,. \tag{2.67}$$

The normalization rules apply as before, forcing the constraint $|\alpha_{0,0}|^2 + \cdots + |\alpha_{1,1}|^2 = 1$.

Tensor product, as mentioned already, obeys a lot of the multiplication rules we're used to. In particular:

- $|\psi\rangle \otimes (|\phi\rangle + |\mu\rangle) = |\psi, \phi\rangle + |\psi, \mu\rangle$ for any $|\psi\rangle$, $|\phi\rangle$, and $|\mu\rangle$ (note that $|\phi\rangle$ and $|\mu\rangle$ have to live in the same underlying vector space). That is, tensor obeys distributive property.
- $\alpha(|\psi, \phi\rangle) = (\alpha\,|\psi\rangle) \otimes |\phi\rangle = |\psi\rangle \otimes (\alpha\,|\phi\rangle)$ for any scalar $\alpha \in \mathbb{C}$ and any two states $|\psi\rangle$ and $|\phi\rangle$. Namely scalars can be absorbed into any of the underlying vectors and the state is the same; equivalently, scalars may be "pulled out" of a tensor.

To demonstrate, consider $|0\rangle \otimes \frac{1}{\sqrt{2}}\,|0\rangle + |1\rangle \otimes \frac{1}{\sqrt{2}}\,|1\rangle$. We may "pull out" the scalar inside the inner system and take advantage of the distributive property to yield the equivalent state $\frac{1}{\sqrt{2}}\,|0, 0\rangle + \frac{1}{\sqrt{2}}\,|1, 1\rangle$.

If the reader likes Cartesian coordinate vectors, then there is a way to describe a tensor product of two vectors as a Cartesian coordinate vector itself. In particular, given:

$$|\psi\rangle = \begin{pmatrix} \alpha_1 \\ \alpha_2 \\ \vdots \\ \alpha_n \end{pmatrix} \qquad |\phi\rangle = \begin{pmatrix} \beta_1 \\ \beta_2 \\ \vdots \\ \beta_m \end{pmatrix}, \tag{2.68}$$

namely two states, one of dimension n and the other of dimension m, then the tensor product $|\psi, \phi\rangle$ may be represented as a Cartesian vector using the Kronecker Product. For this, we essentially multiply the right-most state "into" the left as follows:

$$|\psi,\phi\rangle = |\psi\rangle \otimes |\phi\rangle = \begin{pmatrix} \alpha_1 \\ \alpha_2 \\ \vdots \\ \alpha_n \end{pmatrix} \otimes |\phi\rangle = \begin{pmatrix} \alpha_1 |\phi\rangle \\ \alpha_2 |\phi\rangle \\ \vdots \\ \alpha_n |\phi\rangle \end{pmatrix} = \begin{pmatrix} \alpha_1\beta_1 \\ \alpha_1\beta_2 \\ \vdots \\ \alpha_1\beta_m \\ \alpha_2\beta_1 \\ \alpha_2\beta_2 \\ \vdots \\ \alpha_2\beta_m \\ \vdots \\ \alpha_n\beta_1 \\ \alpha_n\beta_2 \\ \vdots \\ \alpha_n\beta_m \end{pmatrix} \tag{2.69}$$

where, above, each $\alpha_i |\phi\rangle$ represents an m-dimensional vector, namely the scalar product of α_i with $|\phi\rangle$. Note that the dimension of the new system is $n \cdot m$; that is the dimension of multiple quantum states goes up *multiplicatively.*

Note that tensor product, along with Kronecker product, can be applied multiple times which will be used to model three or more systems. The rules apply in the same way. Note that the dimension increases multiplicatively. Thus, if we have one qubit, the dimension of the state space is two; if we have two qubits, the dimension is $2 \times 2 = 4$; with three qubits we need a state space of $2 \times 2 \times 2 = 8$ and so on. The state space of n-qubits is 2^n-dimensional.

2.5.1 Linear operators and tensor product

Tensor product has several nice properties when it comes to linear operators and their application to tensor states. First, let's consider the inner product. Let $|\psi\rangle = |a\rangle |b\rangle$ and $|\phi\rangle = |x\rangle |y\rangle$ be two vectors living in the same tensor space. The adjoint (or in our case, equivalently, the conjugate transpose since everything is finite dimensional) of $|\psi\rangle$ is simply: $\langle\psi| = \langle a| \langle b|$ (which may also be denoted $\langle a, b|$ or $\langle a| \otimes \langle b|$; all of these are equivalent). The interested reader may verify this identity by using the Kronecker product. The inner product, then, is $\langle\psi|\phi\rangle = (\langle a| \langle b|)(|x\rangle |y\rangle)$. Ultimately, we are applying a linear operator (the left $\langle a| \langle b|$) to the state $|x\rangle |y\rangle$. Whenever one has an operator $A \otimes B$ applying to a vector $|x\rangle |y\rangle$, you simply need to remember to apply the left-most operator to the left-most state and the right-most operator to the right most state. In general, if there are multiple tensors, say, $A_1 \otimes A_2 \otimes \cdots \otimes A_n$ applied to $|x_1\rangle \otimes |x_2\rangle \otimes \cdots \otimes |x_n\rangle$, then one will apply A_i to $|x_i\rangle$. That is:

$$A_1 \otimes A_2 \otimes \cdots \otimes A_n |x_1\rangle \otimes |x_2\rangle \otimes \cdots \otimes |x_n\rangle = (A_1 |x_1\rangle) \otimes (A_2 |x_2\rangle) \otimes \cdots \otimes (A_n |x_n\rangle). \tag{2.70}$$

Of course, the map acts linearly on superpositions.

Returning to the inner product, therefore, we find:

$$\langle\psi|\phi\rangle = (\langle a|\langle b|)(|x\rangle|y\rangle) = \langle a|x\rangle\langle b|y\rangle. \tag{2.71}$$

Thus, the inner product of two simple-tensors living in a tensor space is simply the product of the inner product of each individual subspace.

As a more concrete example, let's say $|\psi\rangle = |0,+\rangle$ and $|\phi\rangle = |-,1\rangle$. Then:

$$\langle\psi|\phi\rangle = \langle 0,+|-,1\rangle = \langle 0|-\rangle\langle +|1\rangle = \frac{1}{\sqrt{2}} \times \frac{1}{\sqrt{2}} = \frac{1}{2}. \tag{2.72}$$

Note, be careful when taking the conjugate transpose of a tensor: given the ket $|a,b\rangle$, then its adjoint operator is $\langle a,b|$; the order of the subspaces does not change: "a" is still the left-most, for instance.

What if we have states that are not simple tensors; namely what if we have more complicated states such as: $|\psi\rangle = \alpha|0,1\rangle + \beta|+,-\rangle$ and $|\phi\rangle = \gamma|+,+\rangle + \delta|0,0\rangle$? Then we may simply take advantage of the distributive property and linearity:

$$\begin{aligned}\langle\psi|\phi\rangle &= (\alpha\langle 0,1| + \beta\langle +,-|)(\gamma|+,+\rangle + \delta|0,0\rangle)\\ &= \alpha\gamma\langle 0,1|+,+\rangle + \alpha\delta\langle 0,1|0,0\rangle + \beta\gamma\langle +,-|+,+\rangle + \beta\delta\langle +,-|0,0\rangle\end{aligned}$$

From the above, one may use the earlier rule for simple tensors to solve each individual case. We conclude:

$$\langle\psi|\phi\rangle = \alpha\gamma\frac{1}{2} + \alpha\delta\cdot 0 + \beta\gamma\cdot 0 + \beta\delta\frac{1}{2} = \frac{1}{2}(\alpha\gamma + \beta\delta). \tag{2.73}$$

The above rule of applying linear operators written in tensor form applies to any linear operator not just inner product. For example, if we have two unitary operators U_1 and U_2, each of which act on a single qubit, then given a two qubit state $|\psi\rangle = \sum_{a,b}\alpha_{a,b}|a,b\rangle$, we can construct a unitary operator $U = U_1 \otimes U_2$ and its action will be as follows:

$$U|\psi\rangle = \sum_{a,b}\alpha_{a,b}U|a,b\rangle = \sum_{a,b}\alpha_{a,b}U_1\otimes U_2|a,b\rangle = \sum_{a,b}\alpha_{a,b}U_1|a\rangle\otimes U_2|b\rangle. \tag{2.74}$$

Namely, U_1 will apply to the first qubit while U_2 will apply to the second. Note that, given matrix representations of U_1 and U_2, one may use the Kronecker product to construct a matrix representation of U; however, we advise the reader not to bother and to try and think in terms of Dirac notation as matrix representations, though useful in some applications, quickly become tedious as the number of qubits increases.

2.5.2 Entanglement

Entanglement is one of the fundamental resources in quantum computation and communication, and is a fascinating topic on its own. At a high level, two particles, one owned by Alice and one owned by Bob, are "entangled" if actions on one affect the state of the other, even if they are spatially separated. Even if they are separated by so great a distance so as to rule out any communication between the two particles. Formally, consider a two particle state $|\psi\rangle_{AB}$. We say it is *separable* if we can write $|\psi\rangle_{AB} = |a\rangle_A \otimes |b\rangle_B$; namely, if we can write the two-particle state as a simple tensor of two component systems. As an example, the state $|\psi\rangle_{AB} = \frac{1}{2}(|00\rangle_{AB} + |01\rangle_{AB} + |10\rangle_{AB} + |11\rangle_{AB})$ is separable, since we can write it as $|\psi\rangle_{AB} = |+\rangle_A \otimes |+\rangle_B = |+,+\rangle_{AB}$.

States that are separable are those which admit a very clear disconnection between each party; in particular, for a separable state, any action Alice performs on her system (e.g., a unitary operation on the A system, or a measurement of it) will clearly not affect the state of Bob's particle. Indeed, just try tracing the execution of such an action: the unitary will be of the form $U_A \otimes I_B$ which will not affect Bob; similarly, her measurement operators will apply an identity to Bob's particle and thus not affect his state; regardless, the state will remain $|a'\rangle_A \otimes |b\rangle_B$, where $|a'\rangle_A$ is the altered state after Alice interacts with her particle (i.e., $|a'\rangle = U_A|a\rangle$); Bob's system remains the same.

Opposite to separable states are *entangled states* – i.e., an entangled system is one which is not separable. An important example of this is the so-called Bell state $|\Phi^+\rangle = \frac{1}{\sqrt{2}}(|00\rangle_{AB} + |11\rangle_{AB})$. Why is this state entangled? And why is it the case that now, Alice's operations will affect Bob? Let's answer the first question first. We can show this is an entangled state by contradiction.

Assume, for contradiction, $|\Phi^+\rangle$ is not entangled; therefore it must be separable. If this is the case, we can write it as a simple tensor: $|\Phi^+\rangle = |a\rangle \otimes |b\rangle$ for some qubits $|a\rangle$ and $|b\rangle$. The most general way we can write these two individual qubits is in the form $|a\rangle = \alpha|0\rangle + \beta|1\rangle$ and $|b\rangle = \gamma|0\rangle + \delta|1\rangle$. In this case, we have:

$$|a\rangle \otimes |b\rangle = \alpha\gamma|00\rangle + \alpha\delta|01\rangle + \beta\gamma|10\rangle + \beta\delta|11\rangle. \tag{2.75}$$

For the above to possibly be equal to $|\Phi^+\rangle$, it must be that either $\alpha = 0$ or $\delta = 0$ (to remove the $|01\rangle$ term which never shows up in $|\Phi^+\rangle$). But if $\alpha = 0$, then there is no $|00\rangle$ term! Therefore $\alpha \neq 0$ and it must be $\delta = 0$; but in that case, there is no $|11\rangle$ term. Thus, no solution exists and therefore $|\Phi^+\rangle$ is not separable, but entangled. Note that entangled states cannot be written as a single simple tensor; though, of course, they may be written as the sum of simple tensors!

Exercise 2.15. Consider the following:

1. $\frac{1}{2}(|00\rangle - |01\rangle + |10\rangle - |11\rangle)$
2. $\frac{1}{2}(|00\rangle + |01\rangle + |10\rangle - |11\rangle)$

Show that the first is separable but the second is entangled. To show a state is separable, write it as a simple tensor $|a\rangle \otimes |b\rangle$; to show it is entangled, prove by contradiction that such a separation does not exist.

Now, let's assume Alice applies a unitary operation to her qubit; let's say the Hadamard operator H. Bob does nothing to his; thus, globally, we model this as the operator $H \otimes I_B$

and the state evolves to $H \otimes I_B |\Phi^+\rangle$. We claim that Alice performing her operation and Bob doing nothing, is identical to Bob applying H and Alice doing nothing! Clearly this is not the case with a separable system (please convince yourself of that, if you're not already, by tracing the execution of the operator to, say, $|{+}{+}\rangle_{AB}$; you will get $|0+\rangle$ which is clearly not the same as $|{+}0\rangle$). Let's trace the execution of the operation on the entangled Bell state:

$$H \otimes I_B(\frac{1}{\sqrt{2}}|00\rangle + \frac{1}{\sqrt{2}}|11\rangle) = \frac{1}{\sqrt{2}}|{+}0\rangle + \frac{1}{\sqrt{2}}|{-}1\rangle \tag{2.76}$$

$$= \frac{1}{2}(|00\rangle + |10\rangle + |01\rangle - |11\rangle) = \frac{1}{\sqrt{2}}(|0+\rangle + |1-\rangle) \tag{2.77}$$

$$= I_A \otimes H(\frac{1}{\sqrt{2}}|00\rangle + \frac{1}{\sqrt{2}}|11\rangle) \tag{2.78}$$

Thus, unlike the separable case, actions on one particle affect the other. In this case, Alice applying a Hadamard to her qubit is equivalent to Bob applying it to his, even if he never performed any action on his system. This holds even if the two particles are spatially separated (potentially by billions of miles even in theory).

Exercise 2.16. Show that $H \otimes H |\Phi^+\rangle = |\Phi^+\rangle$. You can do this directly (work out the algebra) or you can use the above identity and the fact that $HH = I$. Also, and related, show that $|\Phi^+\rangle$ may be written in the X basis as $|\Phi^+\rangle = \frac{1}{\sqrt{2}}(|{+}{+}\rangle + |{-}{-}\rangle)$.

2.5.3 Joint and partial measurements

Besides operators, measurements of one particle will affect the other if they are entangled. Let's look again at the Bell state $|\Phi^+\rangle = \frac{1}{\sqrt{2}}|00\rangle + \frac{1}{\sqrt{2}}|11\rangle$. Without going into formal measurements (don't worry, we will!), let's just use our understanding of superposition states to try and determine the possible outcomes of different measurements. First, let's assume we have both particles in the same physical location; this would allow us to measure both systems simultaneously. If we were to measure in the two-qubit Z basis, spanned by $|00\rangle$, $|01\rangle$, $|10\rangle$, $|11\rangle$, it is not difficult to see that we would observe $|00\rangle$ with probability $1/2$; otherwise, we would observe $|11\rangle$. This can be derived from the inner-product formulation of measurements, or using measurement operators, both of which were discussed in Section 2.3. Such a measurement on the entire state is typically called a *joint measurement*, and the rules discussed earlier apply (one must simply work over the new tensor state space).

Now, let's consider *partial measurements*, where only some, but not all, of the underlying systems are measured. For example, maybe two particles are spatially separated, perhaps by a vast distance, thus prohibiting their joint measurement and forcing each particle to be measured individually. Or maybe one just wants to look at part of a system, but not the entire state.

Before going into the formalism, let's think about Alice making a measurement of the first system at a conceptual level. She will see either $|0\rangle$ or $|1\rangle$, each with probability $1/2$.

Note since she is measuring only a single particle, she can only observe a two-dimensional outcome from such a basis measurement and the probability comes from squaring the appropriate coefficient in this specific case (in the general case it will be a bit more involved, but not significantly, as we'll soon discuss). Alternatively, if Bob had measured his particle he would observe the same possible outcomes with the same distribution.

However, and importantly, a measurement, even a partial one of spatially separated systems, causes state collapse of the entire, *global or joint* system! That is, in the first case, if Alice measures and observes $|0\rangle$, the entire system must collapse to a state that could agree with the observed reality. Since Alice saw $|0\rangle$, Bob's particle must collapse to $|0\rangle$ also, causing the joint state to become $|00\rangle$. Any subsequent measurement by Bob will always be $|0\rangle$. Similarly, if Alice observed $|1\rangle$, the entire state collapses to $|11\rangle$ and so Bob will subsequently always observe $|1\rangle$. Thus, even if the two particles are separated by vast distances, Alice and Bob will always get a correlated outcome, randomly either $|0\rangle$ or $|1\rangle$ (for both).

It must be noted that Alice and Bob are *not* communicating faster than light! Indeed, Alice gets a random outcome $|0\rangle$ or $|1\rangle$ same as Bob. While they know their random outcomes will be fully correlated, this does not constitute communication – namely, the distribution or outcomes cannot be "biased" based on some intended message to be sent.

While the above high-level discussion suffices for "simple" states (for example, repeat the above for the state $\frac{1}{\sqrt{2}}|01\rangle + \frac{1}{\sqrt{2}}|10\rangle$ – you should note that, now, Alice and Bob will always get opposite outcomes), it does not work well for more complex states. For instance, consider the *tripartite* state $\alpha\,|00e_0\rangle_{ABC} + \beta\,|10e_1\rangle_{ABC} + \gamma\,|+1e_2\rangle_{ABC}$, where the $|e_i\rangle$ states are sub-normalized. Now, what is the probability of observing $|0\rangle$ if a measurement of the A system, in the Z basis, is performed? What is the probability of observing $|-\rangle$ if a measurement of the B system, in the X basis is performed? What about general measurements (e.g., measurement operators)? We'll actually answer the last question, which answers the first two (since basis measurements can be written as measurement operators).

Consider two-party systems again (this generalizes to additional systems). Let's say Alice wishes to measure her system using some measurement $\{M_x\}_x$, where M_x are measurement operators (see Section 2.3.2), that act on Alice's state space (i.e., on Alice's system, not the entire joint state). Bob does not measure. In this case, we can define new measurement operators $\{\widetilde{M}_x\}_x$, where $\widetilde{M}_x = M_x \otimes I_B$. That these form a valid set of measurement operators is straightforward to show:

$$
\begin{aligned}
\sum_x \widetilde{M}_x^* \widetilde{M}_x &= \sum_x (M_x \otimes I_B)^* \otimes (M_x \otimes I_B) \\
&= \sum_x (M_x^* \otimes I_B^*)(M_x \otimes I_B) \\
&= \sum_x M_x^* M_x \otimes I_B = I_A \otimes I_B = I_{AB}.
\end{aligned}
$$

Above, we used the fact that $(A+B)^* = A^* + B^*$, along with the fact that $I_A \otimes I_B$ is the identity operator on the entire joint space, namely I_{AB}. Since we have a valid measurement operator to represent this partial measurement, all the rules from Section 2.3.2, describing the probability distribution of the measurement and any post-measured state, can be applied. Note that, on observing x, the post measured state becomes:

$$\frac{\widetilde{M}_x \left|\psi\right\rangle_{AB}}{\sqrt{p(x)}} = \frac{M_x \otimes I_B \left|\psi\right\rangle_{AB}}{\sqrt{p(x)}} \tag{2.79}$$

Sometimes, writing out explicit measurement operators is more work than needed. For example, in the simple case of $\alpha\left|0\right\rangle + \beta\left|1\right\rangle$, making a Z basis measurement using measurement operators is more work than simply looking at the probability amplitude and squaring it (though, of course, the end result is identical). When making a basis measurement of one system (i.e., a partial basis measurement), then there is a simpler approach. Let's assume we have a two-party state $\left|\psi\right\rangle_{AB}$ and we want to make a measurement of the A system in some orthonormal basis $\{\left|v_1\right\rangle, \cdots, \left|v_d\right\rangle\}$. We could define measurement operators $M_i = \left|v_i\right\rangle\left\langle v_i\right|_A \otimes I_B$ and work out the algebra. Alternatively, if we write $\left|\psi\right\rangle_{AB}$ in the form:

$$\left|\psi\right\rangle_{AB} = \sum_{i=1}^{d} \left|v_i\right\rangle_A \otimes \left|b_i\right\rangle_B, \tag{2.80}$$

where the $\left|b_i\right\rangle_B$ states are arbitrary (potentially sub-normalized) vectors in the B state space, then it will follow that the probability of observing outcome $\left|v_i\right\rangle_A$ if a partial measurement is made on A, is simply $p(i) = \left\langle b_i | b_i\right\rangle$. The post measured state is $\left|v_i\right\rangle_A \left|b_i\right\rangle_B / \sqrt{p(i)}$. Note that, since $\{\left|v_i\right\rangle\}$ is an orthonormal basis of the A system, any state $\left|\psi\right\rangle_{AB}$ can be written in this form.

Exercise 2.17. Show that the "simple" form of a partial basis measurement, discussed above, will result in the correct distribution and post measured state, as dictated by the measurement rules using measurement operators.

2.5.4 No-cloning theorem

We'll close this section with an important result, namely the *no-cloning theorem*. Specifically, it can be shown that no unitary operator U exists which can perfectly clone an arbitrary quantum state. (Actually, one can prove that no such *quantum operation* exists, and such an operation may consist of measurements, classical post-processing, and additional quantum ancillas.) Formally, let $\left|\psi\right\rangle$ be a quantum state, then to clone this state, we mean apply some quantum circuit (a unitary U) such that:

$$U\left|\psi\right\rangle\left|0\right\rangle = \left|\psi\right\rangle\left|\psi\right\rangle, \tag{2.81}$$

where the input is the original state to clone, $\left|\psi\right\rangle$ and some "blank" ancilla initialized to the $\left|0\right\rangle$ state (this space is where the clone will live, after the operation). The fact that this cannot be done, follows immediately from linearity. Assume, for contradiction, that such an operation exists. Then, it must be that for two different states $|\psi\rangle$ and $|\phi\rangle$, it holds:

$$U|\psi\rangle|0\rangle = |\psi\rangle|\psi\rangle$$
$$U|\phi\rangle|0\rangle = |\phi\rangle|\phi\rangle$$

However, let $|\tau\rangle = \alpha|\psi\rangle + \beta|\phi\rangle$ be some other quantum state, a superposition of both. If U is really a cloning operation, it should hold that $U|\tau\rangle|0\rangle = |\tau\rangle|\tau\rangle$. However, due to linearity, it actually holds that:

$$\begin{aligned} U|\tau\rangle|0\rangle &= U(\alpha|\psi\rangle + \beta|\phi\rangle)|0\rangle = \alpha U|\psi\rangle|0\rangle + \beta U|\phi\rangle|0\rangle \\ &= \alpha|\psi\rangle|\psi\rangle + \beta|\phi\rangle|\phi\rangle \end{aligned}$$

A simple calculation shows that

$$|\tau\rangle|\tau\rangle = \alpha^2|\psi\rangle|\psi\rangle + \beta^2|\phi\rangle|\phi\rangle + \alpha\beta|\psi\rangle|\phi\rangle + \beta\alpha|\phi\rangle|\psi\rangle. \tag{2.82}$$

This leads to a contradiction. Of course, one *can* clone basis states *if* one knows the basis the state is in. However, if the basis is unknown, then no cloner can exist which works for every state. This will be very important, later, when discussing security of various quantum cryptographic protocols.

Exercise 2.18. Let $|\psi\rangle$ and $|\phi\rangle$ be two orthogonal quantum states. Show that there exists a unitary U, specific for these two states, such that:

$$U|\psi\rangle|0\rangle = |\psi\rangle|\psi\rangle$$
$$U|\phi\rangle|0\rangle = |\phi\rangle|\phi\rangle$$

Explain how this is not a contradiction with what we just discussed.

2.6 Quantum information theory

We now turn to the subject of quantum information theory, along with a discussion of some classical information theory to start off with. This is a broad, and truly fascinating field, and we can only scratch the surface here. There are many, more detailed, texts on this subject, see, for instance, [15]. Here, we only cover the tip of the iceberg, introducing topics that will be needed later.

Let's first consider basic Shannon entropy. Imagine the following scenario: there is a source device that will transmit some message with some probability. In particular, let's say the source will choose to send some message m_i, from a set of possible messages $\{m_1, \cdots, m_n\}$, choosing message m_i with some probability p_i. Now, imagine you are the receiver who knows the possible set of messages, and the probability distribution. You then receive a particular message m_i. How much information did you learn? This is the question (amongst others) Claude Shannon asked when developing his entropy function.

As an example, let's say there are two possible messages m_0 and m_1 and that the source sends m_0 with probability 0.99 (most likely) while sending message m_1 with probability .01 (very rarely). Now, let's say you, the receiver, receive message m_0. Well, that really didn't tell you much – after all, you knew it was the most likely message to be sent. On the other hand, if you receive message m_1, now that is very surprising and potentially gives a lot of information as that message was highly unlikely to arrive! This is the basic idea behind Shannon entropy: it will measure how much information you receive, on average (or, alternatively, "how surprised you are") over the messages.

There are a few ways to define and build Shannon entropy – let's work with the *axiomatic approach* (see, for instance, [3]). In particular, we want to consider, first, how much information we get from a message sent with probability p. First, we will note that, since the messages are labeled m_1 through m_n, the message itself is not so important as the index is. If we know the list of messages and send "5" we know we mean m_5. Thus, the only important thing to consider, then, is the probability distribution and, so, the information function will be a function only of probability values p_i. (This may seem strange in practice when the message itself matters – however, if we know the messages and only care about which message we're receiving, then it is fairly clear why we can ignore the message labels and only consider the probability value of a message being sent.)

Let $\mathcal{I}(p)$ be the information you attain on receipt of some message sent with probability p. Axiom (1): we will want information to be non-negative, thus $\mathcal{I}(p) \geq 0$ for all $p \in [0, 1]$. Next, axiom (2), if a message shows up with probability one, then there is no information gained, thus $\mathcal{I}(1) = 0$. Third (3), if two independent events occur, sending message m_i followed by m_j, choosing both independently (thus, their joint distribution is simply $p_i p_j$), the information gained is additive, namely $\mathcal{I}(p_i p_j) = \mathcal{I}(p_i) + \mathcal{I}(p_j)$. Finally, (4) is that the information function should be continuous – namely, small changes to the probability value should not result in drastic changes in information. It can be shown that the above four axioms imply $\mathcal{I}(p) = -\log_b p$, for some base b. The base will be the unit of measurement – here we will almost always have base two as we measure information in bits.

Now, consider the random variable M which chooses message i with probability p_i. The Shannon entropy of this random variable, denoted $H(M)$, is the expected information gain, namely:

$$H(M) = \sum_{i=1}^{m} p_i \mathcal{I}(p) = -\sum_i p_i \log_2 p_i. \tag{2.83}$$

Above, we are simply assuming the base is two since, as mentioned, that is the case we care about in this book. Note that, sometimes, instead of writing $H(M)$, we may write $H(p_1, p_2, \cdots, p_n)$. When $n = 2$ (i.e., there are only two outcomes or messages), then there is only one probability value to consider, since $p_2 = 1 - p_1$. In this case, we often write $h(p)$ to be the *binary Shannon entropy* – that is, the Shannon entropy of the two outcome experiment, with message one being sent with probability p and message two being sent with probability $1-p$. Namely:

$$h(p) = -p\log p - (1-p)\log(1-p). \tag{2.84}$$

Note that $h(p) = h(1-p)$ which may be easily verified. All logarithms in this section will be base two, unless otherwise specified.

While the above axiomatic construction is helpful, it is not immediately clear how this function can be useful for anything. As it turns out, it is highly important in a number of applications. Let's take, for instance, Shannon's *noiseless coding theorem* [12,8]. Let's say a source is outputting a string of symbols (messages) with some probability distribution. Call this random variable M_1. Then, after it chooses a symbol to send, it repeats, choosing another symbol with the same probability distribution, independently of the first. Call this (independent and identical) random variable M_2. And so on producing an infinite series of symbols $M_1M_2\cdots$ each one chosen independently and identically, of the others. Shannon asked, "what are the minimal number of bits that are needed to store the transmitted sequence so that it can be perfectly reconstructed later?" As it turns out the number of bits per symbol is exactly $H(M_i)$. As it turns out, this is the best one can do.

As a concrete example, say the source produces one bit (a zero or one) every round. If the bit chosen is zero with probability 1/2 (otherwise it is one), then you will need $H(\frac{1}{2},\frac{1}{2}) = 1$ bit per round to store the message. Not too surprising. However, now let's say that bit zero is chosen with probability 1/10 (so, on average nine times out of ten, the source is sending a "1"). It should be intuitive that we can "compress" the received message in some way. Shannon's entropy can show exactly how much we can compress by – in this case, we only need $H(1/10, 9/10) = 0.469$ bits per round.

Moving to slightly more complex domains, consider now two random variables X and Y which may, or may not, be independent. We can talk about the *joint entropy*, denoted $H(XY)$, and it is defined in the natural way as:

$$H(XY) = -\sum_{x,y} p(x,y)\log p(x,y), \tag{2.85}$$

where $p(x,y)$ is the joint distribution of X taking the value x and Y taking the value y. If X and Y are independent, then:

$$H(XY) = H(X) + H(Y). \tag{2.86}$$

We can also talk about *conditional entropy*, namely the entropy of X conditioned on knowing the outcome of Y. This is denoted $H(X|Y)$ and is defined to be:

$$H(X|Y) = H(XY) - H(Y). \tag{2.87}$$

You can think of it as the amount of information in both X and Y, subtracted by whatever information is in Y. It measures how much uncertainty we have on X, given that we know the outcome of Y. If X and Y are independent, then $H(X|Y) = H(X)$ (which you can prove in Exercise 2.19). If X and Y are completely correlated, then $H(X|Y) = 0$ since, given Y, you know exactly what X is. Conditional entropy will be very important later, when we discuss the security of QKD. In particular, we will be interested in how much uncertainty an adversary has on Alice or Bob's "key" – even if the adversary has some side information system that may be correlated with Alice or Bob.

Exercise 2.19. Let X and Y be independent random variables. Prove that $H(X|Y) = H(X)$. Similarly, show $H(Y|X) = H(Y)$. Why does this make intuitive sense?

Shannon entropy works well for classical random variables – that is, variables that take classical values based on classical probability distributions. But, now, how do we extend all of this to the quantum setting? For that, we will first introduce *von Neumann entropy*, followed by *quantum min entropy*. We'll first discuss von Neumann entropy as it is the most similar to Shannon entropy (and, in fact, is often called the "quantum analogue" of classical Shannon entropy). Later we'll move on to min entropy which, though more complicated to compute in general, is actually more useful from a cryptographic perspective. Before we can discuss any of these notions, however, we must first discuss *density operators*.

2.6.1 Density operators

Up to this point, we have been working with *pure states* – i.e., systems that are completely self-contained and can be described, entirely, by their state vector. However, there are some instances where this is not the case. As a very basic example, tying well with the discussion above on Shannon entropy, consider a source device that creates quantum states probabilistically. In particular, the source will prepare state $|\psi_i\rangle$ with some probability p_i. Such a system cannot be described, naturally, using state vector formalism – especially when it comes time to describe exactly what the state of the system is when the receiver actually receives something. This is because the system that is sent is no longer "closed" – instead, it depends on the state of an external system (in this case the source device). This is where density operator formalism is useful – it can describe everything we have done so far (namely, pure states), but also much more. In general, a *quantum system* is a density operator which may, or may not, be pure. Note that, a state which is not pure is said to be *mixed* or a *mixed state.*

Density operator formalism allows us to describe scenarios such as the above (along with others – for instance, how do we describe an entangled system where we only have access to one particle? Density operators can do that too!). Given the above example of a source producing some state $|\psi_i\rangle$ with probability p_i, the density operator describing that system is written:

$$\rho = \sum_i p_i |\psi_i\rangle \langle\psi_i|, \tag{2.88}$$

where $|\psi_i\rangle \langle\psi_i|$ is the outer product of $|\psi_i\rangle$ (thus, it is a square matrix or operator).

How do unitary operators work on a density operator? Well, going back to our source producing $|\psi_i\rangle$ with probability p_i (this is called an *ensemble of states*), let's imagine that the state produced gets sent to a "box" which implements some unitary U. Now, recall, the source box really is producing one of the $|\psi_i\rangle$ (it's not producing a superposition of different $|\psi_i\rangle$'s, nor is it producing multiple $|\psi_i\rangle$'s). Thus, at the output end of the "box" we will have state $U|\psi_i\rangle$ with probability p_i. Let $|\phi_i\rangle = U|\psi_i\rangle$. Thus, our density operator, describing the state after the unitary is performed, will be:

$$\sigma = \sum_i p_i |\phi_i\rangle\langle\phi_i| = \sum_i p_i U|\psi_i\rangle\langle\psi_i| U^* = U\rho U^*. \tag{2.89}$$

Above, we used the fact that $\langle\phi_i| = (U|\psi_i\rangle)^* = \langle\psi_i|U^*$. Thus, to apply a unitary operator to a density operator ρ, we multiply U on the left and U^* on the right.

What about measurements? Well, let $\{M_x\}$ be some measurement operators (we can introduce this using projective basis measurements, first, however we might as well go to the end goal and discuss general measurements, which will include the basis measurement case). Let $p(x|i)$ be the probability of observing outcome x given that the input state to the measurement box is $|\psi_i\rangle$, namely:

$$p(x|i) = \langle\psi_i|M_x^* M_x|\psi_i\rangle. \tag{2.90}$$

Now, it can be shown that $tr(M_x^* M_x|\psi_i\rangle\langle\psi_i|) = \langle\psi_i|M_x^* M_x|\psi_i\rangle$ using the *Gram Schmidt process* – see Exercise 2.20. Thus, taking into account the state is $|\psi_i\rangle$ with probability p_i, we may compute the total probability of observing x as simply:

$$p(x) = \sum_i p_i p(x|i) = \sum_i p_i tr(M_x^* M_x|\psi_i\rangle\langle\psi_i|) = tr(M_x^* M_x \rho). \tag{2.91}$$

In general, it can be shown, using the measurement rules from Section 2.3, along with basic probability theory, that the post-measured state, conditioned on observing "x" is the mixed state:

$$\rho_x = \frac{M_x \rho M_x^*}{tr(M_x^* M_x \rho)}. \tag{2.92}$$

Exercise 2.20. Let A be an operator and $|\psi\rangle$ be some vector. Show that:

$$tr(A|\psi\rangle\langle\psi|) = \langle\psi|A|\psi\rangle. \tag{2.93}$$

We introduced density operators using the notion of ensembles of quantum states: namely, using them to model random variables over quantum states. As it turns out, the resulting density operator is a positive Hermitian operator of unit trace (this is easy to verify). Hermitian operators were defined in Definition 2.1.3, while positive operators were defined in Definition 2.1.4.

On the other hand, every positive Hermitian operator is also normal, thus diagonalizable; furthermore, it is easy to check that the eigenvalues in this case will be real (since it is Hermitian) and non-negative (since it is also positive semi-definite) and so may be written in the form $\rho = \sum_i \lambda_i |\psi_i\rangle\langle\psi_i|$, where $\lambda_i \geq 0$ and $|\psi_i\rangle$ are the eigenvalues and orthonormal eigenvectors, respectively. The fact that the trace is one, implies that $\sum_i \lambda_i = 1$. Thus, in this sense, every positive Hermitian operator of unit trace can represent a quantum ensemble of some sort. Thus, we define any *quantum state*, or *density operator*, to be simply a Hermitian positive semi-definite operator of unit trace.

One important density operator that will occur often is the so-called *completely mixed state* and is simply the identity operator, acting on the state space, normalized by $1/d$ (where d is the dimension of the state space). Namely, I/d. For instance, with qubits, which are dimension two, the completely mixed state is $\frac{1}{2}I$. By the completeness relation, it holds that $I/2 = \frac{1}{2}|0\rangle\langle 0| + \frac{1}{2}|1\rangle\langle 1|$. Thus, this density operator represents the following ensemble: with equal, uniform, probability, the qubit is either $|0\rangle$ or $|1\rangle$ and, thus, is completely mixed. Note that, by the completeness relation, we could also write $I/2 = \frac{1}{2}|+\rangle\langle +| + \frac{1}{2}|-\rangle\langle -|$ and, so, it represents the system where the qubit is $|+\rangle$ with one-half probability, otherwise it is $|-\rangle$. We could also have chosen any other basis.

Now, the above brings up an interesting fact: Given an experiment (either a real experiment, or an idealized one), one can write the state as a density operator ρ. However, given ρ, there may be many other experiments that can be represented by the same exact operator! For instance, thinking of the previous paragraph, consider an experiment where a source device prepares a qubit in the state $|0\rangle$ or $|1\rangle$, choosing uniformly at random. The resulting density operator is $\rho = I/2$. Now, consider a completely different experiment where a source device prepares either $|+\rangle$ or $|-\rangle$, uniformly at random; the resulting density operator is $\sigma = I/2$. But, even though these are different experiments, the resulting density operator is the same. Namely, $\rho = \sigma$. How can this be and what does it mean? Keep in mind, for the first experiment, producing ρ, the qubit *really is* $|0\rangle$ or $|1\rangle$; whereas in the second experiment, producing σ, the qubit *really is* $|+\rangle$ or $|-\rangle$, two very different states. How can they produce the same density operator?

Well, let's look at it from a receiver's point of view. Let's say, a receiver device is going to perform some "quantum operations" on the received qubit (maybe some unitary operations, and a measurement, for instance). For the sake of illustration, let's just say a simple Z basis measurement is going to be performed on the incoming qubit. In the first case, with experiment one, producing $\rho = \frac{1}{2}|0\rangle\langle 0| + \frac{1}{2}|1\rangle\langle 1|$, if the receiver measures in the Z basis, then the probability of getting a $|0\rangle$ is, of course, $1/2$ (because with $1/2$ probability the qubit really is a $|0\rangle$). Otherwise, the receiver will observe a $|1\rangle$ with equal $1/2$ probability.

Now, let's consider the same exact measurement for the second experiment, producing $\sigma = \frac{1}{2}|+\rangle\langle +| + \frac{1}{2}|-\rangle\langle -|$. In this case, the qubit really is $|+\rangle$ with probability $1/2$, otherwise it is $|-\rangle$. You may verify that the probability of the receiver observing $|0\rangle$ is, again, $1/2$, same as the first experiment. The reader may readily verify that, even if the receiver applies a unitary operator, and then an arbitrary measurement, the statistics will be identical for both ρ and σ, even though they are two different experiments. Thus, even though these are two different physical experiments on the source side, the resulting outcome as far as the receiver is concerned (which is unaware of which experiment was performed), behaves identically for both systems. Density operators describe this behavior: when two operators are the same, even if they were produced by different systems, as far as the "owner" of the operator is concerned (e.g., the receiver in our case above), they will behave identically.

The above is very important from a cryptographic standpoint. Often, we will want to analyze a cryptographic protocol – doing so will allow us to write out a density operator ρ. However, sometimes, it is easier to consider an alternative protocol – this alternative protocol may not be the one that's actually run, however if it happens to produce the same exact state ρ (even though it did so using completely different operations than the first protocol), we may just as well analyze the second protocol. If two density operators are the same, even if they were produced through different means, there is no way to "distinguish" between them and the resulting output of the system will behave identically in both cases. This is perhaps one of the most subtle, and difficult to understand, part of density operators. It may take some time to really sink in how important this concept is.

Another important, and useful, ability of density operators is their ability to model the classical and quantum output of a measurement. Let ρ_A be some quantum state and let's say we want to measure the A system using some measurement operators $\{M_x\}$ for $x \in \chi$. Measurements, as we know, produce classical data (the measurement outcome) and quantum data (the post measured state). We can capture all of this in a single density operator which will model the classical data (the outcome of the measurement), the post measured quantum state, and finally the probability distribution of observing any particular outcome.

Let $\rho_A^{(x)}$ be the post measured state in the event one observes outcome x. Of course, from the discussion earlier (see Eqs. (2.91) and (2.92)), we know $\rho_A^{(x)} = \frac{M_x \rho_A M_x^*}{p_x}$, where p_x is the probability of observing outcome x. Now, we will create a new space, denoted X, which will store the classical outcome. Let $\{|x\rangle \ : \ x \in \chi\}$ be an orthonormal basis for this space. The process will involve: measure the state resulting in a post-measured state, while saving the outcome in this new register. This can all be represented, compactly, as a single density operator which we'll denote ρ_{XA}, as:

$$\rho_{XA} = \sum_{x \in \chi} p_x \, |x\rangle \langle x|_X \otimes \rho_A^{(x)} = \sum_{x \in \chi} |x\rangle \langle x|_X \otimes M_x \rho_A M_x^*. \tag{2.94}$$

A state in the above form, where one system is "classical" (i.e., written with respect to a single orthonormal basis and representing the storage of classical information), while the other is "quantum" (i.e., storing a quantum state), is often called a *classical-quantum state* or cq-state. They are used all throughout quantum cryptography, as, often, parties will be holding classical data (e.g., a secret key) while the adversary will be holding quantum side information (which may also include classical information depending on the adversary's attack).

Note the notation of the density operators. The subscript XA represents a density operator acting on two subspaces: one for X and one for A. We will use this notation, throughout this text. For instance, a density operator σ_{ABE} represents a system (which may contain classical and quantum information), divided into three groups (or parties or subspaces, depending on the context). These groups are labeled A, B, and E. Such a system may be separable, or entangled; mixed, or pure. This notation will also extend to the concept of *partial trace* where we may "trace out" parties, or systems, in a density operator which will be discussed in the next section.

Exercise 2.21. Consider the pure state $|\psi\rangle_{AE} = \sum_{a=0}^{d-1} |a\rangle_A \otimes |e_a\rangle_E$, where $|e_a\rangle$ are arbitrary, sub-normalized vectors. Of course, we require $\sum_a \langle e_a|e_a\rangle = 1$. Assume a measurement is made of the A register in the $|0\rangle, \cdots, |d-1\rangle$ basis, but the E system is not measured. Answer the following questions:

1. Write out your measurement operators $\{M_0, \cdots, M_{d-1}\}$ explicitly. Namely, what is M_a? Note that it should involve an identity operator on the E system, since we are only measuring the A system.
2. What is the probability of observing a particular $a \in \{0, \cdots, d-1\}$?
3. Using Eq. (2.94), show that we may write the measurement outcome as a mixed state:

$$\rho_{XAE} = \sum_{a=0}^{d-1} |a\rangle\langle a|_X \otimes |a\rangle\langle a|_A \otimes |e_a\rangle\langle e_a| . \tag{2.95}$$

where the X system stores the classical outcome of the measurement. (Note that, using partial trace discussed below, we can actually discard the post-measured "A" system.)

2.6.1.1 Partial trace

One of the powerful abilities of density operators is its ability to model quantum information that one party holds, given only part of a global system. For instance, let's say Alice and Bob share an entangled particle pair in the Bell state $|\Phi^+\rangle = \frac{1}{\sqrt{2}}(|00\rangle_{AB} + |11\rangle_{AB})$. As a density operator, this state is represented by:

$$\rho_{AB} = |\phi^+\rangle\langle\phi^+| = \left(\frac{|00\rangle_{AB} + |11\rangle_{AB}}{\sqrt{2}}\right)\left(\frac{\langle 00|_{AB} + \langle 11|_{AB}}{\sqrt{2}}\right) \tag{2.96}$$

$$= \frac{1}{2}\left(|00\rangle\langle 00|_{AB} + |00\rangle\langle 11|_{AB} + |11\rangle\langle 00|_{AB} + |11\rangle\langle 11|_{AB}\right) \tag{2.97}$$

where, above, we made use of the fact that the "bra" of a state vector is the conjugate transpose of that vector. Note, again, the subscript notation for the density operator, as commented on at the end of the last section.

This is a pure state (it is entangled, but taken together as an AB system, it is pure as it may be written as a state vector and, given the entire system description, there is no uncertainty in the state – it really is the Bell state). However, how do we mathematically represent the state that only Alice holds? Or only Bob holds? How do we "zoom in" on a particular sub-system of a larger state? This is one of the areas where density operators are highly useful as such systems can be represented through a *partial trace*.

Let's say we want to zoom in on Alice (and ignore Bob). To write out the density operator describing only Alice's system, we will write:

$$\rho_A = tr_B \rho_{AB}, \tag{2.98}$$

where tr_B means "trace over" system B (i.e., remove it from consideration but, importantly, don't remove its effect on the other systems). The partial trace is a linear operator, defined by the following action:

$$tr_B(|a_1\rangle\langle a_2|_A \otimes |b_1\rangle\langle b_2|_B) = |a_1\rangle\langle a_2|_A\, tr(|b_1\rangle\langle b_2|_B) = \langle b_2|b_1\rangle\, |a_1\rangle\langle a_2|_A\,. \tag{2.99}$$

Above, we used the fact that $tr(|x\rangle\langle y|) = \langle y|x\rangle$. Note that this is an operator only on the A state space, now.

An easy way to perform a partial trace operation of some density operator ρ_{AB} is to write the system you're tracing out/over in an orthonormal basis. That is, if we're tracing out B, we can write ρ_{AB} in the form $\rho_{AB} = \sum_{b,b'} \rho_A^{(b,b')} \otimes |b\rangle\langle b'|_B$, where $|b\rangle$ is some orthonormal basis for the B state space. You can probably convince yourself that any arbitrary ρ_{AB} can be written in this form for any basis of your choice. Once that's done, using Eq. (2.99) and the linearity property of partial trace, we have:

$$\begin{aligned} tr_B \rho_{AB} &= tr_B\left(\sum_{b,b'} \rho_A^{(b,b')} \otimes |b\rangle\langle b'|\right) \\ &= \sum_{b,b'} tr_B\left(\rho_A^{(b,b')} \otimes |b\rangle\langle b'|\right) \\ &= \sum_{b,b'} \rho_A^{(b,b')} \langle b'|b\rangle = \sum_{b} \rho_A^{(b,b)}. \end{aligned} \tag{2.100}$$

The last equality follows since $\langle b'|b\rangle = 0$ whenever $b \neq b'$ and, otherwise, $\langle b'|b\rangle = 1$.

Let's consider an example, namely the Bell state $\rho_{AB} = |\Phi^+\rangle\langle\Phi^+|$, where $|\Phi^+\rangle = \frac{1}{\sqrt{2}}(|00\rangle + |11\rangle)$. Now, imagine the two particles are separated, so that Alice holds the A particle and Bob holds the B particle. If Alice were to measure in the Z basis, she would see $|0\rangle$ or $|1\rangle$, each with equal probability. If she were to measure in the Hadamard basis, she would also observe either basis vector with equal probability. One can confirm this, using the measurement rules discussed earlier (either for density operators, or for standard pure states).

But what does the partial trace look like (before Alice does anything)? Tracing over Bob reveals the following density operator:

$$\begin{aligned} tr_B(|\Phi^+\rangle\langle\Phi^+|) &= tr_B\left(\frac{1}{2}|00\rangle\langle 00|_{AB} + |00\rangle\langle 11|_{AB} + |11\rangle\langle 00|_{AB} + |11\rangle\langle 11|_{AB}\right) \\ &= \frac{1}{2}|0\rangle\langle 0|_A + \frac{1}{2}|1\rangle\langle 1|_A\,. \end{aligned}$$

For the above, note that:

$$|a,b\rangle\langle c,d| = |a\rangle\langle c|_A \otimes |b\rangle\langle d|_B\,. \tag{2.101}$$

This new density operator describes the system as far as Alice is concerned. Note it is a completely mixed state: measuring in the Z basis yields either $|0\rangle$ or $|1\rangle$ uniformly at random; similarly, measuring in the Hadamard basis also produces a uniform random output. This agrees with our intuition based on a partial measurement of the original pure state. Partial trace constructs a density operator which describes what can happen to a system if you only have that one system to work with and not the entire joint state.

Be careful: while this will correctly model Alice alone, one should keep in mind that Bob's particle has not "disappeared" from the universe. Instead, partial trace is a mathematical tool allowing us to zoom in on a particular party. There is, however, no way to go back to the original state if you don't keep track of the joint system. Partial trace will be very useful later when analyze quantum cryptographic protocols as we will often be interested in how systems behave according to a single party, even if that one party's system is entangled with a larger system. For instance, we will care about how an adversary's system behaves (and how much information it can contain), when it may be partially entangled with Alice and Bob's systems. More on this to come!

2.6.1.2 Trace distance

One last topic we must cover concerning density operators is the *trace distance* which tells us how different (or, how similar) two quantum states are. Let ρ and σ be two density operators acting on the same space. Then, the trace distance between these two systems is denoted $\Delta(\rho,\sigma)$ and is defined to be:

$$\Delta(\rho,\sigma) = \frac{1}{2}||\rho-\sigma|| = \frac{1}{2}tr|\rho-\sigma|, \tag{2.102}$$

where given some normal operator A, we write $|A|$ to mean $|A| = \sqrt{A^*A}$ (see Section 2.1.4 for details on functions such as this).

The above is a somewhat non-intuitive definition. Since our systems are finite dimensional, we can also write the trace distance as:

$$\Delta(\rho,\sigma) = \frac{1}{2}\sum_i |\lambda_i|, \tag{2.103}$$

where λ_i are the eigenvalues of $\rho-\sigma$. Note that if ρ and σ are exactly the same state, then $\rho-\sigma=0$ (the zero operator) and so the eigenvalues are all zero and, thus, $\Delta(\rho,\sigma)$ would be zero. As the distance between ρ and σ increases, Δ will also increase.

Trace distance has several important and useful properties. One is that it is invariant to unitary operations. In particular, let U be a unitary operator. Then:

$$\Delta(\rho,\sigma) = \Delta(U\rho U^*, U\sigma U^*). \tag{2.104}$$

Thus, applying a unitary operator to two quantum states cannot increase (or decrease) their distance from one another. Quantum operations in general (such as, for example, partial trace), cannot increase trace distance. For example, given ρ_{AB} and σ_{AB} it holds that:

$$\Delta(\rho_{AB},\sigma_{AB}) \geq \Delta(\rho_A,\sigma_A). \tag{2.105}$$

Exercise 2.22. Let $\rho_{AB} = |\Phi^+\rangle\langle\Phi^+|$ and $\sigma_{AB} = |\Phi^-\rangle\langle\Phi^-|$, where $|\Phi^\pm\rangle = \frac{1}{\sqrt{2}}(|00\rangle_{AB} \pm |11\rangle_{AB})$. Show that $\Delta(\rho_{AB}, \sigma_{AB}) = 1$ but $\Delta(\rho_A, \sigma_A) = 0$.

Another important property that holds is the *triangle inequality*. In particular for any density operators ρ, σ, and τ (acting on the same state space), it holds that:

$$\begin{aligned}\Delta(\rho, \sigma) &= \frac{1}{2}\,||\rho - \sigma|| = \frac{1}{2}\,||(\rho - \tau) + (\tau - \sigma)|| \\ &\leq \frac{1}{2}\,||\rho - \tau|| + \frac{1}{2}\,||\tau - \sigma|| = \Delta(\rho, \tau) + \Delta(\tau, \sigma).\end{aligned} \tag{2.106}$$

Another important fact of trace distance is that it gives an upper-bound on the probability of distinguishing two systems. Let's say a source outputs either ρ or σ and your goal is to distinguish one from the other. That is, you want to construct some device, which may consist of unitary evolution and measurements, or some combination, which outputs 0 if the device thinks the input is ρ and 1 if the device thinks the output is σ. Let's say the source creates either state with equal probability. The best you can hope for, is to distinguish correctly with probability p_{win}, where:

$$p_{win} = \frac{1}{2} + \frac{1}{2}\Delta(\rho, \sigma). \tag{2.107}$$

Thus, the smaller the trace distance, the more indistinguishable the two systems are (you can only take a random guess at which state was created). When the trace distance is large, the two systems become more distinguishable (you can more likely guess correctly which state was given to you). This is an important concept that we will use often in cryptography.

Trace distance has several other very important properties and uses. We'll see more as we continue in this text.

2.6.2 Quantum entropy

Now that we have described density operators, we can resume our discussion on quantum information theory, turning first to von Neumann entropy. Let ρ_A be a quantum system (a density operator), acting on some Hilbert space $\mathcal{H}_A$ of dimension n. Then, we write $H(A)_\rho$ to be the von Neumann entropy of ρ_A (note the ρ subscript, to indicate this is von Neumann entropy, and not Shannon entropy; note that some authors write $S(\rho_A)$ instead). This is defined to be (see, for instance, [10]):

$$H(A)_\rho = -tr\rho\log\rho. \tag{2.108}$$

Since ρ_A is a density operator, it is diagonalizable. Thus, we can write $\rho_A = \sum_i \lambda_i |v_i\rangle\langle v_i|$, where $|v_i\rangle$ form an orthonormal basis (the eigenvectors) and λ_i are real, non-negative, eigenvalues which sum to unity. Using matrix functions, discussed in Section 2.1.4, we have:

$$H(A)_\rho = -tr\rho\log\rho = -\sum_i \lambda_i \log_2 \lambda_i = H(\lambda_1, \cdots, \lambda_n). \tag{2.109}$$

That is, von Neumann entropy is the Shannon entropy of the eigenvalues of the density operator.

Let's do a few examples. Consider the pure state $\rho_A = |0\rangle\langle 0|$. Intuitively, it seems like this state has no uncertainty to it (it is, after all, no different mathematically from a classical "0" with probability one). And, indeed, we see that, since we can write $\rho_A = 1\,|0\rangle\langle 0| + 0\,|1\rangle\langle 1|$, we have $H(A)_\rho = H(1,0) = 0$.

Now, what if $\rho_A = |+\rangle\langle +|$. Perhaps one might think that there is some uncertainty here, since this is a superposition state. However, we can write $\rho_A = 1\,|+\rangle\langle +| + 0\,|-\rangle\langle -|$ and, so, once again we find $H(A)_\rho = H(1,0) = 0$. That is, there is no uncertainty in this state, despite the fact that it is a superposition state of the form $\frac{1}{\sqrt{2}}|0\rangle + \frac{1}{\sqrt{2}}|1\rangle$. However, this is perhaps not too surprising once one thinks about it for some time: indeed, the basis label really makes no difference. If we have a pure state in the form $|0\rangle$, and we know it's a $|0\rangle$, then there is no uncertainty. Similarly, if we have a pure state $|+\rangle$ and we know it is $|+\rangle$, then there is, again, no uncertainty. Pure states, such as this, represent states on which we have complete knowledge – they really are in a particular state. It is only once we get to a mixed state where there can be some uncertainty.

Exercise 2.23. Let ρ_A be a pure state (i.e., $\rho_A = |\psi\rangle\langle\psi|$ for some vector $|\psi\rangle$ of arbitrary, but finite, dimension). Show that $H(A)_\rho = 0$. Similarly, show that, if $H(A)_\rho = 0$, then ρ_A must be pure. That is, prove the following:

$$H(A)_\rho = 0 \iff \rho_A \text{ is a pure state.} \tag{2.110}$$

As another example, now consider the mixed state $\rho_A = \frac{1}{2}|0\rangle\langle 0| + \frac{1}{2}|1\rangle\langle 1|$. In this case, we find $H(A)_\rho = H(\frac{1}{2},\frac{1}{2}) = 1$. Again, not very surprising, since this density operator represents a "classical coin flip" – with one half probability the state is a "0" otherwise it is a "1." But, as it turns out, the mixed state $\sigma_A = \frac{1}{2}|+\rangle\langle +| + \frac{1}{2}|-\rangle\langle -|$ also has von Neumann entropy equal to one.

Now, let's consider $\rho_A = \frac{1}{2}|0\rangle\langle 0| + \frac{1}{2}|+\rangle\langle +|$. This state is not in a diagonal form (since $|0\rangle$ and $|+\rangle$ are not orthogonal). Thus, to compute the von Neumann entropy, we must diagonalize it. We can do so, numerically, and find that the eigenvalues of ρ_A are approximately 0.854 and 0.146 implying that the von Neumann entropy is 0.599.

Von Neumann entropy has several useful properties. In particular, it holds that $H(A)_\rho \geq 0$ and $H(A)_\rho \leq \log_2 \dim \rho_A$. The von Neumann entropy is maximal only on the purely mixed state I/d, where d is the dimension of the system. Von Neumann entropy is also concave. In particular, if $\rho_A = \sum_i p_i \rho_A^{(i)}$, then:

$$H(A)_\rho \geq \sum_i p_i H(A)_{\rho^{(i)}}, \tag{2.111}$$

where, by $H(A)_{\rho^{(i)}}$, we mean the entropy in the state $\rho_A^{(i)}$. Other properties will be brought up as needed in this book.

As with Shannon entropy, we can also talk about joint and conditional entropies. Let ρ_{AB} be a density operator acting on joint Hilbert space $\mathcal{H}_A \otimes \mathcal{H}_B$. Then, we write $H(AB)_\rho$ to mean the entropy in the entire system. This will be the Shannon entropy of the eigenvalues of ρ_{AB}. If we were to write $H(A)_\rho$, for a given ρ_{AB}, we would mean the entropy in the state $\rho_A = tr_B \rho_{AB}$. Finally, the conditional von Neumann entropy, denoted $H(A|B)_\rho$ is defined to be:

$$H(A|B)_\rho = H(AB)_\rho - H(B)_\rho, \tag{2.112}$$

where, similar to before, the last entropy is the entropy of $\rho_B = tr_A \rho_{AB}$.

In QKD security proofs (and other proofs of security), at least in the so-called *asymptotic setting*, we will need to find bounds on $H(A|E)_\rho$ for some state ρ_{AE}. Such a state is shared between Alice (the A register) and an adversary Eve (who holds the E register). Typically, such states are classical-quantum and may be written as:

$$p_0 |0\rangle \langle 0|_A \otimes \rho_E^{(0)} + p_1 |1\rangle \langle 1|_A \otimes \rho_E^{(1)}. \tag{2.113}$$

Above, $\rho_E^{(a)}$ is Eve's conditional state, namely her quantum state in the event Alice's bit register is $a \in \{0, 1\}$. Since these are quantum states, i.e., they are density operators, they are diagonalizable. Furthermore, they have real, non-negative, eigenvalues (see Section 2.6.1). In this case, we can write the above in the following form:

$$p_0 |0\rangle \langle 0|_A \otimes \left(\sum_{i=0}^{N} \lambda_i^0 |E_i^0\rangle \langle E_i^0|_E \right) + p_1 |1\rangle \langle 1|_A \otimes \left(\sum_{i=0}^{N} \lambda_i^1 |E_i^1\rangle \langle E_i^1|_E \right). \tag{2.114}$$

By "absorbing" all the scalars into Eve's state vectors, we may actually write the above more simply as:

$$\frac{1}{M} |0\rangle \langle 0|_A \otimes \left(\sum_{i=0}^{N} |F_i^0\rangle \langle F_i^0|_E \right) + \frac{1}{M} |1\rangle \langle 1|_A \otimes \left(\sum_{i=0}^{N} |F_i^1\rangle \langle F_i^1|_E \right) \tag{2.115}$$

where M is some normalization term (see Exercise 2.24 below).

The question, now, is what is $H(A|E)_\rho$? To compute this, we would need the eigenvalues of ρ_{AE} and ρ_E, however the vectors $|F_i^j\rangle$ are not completely known. It can be shown that if some information is available on them, in particular if one knows (or can bound) the inner products $\langle F_i^0|F_i^0\rangle$, $\langle F_i^1|F_i^1\rangle$, and also $\langle F_i^0|F_i^1\rangle$, then we can bound the von Neumann entropy. This is stated in Theorem 2.1 below which is easily proven (a proof may be found in [6] which is where this particular wording of the theorem also comes from).

Theorem 2.1. *(From [6]): Given a state ρ_{AE} of the form in Eq. (2.115), then it holds that:*

$$H(A|E)_\rho \geq \sum_{i=1}^{N} \left(\frac{\langle F_i^0|F_i^0\rangle + \langle F_i^1|F_i^1\rangle}{M} \right) \times S_i, \tag{2.116}$$

where:

$$S_i = h\left(\frac{\langle F_i^0|F_i^0\rangle}{\langle F_i^0|F_i^0\rangle + \langle F_i^1|F_i^1\rangle} \right) - h(\lambda_i), \tag{2.117}$$

and, finally:

$$\lambda_i = \frac{1}{2}\left(1+\frac{\sqrt{(\langle F_i^0|F_i^0\rangle - \langle F_i^1|F_i^1\rangle)^2 + 4Re^2\langle F_i^0|F_i^1\rangle}}{\langle F_i^0|F_i^0\rangle + \langle F_i^1|F_i^1\rangle}\right). \tag{2.118}$$

This theorem will be helpful later, in Chapter 5, when we prove security of the BB84 QKD protocol against so-called "collective" attacks. Above, $h(x)$ is the binary Shannon entropy function (Eq. (2.84)).

Exercise 2.24. Show that Eq. (2.114) may be written in the form depicted in Eq. (2.115).

2.6.2.1 Quantum min entropy

Now, von Neumann entropy is highly useful when analyzing the security of a QKD protocol, in asymptotic scenarios. However, one other very important entropy will be *conditional quantum min entropy,* denoted $H_\infty(A|E)_\rho$. This is a "worst case" entropy measure that we will use a lot, later, when proving the security of various quantum cryptographic systems, in the *finite key* setting (the distinction between the asymptotic scenario and finite key scenario will be made clear in Chapter 5).

Let ρ_{AE} be a quantum state, then the quantum min entropy is defined to be [11]:

$$H_\infty(A|E)_\rho = \sup_{\sigma_E} \max\{\lambda \in \mathbb{R} : 2^{-\lambda} I_A \otimes \sigma_E - \rho_{AE} \geq 0\}, \tag{2.119}$$

where the above supremum is over all σ_E acting on the E state space (Hilbert space) and where, by $A - B \geq 0$, for operators A and B, we mean $A - B$ is *positive semi-definite,* defined in Definition 2.1.4. Compared to von Neumann entropy, this is, perhaps, not as intuitive a function. As it turns out, when ρ_{AE} is a cq-state (see Section 2.6.1) with $\rho_{AE} = \sum_x p_x |x\rangle\langle x|_A \otimes \rho_E^{(x)}$, then quantum min entropy has a surprisingly nice operational definition, as proven in [5]. In this case, it turns out that the min entropy is related to the maximal guessing probability of Alice's state, given Eve's side information, namely:

$$H_\infty(A|E)_\rho = -\log p_g(A|E), \tag{2.120}$$

where $p_g(A|E)$ is the guessing probability:

$$p_g(A|E) = \max_{\{\mathcal{E}_x\}}\left(\sum_x p_x tr\left(\mathcal{E}_x \rho_E^{(x)}\right)\right). \tag{2.121}$$

The maximization, above, is over all POVMs acting on Eve's Hilbert space. Recall that $tr\mathcal{E}_x\rho_E^{(x)}$ is the probability that Eve's measurement produces outcome x, given that the state actually is $\rho_E^{(x)}$. Thus, the above represents the probability that Eve's measurement matches Alice's register, maximized over all possible measurements that Eve could perform on her ancilla system.

Sometimes, we will want to "discard" bad events which happen with a small probability. In this case, it is often useful to switch to *smooth min entropy*, denoted $H_\infty^\epsilon(A|E)_\rho$, and defined to be [11]:

$$H_\infty^\epsilon(A|E)_\rho = \sup_{\sigma_{AE}} H_\infty(A|E)_\sigma, \tag{2.122}$$

where the supremum is over all density operators σ_{AE} which are ϵ-close to ρ_{AE} in trace distance, namely $||\rho_{AE} - \sigma_{AE}|| \leq \epsilon$. This will be very useful, later, when analyzing QKD protocols as there may be certain events which, though occurring with small probability, would cause the min entropy to be very low, or zero, were they to occur. By switching to smooth min entropy, we may "discard" these events, though it must be done carefully, as we will see as this book progresses.

Let's take a look at some examples. If $\rho_{AE} = \sum_x p_x |x\rangle\langle x|_A \otimes |x\rangle\langle x|_E$, then it can be shown that $H_\infty(A|E)_\rho = 0$ since Eve is fully correlated, Eve will know exactly what data Alice has. On the other hand, consider: $\sigma_{AE} = \sum_x p_x |x\rangle\langle x|_A \otimes |0\rangle\langle 0|_E$. In this case, Eve's system is fully independent (separable) of Alice's system, and it can be shown that $H_\infty(A|E)_\sigma = -\log \max_x p_x$.

Quantum min entropy is, in general, a difficult quantity to bound. However, there are some cases where it is easy, or "easier." Besides entropic uncertainty, which we discuss below in Section 2.6.2.2, there are certain types of states where min entropy may be readily calculated or bounded. One interesting case is when we condition on a classical system. In particular, consider the state:

$$\rho_{CAE} = \sum_{c=0}^{N} p_c |c\rangle\langle c|_C \otimes \rho_{AE}^{(c)}. \tag{2.123}$$

This state has a classical register C where, if $C = c$, then the AE system is in the state $\rho_{AE}^{(c)}$. In this case, we have:

$$H_\infty(A|E)_\rho \geq H_\infty(A|EC)_\rho \geq \min_c H_\infty(A|E, C = c)_\rho \tag{2.124}$$

The first inequality actually applies to any state: namely, conditioning on another system can only decrease uncertainty. The second inequality applies for the specific state in question. Here, we write $H_\infty(A|E, C = c)_\rho$ to mean the min entropy in the conditional AE state when $C = c$, namely $H_\infty(A|E, C = c)_\rho = H_\infty(A|E)_{\rho^{(c)}}$. The above states that min entropy may be bounded by the "worst case" event. Von Neumann entropy has significantly better lower bounds in these cases; however, min entropy is, in a way, a "worst case" entropy measure ("worst case" in the terms of cryptography, where we would ideally want an adversary E to have a lot of uncertainty in general).

Another interesting case for min entropy is when we have a "small" superposition. Namely, consider the following (pure) state where the A system consists of n qubits and the E system is arbitrary, and potentially entangled:

$$|\psi\rangle_{AE} = \sum_{i \in J} \alpha_i |i\rangle_A |E_i\rangle_E , \tag{2.125}$$

where $J \subset \{0, 1\}^n$. Now, let's assume a measurement of each qubit of A is made in some basis; let X be the random variable denoting this measurement and let the post measured system be ρ_{XE} (see Section 2.6.1). What is $H_\infty(X|E)_\rho$? This is not an easy question to answer. Now consider the following *mixed* state:

$$\sigma_{AE} = \sum_{i \in J} |\alpha_i|^2 |i\rangle\langle i| \otimes |E_i\rangle\langle E_i| . \tag{2.126}$$

Recall the difference between $|\psi\rangle_{AE}$ and σ_{AE}. The former is a superposition of states, while the latter is a mixed state (the A system really is $|i\rangle$ with probability $|\alpha_i|^2$). Now, let's say a measurement is made of the A system, using the same basis as before, of this mixed state, resulting in a new density operator σ_{XE}. In general, computing $H_\infty(X|E)_\sigma$ is straightforward: one must consider the "worst case" $i \in J$ (using Eq. (2.124)) and follow the measurement on that. As it turns out, we can use this, along with $|J|$ to bound the entropy in the case the state was pure. In particular, it was shown in [2] (based on a proof in [11]), that:

$$H_\infty(X|E)_\rho \geq H_\infty(X|E)_\sigma - \log_2 |J|. \tag{2.127}$$

If $|J|$ is small, the entropy of the pure-state case cannot be too different from the mixed state case. This property will be useful, later, when analyzing the security of various quantum cryptographic systems.

2.6.2.2 Entropic uncertainty

In general, we will be dealing with systems of high dimensions, where Alice's system is of dimension 2^n and Eve's system is of arbitrary, but finite, dimension. In this case, computing the quantum min entropy is difficult, often more so than computing the von Neumann entropy (the latter of which admits several nice simplifications, and obeys some more helpful properties than min entropy). However, there are several techniques that we can employ to at least bound min entropy, one of which is *entropic uncertainty*.

Before discussing entropic uncertainty for quantum min entropy, we'll discuss some basic uncertainty relations for von Neumann entropy (which are also very useful, especially in cryptographic settings). Let's assume we have a quantum state ρ_A and, furthermore, let's make it easy on ourselves and say ρ_A is a qubit (so, two dimensional). Let's also say we have two orthonormal bases $N = \{|n_0\rangle, |n_1\rangle\}$ and $M = \{|m_0\rangle, |m_1\rangle\}$. If we measure ρ_A in basis N we can consider the outcome to be a random variable and call the post measured (mixed) state ρ_N. Otherwise, if we measure in M we can call the post measured state ρ_M. Then, we can talk about the uncertainty contained in these two states (or, equivalently, the uncertainty in the final measurement outcome). As it turns out, the following expression always holds [7]:

$$H(M) + H(N) \geq -\log_2 c, \tag{2.128}$$

where:

$$c = \max_{i,j} |\langle m_i | n_j \rangle|^2 \tag{2.129}$$

Above $H(M)$ is the Shannon entropy of the classical measurement outcome after measuring in basis M (similar for $H(N)$). Essentially, the above states that, given *any* quantum state, one cannot be certain of the outcome of two different measurements – there must always be some uncertainty in the system and the amount of uncertainty will depend on the "overlap" of the two bases, or, ultimately, how different they are.

Let's consider some basic examples. Let's say $M = N = Z$, that is, both measurement bases are the computational Z basis. In this case, $c = 1$ and, indeed, we can readily find quantum states where $H(Z) + H(Z) = 0$. Just consider $\rho = |0\rangle\langle 0|$. On the other hand, let's say $M = Z$ and $N = X$. Now, $c = 1/2$ and so, for every state it should hold that $H(Z) + H(X) \geq 1$. We can check this at least for some states. Consider $\rho = |0\rangle\langle 0|$ as before. In this case, $H(Z) = 0$ (why?) but $H(X) = 1$ since an X basis measurement will result in a completely mixed state with maximal entropy. In this case, the bound above holds. You may check this for other states of your choosing; as it turns out, it always works.

Now, the above is interesting, however from a cryptographic standpoint, where there exists an adversary system, it is not as useful as it might be. However, there are many different entropic uncertainty relations (see, for instance, [14,4] for an excellent survey). One in particular was discovered in [1], and works for tripartite systems of the form ρ_{ABE}. Let's assume Alice makes a measurement of her system in either the M or N basis (as before), leading to density operator $\rho_{A_M BE}$ or $\rho_{A_N BE}$ respectively (i.e., A_M is the new system representing Alice's outcome in the M basis). This uncertainty relation from [1] reads:

$$H(A_M|E)_\rho + H(A_N|B)_\rho \geq -\log_2 c, \tag{2.130}$$

where c is the same as in Eq. (2.129). Now, we have a conditional entropy expression. This will be very useful in future chapters, when we analyze QKD systems and other cryptographic systems. In particular, we can allow Eve (E) to create a system, perhaps entangled with herself, and send part of it to Alice and part of it to Bob. So long as Alice and Bob have low uncertainty, then Eve will be very uncertain of Alice's measurement outcome. The greater the difference between the two bases (as computed by c), the more uncertain she will be.

The above relations are suitable for von Neumann entropy, however similar relations exist for quantum min entropy. Let ρ_{ABE} be a quantum state where, now, the A and B system consist of n registers each of dimension d (e.g., if dealing with qubits, as is often the case, $d = 2$ and so the dimension of A and B is 2^n). Let M and N be d-dimensional orthonormal bases with overlap function given by c in Eq. (2.129). Then, it was shown in

[13] that:

$$H^{\epsilon}_{\infty}(A_N|E)_\rho + H^{\epsilon}_{max}(A_M|B)_\rho \geq -n\log_2 c, \tag{2.131}$$

where $H_{max}(A|B)$ is the so-called *max-entropy* defined by:

$$H_{max}(A|B)_\rho = -H_\infty(A|C)_\rho, \tag{2.132}$$

where the above min entropy is evaluated for a *purification* ρ_{ABC} of ρ_{AB}. (To *purify* a mixed state ρ_{AB} means to add an additional system C that turns the entire joint state, ABC into a pure state.) $H^{\epsilon}_{max}(A|B)$ is a *smoothed* version of max entropy. We will not go into great detail on max entropy, here, but will recall necessary details, later, when required.

Importantly, the above uncertainty relation will be very useful in future chapters. In general, max entropy can be estimated by Alice and Bob directly; if it is small (i.e., Alice and Bob have a low degree of uncertainty between their outcomes), then the min entropy, between Alice and the Eve system, must be large.

References

[1] Mario Berta, Matthias Christandl, Roger Colbeck, Joseph M. Renes, Renato Renner, The uncertainty principle in the presence of quantum memory, Nature Physics 6 (9) (2010) 659–662.

[2] Niek J. Bouman, Serge Fehr, Sampling in a quantum population, and applications, in: Annual Cryptology Conference, Springer, 2010, pp. 724–741.

[3] Tom Carter, An introduction to information theory and entropy, Complex Systems Summer School, Santa Fe, 2007, pp. 15–34.

[4] Patrick J. Coles, Mario Berta, Marco Tomamichel, Stephanie Wehner, Entropic uncertainty relations and their applications, Reviews of Modern Physics 89 (1) (2017) 015002.

[5] Robert Konig, Renato Renner, Christian Schaffner, The operational meaning of min-and max-entropy, IEEE Transactions on Information Theory 55 (9) (2009) 4337–4347.

[6] Walter O. Krawec, Quantum key distribution with mismatched measurements over arbitrary channels, Quantum Information & Computation 17 (3&4) (2017) 209–241.

[7] Hans Maassen, Jos B.M. Uffink, Generalized entropic uncertainty relations, Physical Review Letters 60 (12) (1988) 1103.

[8] David J.C. MacKay, Information Theory, Inference and Learning Algorithms, Cambridge University Press, 2003.

[9] Michael A. Nielsen, Isaac L. Chuang, Quantum Computation and Quantum Information, Cambridge University Press, 2010.

[10] Dénes Petz, Entropy, von Neumann and the von Neumann entropy: dedicated to the memory of Alfred Wehrl, in: John von Neumann and the Foundations of Quantum Physics, Springer, 2001, pp. 83–96.

[11] Renato Renner, Security of quantum key distribution, International Journal of Quantum Information 6 (01) (2008) 1–127.

[12] Claude E. Shannon, A mathematical theory of communication, The Bell System Technical Journal 27 (3) (1948) 379–423.

[13] Marco Tomamichel, Renato Renner, Uncertainty relation for smooth entropies, Physical Review Letters 106 (11) (2011) 110506.

[14] Stephanie Wehner, Andreas Winter, Entropic uncertainty relations—a survey, New Journal of Physics 12 (2) (2010) 025009.

[15] Mark M. Wilde, Quantum Information Theory, Cambridge University Press, 2013.

3 Modern cryptography

Key distribution is just one of many cryptographic systems studied in *modern cryptography.* It is, however, a vital primitive as it allows for the establishment of secret keys between parties; secret keys being one of the fundamental requirements of many other cryptographic systems. Quantum key distribution (QKD) is just one way to establish a secret key, using quantum communication at its heart. There are other methods and, as with anything, each comes with their own advantages and disadvantages. If one wishes to study QKD it is important to be aware of alternative solutions and where, exactly, QKD sits within the field of cryptography. This chapter serves as a very high-level introduction to the field of modern cryptography. It is hoped that this chapter will provide some glimpse as to this fascinating field, while also providing the reader with a baseline foundation to better understand the security definitions and also how QKD can be used as a sub-protocol within more advanced cryptographic systems. This chapter will be brief – for more information, the reader is referred to the excellent references of [4,3].

Modern cryptography, as distinguished from *historical cryptography,* is a mathematical science based on several key principles. Most importantly, perhaps, is the principle of the mathematical proof of security, which is the gold standard of any cryptographic primitive. Often the case, this proof requires some underlying assumptions leading us to another important principle, namely that any assumptions used for security must be as minimal as possible, and be clearly stated. Some cryptographic systems, such as RSA encryption, require *computational assumptions* (for instance, that it is hard to factor certain numbers – though, actually, a slightly more involved assumption is required). Others, such as QKD, require *physical assumptions* (namely, that the laws of quantum mechanics, discussed in Chapter 2, hold, along with assumptions on the security of the devices used). Classical crypto systems (those relying on classical communication and computation only), are typically proven through a *proof by reduction* – namely, proofs of security require reducing the problem to breaking the assumption (a proof method inspired by complexity theory). QKD proofs of security typically, though not always, rely on information theoretic methods and are more direct.

Of course, any mathematical proof requires rigorous definitions. For instance, to prove a system is secure, we need to know what it actually means to *be* secure. This is not always so obvious and, in fact, a good portion of cryptographic research deals with finding good definitions of security. For QKD, our security definition will involve determining the distance from an "ideal" secret key to the real one produced. Other systems, for instance, *encryption,* involve mathematical games played between an adversary and an oracle.

Quantum Communication and Cryptography. https://doi.org/10.1016/B978-0-44-321532-2.00010-3

3.1 Encryption

We'll begin with some discussion on historical cryptographic systems which will provide some insight into the challenges of designing and proving security of systems. In particular, we will look into some historical *encryption protocols*. Let's assume we have two parties, Alice and Bob (recurring characters in this book!) who have some communication channel connecting them, and they wish to communicate in such a way as to prevent a third-party from intercepting and understanding their messages. Exactly what capabilities the adversary has, in terms both of computational abilities and also how the adversary is allowed to interact with the communication channel, is determined by the attack model. For the purposes of this section, let's assume the adversary has unlimited computational power, but is restricted to an *eavesdropping* attack – namely, the adversary can intercept any transmission between the two parties, but she cannot send messages of her own, or tamper with messages sent. This is the EAV security model. See Fig. 3.1.

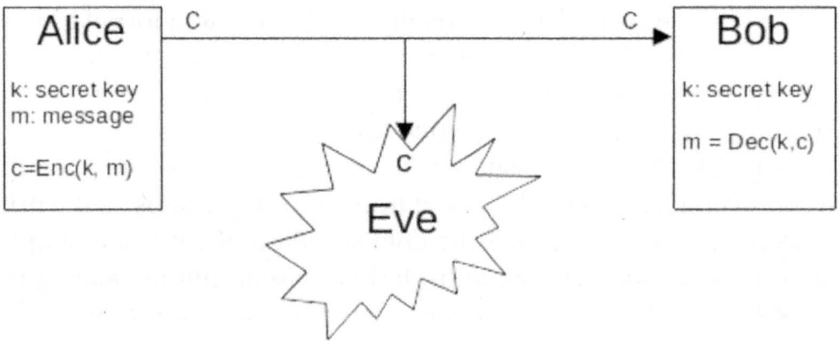

FIGURE 3.1 Eavesdropping (EAV) Security in the private key setting. Here Alice and Bob share a secret key k. Alice (or equivalently Bob) sends a message to Bob (or Alice), encrypting the message m using the secret key k. The resulting ciphertext c is also captured by the adversary Eve. In the EAV model, the adversary can only listen to the communication, but not tamper with the communication. Stronger models of security are discussed later.

Perhaps the most natural place to start would be discussing so-called *Shift Ciphers* (the well-known Caesar cipher is an example of a Shift cipher). The message is a string of characters $m = m_1 m_2 \cdots m_n$, with each m_i from some d-character alphabet (e.g., $d = 26$). Alice and Bob begin with a *secret key k* which is a randomly chosen number between zero and $d - 1$ inclusive. Next, to encrypt a message (also known as a *plain text*), they run the following algorithm:

$$\mathsf{Enc}(k, m) = (m_1 + k)||(m_2 + k)|| \cdots ||(m_n + k), \tag{3.1}$$

where all arithmetic, above, is done modulo d. Above, we use "||" to mean bit-string concatenation. The output of any encryption protocol is called a *ciphertext*. Decryption is straightforward: simply subtract each character of the ciphertext by the key, modulo d.

Of course, the above is clearly not secure. Two issues in particular stand out: first, there are only d possible keys, where d is the size of the alphabet (which is typically very small, or at least too small). The *key space* (or the set of all possible keys) should be so large, as to rule out a *brute-force search attack* (an attack where the adversary will simply try all possible keys). Today we typically talk about key-spaces that are at least 2^{128} in size or more.

The second glaring issue with this system is its inability to hide underlying structure in a ciphertext. Think of yourself, now, as the adversary and you receive a cipher text $c = c_1c_2\cdots c_6$ (a six character ciphertext). Now, you were probably not born yesterday, so you have some prior information on the situation – in particular, you may know something about the messages that may be sent. For instance, you may know the message is "attack" or "defend." (A more modern example may be you know the message is a data packet with its well-defined header.) You may even know that the message "attack" is more likely than "defend." Now, let's say the ciphertext you observe looks like the following:

```
FYYFHP
```

Which message did Caesar send? Well "attack" of course! You can figure that out by going through all possible keys; however, you can also readily see the message *has* to be attack due to the double-up `Y`'s in the ciphertext, which correspond to the doubled-up *t*'s in the original message. The message "defend" would not have that double character in the ciphertext. Clearly, this is not a secure system to use by today's standards.

But what do we even mean by *secure*? Taking some candidate definitions of security from [4], perhaps we can define security in the following way: Given a ciphertext c, which is the encryption of some message m using a secret key k, then it is impossible to learn the secret key k. This seems like a good start – and not being able to learn the secret key is certainly *necessary* for security; however it is not *sufficient*. Indeed, consider the function $\texttt{Enc}(k, m) = m$. Here, the ciphertext is the message – clearly insecure from any reasonable point of view, yet it is also clearly impossible to learn the key k given the ciphertext $c = m$. Thus, this would be a secure encryption system according to this definition, but clearly it is not actually secure; thus the definition is flawed. For more discussion on this, and other candidate (flawed) definitions the reader is encouraged to read the excellent text [4].

So, on the one hand, we have as a core principle: the need to prove security, mathematically. On the other hand, to do so, we need good definitions of security. The "gold standard," at a high level, states that, given a ciphertext, an adversary cannot compute any function of the underlying encrypted message [4]. Think about why this is a good definition: functions can be comparisons (is the message equal to something or is the message greater than a value?). They can be relational (is the message related to another). And so on. None of these functions can be computed on the message given c. That is a powerful notion! (With any definition, sometimes the reader must simply pause and let the definition soak in – the only way to truly grasp some concepts is to take a step back, write it down, and let it soak, as many definitions, especially in cryptography, are subtle.)

There's one last historical tidbit to discuss, namely "security through obscurity." For much of history (and indeed, still today sometimes), the notion of keeping the cryptographic algorithm secret had a strong following. That is, if the algorithm itself is obscured from view, then it should hold that the ciphertext should also be secure. However, in 1888, Kerckhoffs, a Dutch cryptographer, proposed several principles for the design of strong cryptographic systems. What is known, today, as Kerckhoffs's Principle, states that security of the system should not require secrecy of the algorithm – namely, even if the algorithm or protocol falls into enemy hands, the system can still be used. Anything that fits this principle, then, relies on the secrecy of a "key" as opposed to the secrecy of an algorithm or protocol. Clearly keeping a key secret is easier than an entire algorithm; also, if the key ever "leaks," well sure that's very bad, but it's easier to get running again by producing a new key – if you had to produce an entire new algorithm, that would be a significantly greater challenge. In general, Kerckhoffs's Principle is followed by most cryptographers, at least as far as this author is aware. In this book, we will certainly always assume this principle holds – so when discussing security of anything, the algorithm or protocol, in its entirety, will be known to the adversary.

3.1.1 Definitions of security for encryption

The above was a high-level explanation of the definition of security. How about something more formal. Let's return to standard private key encryption. We'll introduce the Shannon definition of perfect secrecy first, followed by a computational notion. Let's define three sets and three probability distributions over them:

1. *Message Space* $\mathcal{M}$ is the set of all possible messages that might be encrypted. Let M be the random variable choosing a message at random according to some distribution.
2. *Key space* $\mathcal{K}$ is the set of all possible keys that might be used and K is the random variable which picks a key at random for a user pair. This is usually the uniform distribution when dealing with symmetric key encryption.
3. *Cipher space* $\mathcal{C}$ is the set of all possible ciphertext outputs that might occur given $\texttt{Enc}$. The random variable, C, is induced by the encryption function, along with the message and keys that are chosen. Namely $C = \texttt{Enc}(K, M)$, where the key and message is chosen randomly according to K and M.

In Shannon's definition of security [7], we assume the adversary knows M (and, of course, the encryption protocol according to Kerckhoffs's Principle). Then, the users encrypt a message, choosing according to M, and send a ciphertext c. This ciphertext is sent across a communication channel over which an adversary has unlimited computational power, but only has *eavesdropping* capability on the channel. Such a channel allows any transmitted data to be copied, exactly, to the adversary; however the adversary is not allowed to tamper with a transmission (i.e., launch a so-called "man in the middle" attack). See Fig. 3.1.

On receiving the ciphertext, c, does the adversary learn anything about the message? Namely, we assume at the start of the "experiment," the adversary knew the distribution M of likely messages to be sent. Does this distribution change, even a tiny amount, after learning c? If it does, then the system is not secure.

Definition 3.1.1. An encryption protocol is said to have *perfect secrecy* if, given the above setup, and for any distribution M, it holds that:

$$Pr(M) = Pr(M|C). \tag{3.2}$$

Namely, the adversary's understanding of the message distribution does not change even after seeing C. Note that, this equivalently says that M and C are independent distributions. Thus, C cannot leak any information on M.

The above is a powerful definition of security as it does not assume anything about the adversary's computational power. Thus, a perfectly secret encryption scheme would be secure even against quantum computers (or some future, undiscovered, technology). Do such systems exist? Clearly, Caesar cipher does not work: indeed, assume M is the distribution picking the message "attack" with probability 1/2 or "defend" with probability 1/2. On receiving ciphertext `FYYFHP`, then the conditional probability distribution has message "attack" being unit probability. Note that to be perfectly secret, Eq. (3.2) must hold for any distribution M; thus, to prove something is not perfectly secret, one need only find one distribution where this is not the case.

A scheme that is perfectly secret is *one time pad* (OTP). This encryption protocol has a message, key, and cipher space consisting of n-bit strings, namely $\mathcal{M} = \mathcal{C} = \mathcal{K} = \{0,1\}^n$. Then, to encrypt a message m, one simply performs a bitwise XOR operation with k, namely $\texttt{Enc}(k, m) = k \oplus m$. Here, $\oplus$ is the bitwise XOR operator (modulo two arithmetic).

To see why this might be secure, let's consider a specific example. Let's say we have three bit messages (so $n = 3$). And let's say that message $m = 101$ is chosen with 1/2 probability, otherwise message $m = 000$ is chosen. Now, you, the adversary, know this distribution. Let's say you receive ciphertext $c = 010$. What message was sent? Well, it could have been the first message $m = 101$ if the key happened to be $k = 111$ (then $m \oplus k = 010$). On the other hand, it could also have been the other message, namely $m = 000$ if the key happened to be $k = 010$. The probability that either key was chosen is exactly equal and there are no other possibilities to account for. Thus, given $c = 010$, it is just as likely the message was 000 or 101 and, so, $Pr(M|C) = Pr(M)$. This, of course, is one specific example, however you can try to write out a full proof (see Exercise 3.1) using this logic.

Exercise 3.1. Prove that OTP is perfectly secret.

If OTP is perfectly secret, meaning no matter how advanced the adversary's computational power is, she can never crack the encryption, why don't we use it? Well, it actually was used in the past, though it has since been replaced by modern systems relying on short keys and computational assumptions. The main disadvantage to OTP is that the key must be the same size (or larger) than the message being encrypted. If you think of this from a practical standpoint, if you ever wanted to encrypt, say, a 1 TB hard drive using OTP, you would need a 1 TB secret key. You would also need a way to store that and, so, you would need a second 1 TB hard drive just to save the key. Of course the key must be kept secret, so you must also protect that second hard drive and, if you wanted to allow another party to decrypt, you would need to somehow get that hard drive storing the key securely to the decryptor. Clearly this is not very practical when we can have systems, today, that can encrypt a 1 TB hard drive using a 256-bit secret key! The other disadvantage to OTP is that the key can only be used once. If the key is ever reused, an adversary can learn something about the underlying messages. Thus, going back to our hard drive example, if you ever changed something on the disk, you would need to recreate and re-distribute that key!

Perhaps the reader is thinking that this is an artifact of OTP. Maybe there is a system that is perfectly secret but with the $|\mathcal{K}| < |\mathcal{M}|$. However, such a protocol cannot exist, which can be mathematically proven. For details, we refer the reader, again, to [4].

Today, we use systems such as AES (the "Advanced Encryption Standard") [5] which can encrypt any size file with a relatively small key of 256-bits. (Technically, AES can encrypt using 128-bit keys also, however due to Grover's algorithm, such key sizes are not post-quantum secure.) However, security of any such system must, by mathematical necessity, rely on computational assumptions and cannot be perfectly secret. However, quantum key distribution (QKD), discussed in detail in this book, may pave the way for OTP to someday be feasible again. If key management and distribution is one of the primary issues with OTP's practical usage, QKD is a technology that can, someday hopefully, stream secret keys between parties at a rate fast enough to allow for encryption and decryption of data. QKD solves the key distribution problem (though it does not solve the key management problem – one will still have a massive key at the end of the day). How QKD operates and how it can be used in classical cryptographic systems such as OTP (or with AES and other classical symmetric key cryptographic systems), will be discussed throughout this book.

3.1.2 Computational EAV security and PRGs

Perfect secrecy is a strong notion, but it is hard to achieve in practice due to the need for large one-time keys. However, the assumption that an adversary has unlimited computational power and time is rather strong. Thus, much of modern cryptography involves *computational security* where, now, security is based on computational assumptions. Defining security in this sense usually involves setting up a game played between an adversary and an oracle. For encryption, the oracle is going to simulate Alice and Bob encrypting a message to one another, leaking information to the adversary based on information the adversary might have learned in the real scenario (e.g., leaking a ciphertext). From this, the adversary must "guess" which message was encrypted. If the adversary can guess with higher (or different) probability which message was chosen than it might have before the game started, we will say the system is insecure (as, ultimately, the adversary is learning something about the message sent given the ciphertext). We'll discuss security against eavesdropping attacks now, and then briefly mention stronger notions of security including CPA security and CCA security.

Let's recall the general eavesdropping scenario, shown in Fig. 3.1. Let's assume the simplest scenario we can think of: a key is generated and distributed (somehow) so Alice and Bob share a secret key. Next, Alice sends a message to Bob after encrypting it. Eve, the eavesdropping adversary, intercepts the ciphertext and wants to learn something about the underlying message. Let's try to simplify the scenario as much as possible in favor of the adversary. When coming up with a definition of security, one always wants to give as much advantage as possible to the adversary, while still making it possible for secure systems to exist.

In general, Alice may choose to send one of many potential messages from which the adversary must guess at. Clearly, if Alice only chooses one message, all the time, then there is no point in discussing security (if there is only one message that can ever be sent, and everyone knows that is the message you will send, there is no point in encrypting it really). On the other hand, if there are five million potential messages Alice may choose from and Eve must guess which one was chosen, given the ciphertext, this is a hard challenge. Thus, to make it as "easy" as possible for the adversary, while still having something interesting to talk about, we'll assume there are only *two* messages Alice may send to Bob. Namely, Alice will choose one message of the two, encrypt it resulting in ciphertext c, and send the ciphertext to Bob. Eve will learn c and must guess which message was encrypted. If the probability of choosing either message is $1/2$ initially, then we will say the system is insecure if Eve's probability of guessing correctly is "different from" $1/2$ (how different it must be, we will discuss momentarily).

Now, the question is, what should the two messages be? Clearly, some messages, even for an insecure crypto system, will be secure (e.g., try distinguishing the encryption of the message "aa" versus "bb" in a shift cipher – it's impossible, yet shift cipher is clearly insecure). Instead of trying to pin down a good message pair, let's just let the adversary choose the two messages! This gives a great advantage to the adversary. In real life, where an adversary most likely does not have this ability, things can only be better.

Thus, we come to our final definition of security for the eavesdropping model. Fix some encryption protocol $\mathtt{Enc}$ and let $\mathcal{A}$ be some algorithm (the adversary) and it will interface with an oracle algorithm $\mathcal{O}$. The game starts with the oracle choosing a random secret key for the encryption protocol being analyzed. This simulates Alice and Bob starting the day with a secret key (again, exactly *how* they get this key is another story – one that can be solved with QKD for instance). Now, the adversary algorithm is allowed to run and it must output two messages m_0 and m_1. These are called the *challenge messages*. It must hold that $|m_0| = |m_1|$ (i.e., the two messages are the same size – as it turns out, we cannot hide the size of a message, if the message can be arbitrarily long). It may choose these messages randomly, and it may think about the choice for a while (keep in mind, the adversary algorithm should know the $\mathtt{Enc}$ protocol being used). The oracle will then choose a random bit b, uniformly at random (i.e., flip a coin) and send to the adversary algorithm the encryption of the b'th message. Namely, the adversary receives:

$$c = \mathtt{Enc}(k, m_b). \tag{3.3}$$

This is called the *challenge ciphertext*. The adversary is now allowed to think some more, but at the end of the day it must output a guess as to what b is. Namely, the adversary outputs a bit b'. We say the adversary "wins" if $b = b'$, otherwise it loses.

Let's call the above experiment the EAV-Privacy experiment. We are interested in analyzing the probability of an adversary winning, where the probability is over the choice of k, the choice of b, and also any internal randomness that $\mathtt{Enc}$ and $\mathcal{A}$ use. Now, what should this probability be for an encryption protocol to be secure? It should not be 0 (i.e., the adversary always loses). Because, if you have an adversary algorithm $\mathcal{A}$ that always loses, you can build a new adversary $\mathcal{A}'$ that always wins! This new algorithm will simply run $\mathcal{A}$ and output the opposite answer. Instead, we need the probability of winning to be 1/2, or at least very, very close to 1/2.

Why is this good? Well, if you think about it – if the adversary algorithm never receives the ciphertext c from Eq. (3.3), what is the probability of winning? Well, in this case, it is simply 1/2 because the adversary can only guess at b (it has no other information). Now, the adversary receives c, but if the probability of winning remains nearly 1/2, then clearly the ciphertext is not leaking that much information, if any. Perhaps the reader can see the similarities to Definition 3.1.1.

Now, how close to 1/2 should the probability be? Well, it should be no larger, or smaller, than a *negligible* amount:

Definition 3.1.2. A function $f : \mathbb{N} \to \mathbb{R}$ is said to be *negligible* if for every polynomial $p(n)$, there exists a value N_0 such that for every $n > N_0$ it holds that:

$$f(n) < \frac{1}{|p(n)|}. \tag{3.4}$$

Namely, a negligible function goes to zero faster than any inverse polynomial (i.e., pretty fast!).

An example of a negligible function is $f(n) = 2^{-n}$, though there are many others. But what should "n" be? Well, that will be our *security parameter* – a setting the user may set to a desirable setting to ensure the adversary's advantage (namely, the distance from 1/2) is sufficiently small. In general, n is the size of the secret key.

Next, we must consider the computational abilities of the adversary algorithm. In the computational setting, we will assume the adversary is bounded in running time by a polynomial $p(n)$. Namely, we will say that the adversary is a probabilistic polynomial time (PPT) algorithm (later in this book, we will have QPT adversaries: quantum polynomial time). Given all of this, we can now define computational EAV security:

Definition 3.1.3. An encryption protocol is said to be EAV-secure if, for every PPT algorithm $\mathcal{A}$, there exists a negligible function $f(n)$ such that:

$$Pr(\mathcal{A} \text{ wins the EAV-game}) \leq \frac{1}{2} + f(n). \tag{3.5}$$

Note that $f(n)$ depends on the adversary. Also note that it is sufficient to write the probability must be "less than" $1/2 + f(n)$ since, if there is an adversary $\mathcal{A}$ that wins with probability substantially lower than 1/2 (e.g., with probability 0), then there is another adversary that wins with probability much larger than 1/2 (e.g., with probability 1). Since the above definition says "for every" adversary, it is sufficient to write "less than" in the probability expression.

Note that OTP is EAV secure in the computational sense (this should be obvious, since it is perfectly secret in the information theoretical sense where the adversary has no limits). Are there other interesting things we can do in the EAV model? Well, it turns out that we can define secure encryption systems in the EAV model that can encrypt messages larger than the key. (Recall, this is impossible in the perfect secrecy model). For that, we need a cryptographic primitive known as a *Pseudo Random Generator (PRG).*

3.1.2.1 Pseudo random generators (PRG)

A PRG is a deterministic algorithm that, given a small random input (called a *seed*), produces a large output that "looks" random. Let $G : \{0,1\}^n \rightarrow \{0,1\}^N$ be some deterministic function, with $N > n$. Then G is a PRG, at a high level, if its output, given a random input, is indistinguishable from a truly uniform random N-bit output.

But what does it mean to "look random?" We can define security of such a system, rigorously, through the use of *distinguishers.* Let D be some PPT algorithm which takes, as input, an N-bit string and outputs a single bit. We will call D our distinguisher, as it will try to distinguish the output of G from a truly random string. In particular, we look at the following difference in probability values:

$$|Pr(D(G(U_n)) = 1) - Pr(D(U_N) = 1)|, \tag{3.6}$$

where, by U_m, we mean a random variable which chooses a random m-bit string, uniformly at random (i.e., each possible string appears with probability 2^{-m}). The probability in the first is over U_n, along with any randomness internal to D. *Note there is no randomness within G.* The second probability is over the random N-bit string (a much larger string potentially), and also any randomness internal to D.

Now, looking at Eq. (3.6), if G really is a good PRG, then it should be that $G(U_n)$ "looks like" U_N. Thus, if G really is a PRG, these two probability values should be very, very close to one another. If there is a large separation in the two probabilities, then D is doing *something* that is distinguishing between the two – it is behaving differently on the two resources. Thus, we have the following definition:

Definition 3.1.4. Let $G : \{0,1\}^n \rightarrow \{0,1\}^N$ be a deterministic algorithm with $n < N$. Then, we say G is a PRG if for every PPT algorithm D, there exists a negligible function $f(n)$, such that:

$$|Pr(D(G(U_n)) = 1) - Pr(D(U_N) = 1)| \leq f(n). \tag{3.7}$$

PRGs can be used to build EAV secure encryption protocols, allowing for relatively short keys to encrypt long messages. Let G be a PRG, then consider the following encryption protocol:

$$\mathsf{Enc}(k, m) = G(k) \oplus m, \tag{3.8}$$

where the key k is of size n-bits and the message, m, is of size $N > n$ bits. This is an EAV-secure scheme – we won't prove it rigorously, but if you think that $G(k)$ "looks random," at least from a computational distinguisher perspective, then $G(k) \oplus m$ should look like $U_N \oplus m$. That, is exactly OTP which is perfectly secret. Thus, $G(k) \oplus m$ is computationally secure.

We comment that PRGs exist only by making other computational assumptions (e.g., if so-called *one-way functions* exist). Another interesting fact is that, if a PRG exists which can stretch the input by just one bit (i.e., if $N = n + 1$), then you can build a PRG H that stretches by any polynomial amount. Thus, one only needs to expand the seed input by a single bit and one can build another PRG, using the first as a building block, to stretch up to any polynomial amount. There are many other fascinating results and properties of PRGs – the interested reader is referred to [4,3].

We comment that PRGs are used often in practice to derive cryptographically secure random strings. However, later, when discussing QKD, we will need true random number generators – QKD will be proven secure from an information theoretical perspective, much like perfect security, and unlike the computational notions of PPT algorithms. QKD will require a large amount of randomness to operate and this randomness must be true randomness. If one were to use a PRG, one will no longer have information theoretic security, and the QKD system may only be computationally secure (at which point, you might as well use other systems). That being said, we will see the definition of PRGs again in Chapters 8 and 10, when we discuss new notions of pseudo-random states, and other quantum cryptographic protocols. There, we will also be in the computational security realm – however, there, our adversary will be polynomial time quantum circuits.

3.1.3 CPA security and PRFs

Note that the definition above only involves the transmission of one message between Alice and Bob. If these parties were to ever reuse the key to send a second message, the security definition would be invalidated. EAV-security, as presented earlier, really is a one time process. Clearly we would like something stronger, allowing us to distribute a key once, and use it multiple times. This leads us to the stronger definitions of security known as CPA security (Chosen Plaintext Attack) and CCA security (Chosen Ciphertext Attack). The first gives us security if the key is used multiple times; the second gives us additional security in the face of a man-in-the-middle adversary (one who is allowed to tamper with the channel connecting Alice and Bob and even send false ciphertexts). Clearly EAV is the "weakest" while CCA is the "strongest" of these three (CPA sits in the middle).

To define CPA security, we will give the adversary the additional ability to choose plaintext messages and receive an immediate response in the form of a ciphertext, namely the encryption of that plaintext. We say the adversary has *oracle access* to the encryption function: namely, the oracle will compute $\mathtt{Enc}(k, \cdot)$, replacing the "$\cdot$" with any message of the adversary's choice. This process can be repeated multiple times (technically a polynomial number of times). Eventually, the adversary must output two challenge messages, as with EAV, and it will receive one ciphertext, the encryption of one of the two messages. The adversary is allowed to make additional queries to the encryption oracle (including asking for the encryption of the two challenge messages!). Finally, the adversary must output a guess as to which of the two challenge messages was encrypted. We require that the probability of winning be negligibly close to 1/2 (same as with Definition 3.1.3 for EAV security).

Why does this guard against the encryption of multiple messages? Well, if we think about how the oracle-adversary game is supposed to simulate the real world scenario, imagine the case of an adversary listening to the communication channel for some time. Over this time, multiple messages have traveled from Alice to Bob, all encrypted with the same key k. Now, over time, maybe Eve has somehow managed to learn something about what these messages were (i.e., learn something about the correlation between messages and ciphertexts sent). Or maybe even Eve can somehow "trick" Alice or Bob into sending an encrypted version of a particular message that Eve can influence (something that has happened often in history, actually). As we always want to give the adversary as much advantage as possible, we might as well just let the adversary pick whatever message it wants, and receive an immediate encryption of that message.

But, now, we have a potential problem: the adversary is allowed to ask for the encryption of the challenge messages! Clearly OTP (which is EAV-secure) is not CPA secure. Indeed, consider the following adversary:

1. First, the adversary sends two challenge messages, $m_0 = 00\cdots 0$ and $m_1 = 11\cdots 1$.
2. The adversary receives the challenge ciphertext $c = \mathtt{Enc}(k, m_b) = m_b \oplus k$.
3. If this were EAV security, we would be done here, and have to output a guess – since OTP is perfectly secret, the above ciphertext leaks no information, and so we would only win half the time. However, since this is the CPA security game, we are allowed oracle access to $\mathtt{Enc}(k, \cdot)$. Let's ask (politely of course!) for the encryption of m_0. The oracle will immediately respond with $c_0 = \mathtt{Enc}(k, 0\cdots 0) = k$.
4. We can now check if $c_0 = c$; if it does, we output our guess that message m_0 was encrypted. Otherwise it must be message m_1 that was encrypted. (We actually have the secret key at this point, so we can probably run wild, but we'll stop the attack here.)

The above attack will always win, thus OTP is not CPA secure. In fact, as an exercise, you should try the following:

Exercise 3.2. Show that *any* deterministic encryption scheme cannot be CPA secure. By deterministic, we mean that if you encrypt the same message twice, with the same key, you receive the same ciphertext.

The above exercise shows that, to be CPA secure, we now need *randomized ciphertexts*. That is, if we encrypt the message, say, $0\cdots 0$ we receive some ciphertext c. If we encrypted the same exact ciphertext, with the same exact key, once again, we will receive some other ciphertext c' which, with very high probability, will not be equal to c (and should look independent of c in some way). To build a CPA secure system, we will need a new building block: a Pseudo Random Function (PRF).

Let's consider an arbitrary function $f:\{0,1\}^n \to \{0,1\}^n$ (it need not have any particular properties). The most general way to describe a function is as a lookup table with input/output pairs, as shown in Table 3.1.

Table 3.1 An arbitrary function $f:\{0,1\}^n \to \{0,1\}^n$ may be described as a lookup table, mapping each input $i \in \{0,1\}^n$ to some bit string $y_i \in \{0,1\}^n$. It takes $n2^n$ bits to represent a function, completely, in this way.

Input	**Output**
$0\cdots 00$	y_0
$0\cdots 01$	y_1
$\vdots$	$\vdots$
$1\cdots 11$	y_{2^n-1}

However, instead of a table, we may also write the table in terms only of the outputs by concatenating them all into a single, large, bit string. Namely, $w = y_0 || y_1 || \cdots || y_{2^n-1}$, where w is a bit-string of size $n2^n$-bits. Given the string w, if we want to know what $f(x)$ is, we simply break w into blocks of n-bits, and look at the x'th block for the output. Thus, in this sense, we can actually think of the function as a bit string, i.e., $f \equiv w$.

Now, consider the set of all possible functions from n-bits to n-bits. (Again, these need not be injective or surjective – just any function whatever.) That is, consider the set:

$$\mathcal{F}_n = \{f:\{0,1\}^n \to \{0,1\}^n\}. \tag{3.9}$$

What is $|\mathcal{F}_n|$? Well, if every function can be considered a bit string of size $n2^n$, and if, equivalently, every bit string of that size can be used to represent a function, then $|\mathcal{F}_n| = |\{0,1\}^{n2^n}| = 2^{n2^n}$. This is a huge, huge set, even for very small n; i.e., there are many functions from n bits to n bits.

Now, let's think about what it means to pick a random function f from the set $\mathcal{F}_n$. Well, this is equivalent to choosing a random $n2^n$ bit string or, alternatively, picking random values for all the y_i's that appear in its look up table (there are 2^n such y_i's). Thus, when we say we have a *random function*, we mean we have a deterministic function f that was *chosen* randomly. A random function itself is deterministic – it is the choice that is random, but once it is chosen, it is "set in stone" and cannot change. For example, if f is a random function, then $f(0101\cdots 01)$ "looks" like a randomly chosen value z; however if we evaluate $f(0101\cdots 01)$ again (i.e., run it on the same input), we will get the same exact z over and over again. On the other hand, any two or more distinct inputs will have independent outputs (the outputs for each input are chosen randomly and independently). Thus, $f(0\cdots00)$ and $f(0\cdots01)$ are two randomly chosen, independent, n-bit strings. Even if the inputs are related, the outputs will be independent. But the function is deterministic. This is a random function.

Choosing a random function, in practice, is impractical. Even if $n = 128$, you would need 128×2^{128} bits to write the function down. A PRF is a *deterministic* polynomial time algorithm that "looks like" a random function. Let $F : \{0,1\}^n \times \{0,1\}^n \to \{0,1\}^n$ be a deterministic algorithm with two inputs and one output. The first input will be called the key input and is chosen randomly then fixed, while the second input can be anything. In general, we will write $F(k,x)$ as $F_k(x)$, making the key input a subscript. Choosing a key will "simulate" choosing a random function, at least as far as a computational distinguisher is concerned. Since the key is only n-bits long, this is much easier to write down and choose. The fact that such a thing can be similar to choosing a random function (of size $n2^n$ bits), as far as a computational adversary is concerned, is truly amazing, at least in this author's opinion!

Now, to define security of a PRF, we will again turn to distinguishers. However, unlike the PRG case (Definition 3.1.4), we cannot give the distinguisher a description of the function (since it is too large for a polynomial time D to even process). So, instead, we will give D *oracle access* to a random function, or a PRF function.

Consider a PPT algorithm D and a random function f. The distinguisher receives no input (unlike the PRG case), however it is allowed to "ask questions" or "make queries" of its oracle. Whenever D asks for the function at x (by sending x to the oracle), the oracle will compute $y = f(x)$ and send y to the distinguisher. The distinguisher is allowed to make as many questions as it wants (up to a polynomial number since it is a PPT algorithm). At the end, however, it must output a single bit. Let's call this experiment D^f.

Consider, now, the same PPT D, but now a random key k is chosen. The distinguisher (the same algorithm) asks questions x, but now the oracle answers with $F_k(x)$. Again, the distinguisher must output a single bit. We'll call this experiment D^{F_k}.

We now look at the difference in probabilities of the distinguisher outputting one in the first experiment versus the second. Namely, consider:

$$|Pr(D^f = 1) - Pr(D^{F_k} = 1)|, \tag{3.10}$$

where the first probability is over the choice of random function f, and any internal randomness within D; while the second probability is over the choice of key k, and, again, any internal randomness to D. For F to be a PRF, these two probabilities must be negligibly close to one another.

Definition 3.1.5. Let F_k be a polynomial time deterministic algorithm with key size n-bits. Then, F_k is a PRF if for every PPT distinguisher D, there exists a negligible function $g(n)$

such that:

$$|Pr(D^f = 1) - Pr(D^{F_k} = 1)| \leq g(n) \tag{3.11}$$

In general, D is asking questions and getting answers – in the first case (with random function f), the answers are all independently chosen random numbers; in the second case (the deterministic polynomial time algorithm F_k), the answers are just the output of a deterministic polynomial time algorithm. A PPT algorithm cannot tell the difference between the two answers. Note that D knows the algorithm used, just not the key k.

The output of a PRF "looks" random, even if the input is known. Such a system can be used to build a CPA-secure encryption system: Let k be the secret key, which is also a valid secret key for a PRF. Then, to encrypt message $m \in \{0, 1\}^n$, we first choose a random $r \in \{0, 1\}^n$ and evaluate $F_k(r)$. Note that this should be computationally indistinguishable from a randomly chosen number – thus, it may be used as a pad for a computationally secure version of OTP. Indeed, the output of the encryption protocol will be:

$$\mathtt{Enc}(k, m) = (r, F_k(r) \oplus m). \tag{3.12}$$

Note that the randomly chosen token r is part of the cipher text. However, that's OK since PRFs look random even if the input is known (even if the adversary chooses the input!). So long as the key k stays secret, $F_k(r)$ will appear random. Each time you encrypt the same message, an independent random r' is chosen. With high probability this will be a unique number and, thus, $F_k(r')$ will look like it was randomly chosen and independent of the others (even though it is actually not since it is the output of a deterministic function, it's assumed a computational distinguisher cannot determine that it is not random). A proof by reduction can be used to prove that this is a CPA secure scheme, essentially by showing that if it is *not* secure, then the underlying PRF is not secure. If it is not a secure encryption protocol, one may use an adversary that wins the CPA security game, discussed earlier, as a subroutine in a distinguisher algorithm D that is able to distinguish the PRF from a truly random function.

In slightly more detail, let's assume the above CPA secure encryption scheme is actually insecure. Thus, there exists some adversary $\mathcal{A}$ which wins the game with some non-negligible advantage greater than 1/2. Let $p_{win}^{F_k}(\mathcal{A})$ be the probability that this adversary wins the CPA game, for this particular encryption protocol (which uses a PRF). Now, let's consider a toy version, where we replace the PRF with a real random function. The secret key is the random function choice f, and the encryption outputs $(r, f(r) \oplus m)$. Clearly this is not practical, but it's fine to think of these things in a proof. Let $p_{win}^{f}(\mathcal{A})$ be the probability that this same adversary wins the CPA game, but with this toy protocol, using a real random function. It can be shown that $p_{win}^{f}(\mathcal{A}) = \frac{1}{2} + negl.$, where $negl.$ is a negligible function (in the security parameter n).

To see why $p_{win}^{f}(\mathcal{A})$ is as claimed, consider the CPA experiment. This adversary is going to make several queries of messages $m_1, \cdots, m_p$ and receive valid ciphertexts $c_1, \cdots, c_p$, where each $c_i = (r_i, f(r_i) \oplus m_i)$, where each r_i is an n-bit string, each chosen uniformly at random. At some point the adversary must output two *challenge messages*, which we'll denote m_0^* and m_1^* (the star superscripts will be used to indicate the special challenge messages and ciphertexts). One of these is encrypted and returned as $c^* = (r^*, f(r^*) \oplus m_b^*)$ (where b is a randomly chosen bit; the goal of $\mathcal{A}$ is to guess what b is). The adversary is allowed to make q additional message queries and receive additional responses. These queries can even be one or both of m_i^*. Note that the total number of queries, $p + q$ is polynomial in n.

There are two things that can happen. First, a "bad" event where $r^* = r_i$ for one (or more) of the $p+q$ queries made. Since the adversary knows m_i, the adversary now knows $f(r_i)$ and, thus $f(r^*)$, allowing the adversary to decrypt c^* and learn b. However, since $p+q$ is polynomial, and since there are 2^n possible r values, the probability that this happens is negligible. The second "good" event is where $r^* \neq r_i$ for all queries i. In this case, $f(r^*)$ is essentially a uniform, independent, random bit string and, thus, the ciphertext is OTP encryption. The probability of winning against OTP encryption is simply 1/2.

Exercise 3.3. Prove formally that $p_{win}^f(\mathcal{A}) = \frac{1}{2} + negl(n)$, for any adversary $\mathcal{A}$, where $f : \{0,1\}^n \rightarrow \{0,1\}^n$ is a random function. The probability is over the choice of random function f, b, and any randomness internal to $\mathcal{A}$.

So the toy protocol, using f, is secure, but it's not practical; we care about $p_{win}^{F_k}(\mathcal{A})$. Here is where we will use the assumption that F_k is a secure PRF. To do so, we'll use the adversary $\mathcal{A}$ as a subroutine in a PRF distinguisher. Then, we'll show that, since we're assuming the PRF is secure, the subroutine $\mathcal{A}$ cannot actually be winning the real CPA experiment substantially greater than 1/2.

Consider the following distinguisher algorithm D which will use $\mathcal{A}$ as a subroutine. First, it runs the $\mathcal{A}$ algorithm; every-time $\mathcal{A}$ asks for the encryption of some message m_i, the distinguisher will simulate the CPA oracle. First, D will choose a random r_i; it will then send r_i to its oracle. The oracle returns either $f(r_i)$ (if we're running D^f), or $F_k(r_i)$ (if we're running D^{F_k}). Let y_i be the value returned and D will now compute $c_i = (r_i, y_i \oplus m_i)$ and send it to $\mathcal{A}$. This repeats every-time $\mathcal{A}$ asks for the encryption of some message.

Eventually, $\mathcal{A}$ outputs the challenge messages m_0^* and m_1^*. The distinguisher D will choose r^* and, again use its oracle to simulate the encryption. Namely, it asks for the function evaluation at r^* and receives y^*. The distinguisher then chooses a random bit b and sends back $c^* = (r^*, y^* \oplus m_b)$.

At the end, $\mathcal{A}$ will output a guess b'. If $b' = b$ (i.e., if the adversary wins the CPA game which the distinguisher is simulating), then the distinguisher outputs one. Otherwise the distinguisher outputs zero. Note that, if D is interacting with f, then the probability of outputting one is $Pr(D^f = 1) = p_{win}^f(\mathcal{A})$ since D was simulating, exactly, the CPA oracle for $\mathcal{A}$. Similarly, $Pr(D^{F_k} = 1) = p_{win}^{F_k}(\mathcal{A})$. Since F_k is assumed to be a PRF, we have that there exists some negligible function $g(n)$ such that:

$$\left|p_{win}^f(\mathcal{A}) - p_{win}^{F_k}(\mathcal{A})\right| = \left|Pr(D^f = 1) - Pr(D^{F_k} = 1)\right| \leq g(n). \tag{3.13}$$

But, combined with our analysis of $p^{f}_{win}(\mathcal{A})$, this implies that $p^{F_k}_{win}(\mathcal{A}) \leq \frac{1}{2} + negl.$ and we're done!

The above is a mostly completely proof of security of the above encryption protocol and it illustrates how many cryptographic protocols are proven secure, by making some underlying assumption. Similar proofs will be used in quantum cryptography when making some form of computational assumption.

Exercise 3.4. Let F_k be a secure PRF with n-bit keys, inputs, and outputs. Consider the following encryption scheme for $2n$ bit messages. Namely, $\texttt{Enc}(k, m) = \texttt{Enc}(k, m_L || m_R)$, where m is $2n$ bits, divided into m_L (the left n bits) and m_R (the right n bits). The encryption function will choose a random $r \in \{0, 1\}^n$ and output the ciphertext:

$$c = (r, F_k(r) \oplus m_L || F_k(r + 1) \oplus m_R) \tag{3.14}$$

where, by $r + 1$ we mean standard integer addition. Note that "$||$" is bit-string concatenation, here. Prove, using a cryptographic proof by reduction, as above, that this is a CPA secure encryption scheme. Can you think of a way to extend this to $3n$ bit messages? Or $\lambda \cdot n$ bit messages, for some polynomial λ?

What happens if we change the protocol to:

$$c = (r, F_k(r) \oplus m_L || F_k(r) \oplus m_R), \tag{3.15}$$

(now there is no "$r + 1$" term.) Show that this modified version is not even EAV secure, let alone CPA secure.

3.2 Public key encryption

For most of history, cryptography was restricted to the so-called *private key* or *symmetric key* setting. Here, Alice and Bob start the day with a shared secret key and then perform some cryptographic task. It wasn't until the 1970s that the *public key* cryptography was publicly introduced [2] (there were some classified constructions, before then). Here, one party (say, Alice) has a so-called public key and a private key pair. These two keys are related to one-another mathematically somehow. Bob (and anyone else, including the adversary) has a copy of the public key. How Bob comes by the public key is a *public* key distribution problem. However, it should be easy to see, that distributing a public key is easier than distributing a secret key. Today, we often distribute public keys over the Internet through a certificate authority system, though exactly how this works is outside the scope of this book. This notion that one party has a secret key and all others have a public key creates an asymmetry between parties and, thus, public key protocols are also called *asymmetric key* systems.

We will discuss, here, briefly public key encryption systems. These consist of three algorithms: A key generation algorithm `Gen`, an encryption function `Enc`, and a decryption function `Dec`. The key generation algorithm will output a secret key/public key pair which we will denote throughout this section by sk (for the secret key) and pk (for the public key). The encryption algorithm takes as input the public key pk, and a message m, and produces a ciphertext c, namely $\texttt{Enc}(pk, m) = c$. Finally, the decryption algorithm takes the secret key sk and a ciphertext c and produces a message m. The following *correctness* guarantee should hold:

$$\texttt{Dec}(sk, \texttt{Enc}(pk, m)) = m. \tag{3.16}$$

Thus, in public key encryption, the public key is used to encrypt a message, while the secret key is used to decrypt. In general, we prove security of such systems assuming everyone holds a copy of the public key, *including the adversary!*

This is a rather remarkable prospect: consider the following scenario, in a large lecture hall, the instructor computes a secret key and public key pair and writes them down in secret. Next, he or she shouts out the public key to all students in the audience. Everyone in the audience now has exactly the same information as one another – the only unknown element is the secret key which only the instructor holds. Now, anyone in the audience may take a message and compute a ciphertext, using the given public key (which everyone knows). That person may now shout out the ciphertext to the instructor – all members of the audience have the ciphertext (encrypted with the public key), and the public key itself. However, only one other person in the room will be able to decrypt that message: the instructor, with the secret key! Of course, in practice, the numbers required are so incredibly large, that these computations cannot be done, realistically, using pen and paper, of course.

To illustrate how this may be achieved, we will look at the RSA crypto system (which is technically broken if quantum computers ever scale large enough, due to Shor's Algorithm). RSA depends on the so-called RSA assumption which, itself, requires factoring of certain numbers to be computationally difficult. Namely, given a composite number N, the product of two large primes p and q, it is difficult to find p or q (difficult meaning a PPT adversary cannot do so). Technically the RSA assumption is a bit more subtle than this, however factoring will definitely break RSA.

We will first discuss the RSA `Gen` algorithm. Note that, for symmetric (private) key systems, we didn't even bother discussing key generation algorithms – this is due to the fact that the systems we discussed in the previous chapter had keys consisting of simple bit strings; choosing a key meant choosing a random bit string. This is fairly trivial (though, in practice, perhaps not too trivial since one must have a good source of randomness – something quantum cryptography can help with as we discuss in Chapter 8) and so we didn't bother to discuss it. Public key systems, however, have complicated key generation algorithms due to the necessity of the public key and secret key "pairing" together in such a way that (1) the secret key "unlocks" information the public key encrypted; and (2) given the public key, one cannot find the secret key in a computationally efficient way (e.g., through a PPT algorithm).

For RSA, `Gen` will first choose two large, random primes p and q. In practice these primes can be thousands of bits long. The value $N = pq$ is computed, along with the value $t = (p-1)(q-1)$. For those interested, t is Euler's totient function, which counts the number of relatively prime numbers less than N in this case. One may actually use Carmichael's totient function to produce a more efficient secret key, however these details are outside the scope of this book; we are only presenting the basic, original, version of RSA key generation and encryption, here [6] often called the "text book version."

Next, a number e is chosen, such that $\gcd(e,t) = 1$ where $\gcd(a,b)$ is the *greatest common divisor* function of a and b. Namely, in this case, no number, other than one, will divide both e and t. This number, e will be the public key. Finally, a number d is found such that:

$$ed \equiv 1 \bmod t. \tag{3.17}$$

Above, $a \equiv b \bmod m$ means that m divides into $a - b$ (or, equivalently, b is the remainder of a divided by m). Such a number d is guaranteed to exist since $\gcd(e,t) = 1$. Furthermore, such a number is easily found (easy, here, meaning in poly time), using the *Extended Euclidean Algorithm*. However, to do so, one must know t and e. The public key will be $pk = (e, N)$ while the secret key is $sk = (d, N)$. Note that the number N is in both keys; the "secret" part of the secret key is the number d which satisfies the above equation.

We will now discuss the "text book" version of RSA encryption. As it turns out, this version is insecure; however it can be augmented to create a secure public key encryption system (one that is actually CCA secure, a notion stronger than CPA security and one which we will discuss below, in Section 3.3).

Let m be a message. In public key systems, messages are usually no longer considered arbitrary bit strings, as they were in the symmetric case, but are now elements of some group; in our case we think of them as integers modulo N (which, recall, is $N = pq$). Given an m, the algorithm simply computes:

$$\mathsf{Enc}(pk, m) = m^e \mod N. \tag{3.18}$$

To decrypt, we take a ciphertext and raise it to the power d (modulo N).

Why does this system work? Well it comes down to a number theoretic fact known as Fermat's Little Theorem. We won't go through the details here, however it can be shown that $(m^e)^d \mod N = m^1 \mod N = m$.

However, the above version, the so-called "Text Book" version, is not even EAV secure. EAV security for public key encryption is identical to the symmetric key definition, discussed earlier, except that here the adversary is given a copy of the public key.

The reason Text Book RSA is not secure is that it is deterministic. If an adversary outputs two challenge messages m_0 and m_1 and receives a ciphertext back $c = \mathsf{Enc}(pk, m_b) = (m_b)^e \mod N$, where b is the message chosen by the oracle to encrypt, the adversary can simply compute, locally, m_0^e and m_1^e to see which one matches. Note that the adversary always is assumed to have a copy of the public key. The adversary will always win with this strategy; this "proof" shows also that any public key encryption system must have randomized ciphertexts, even in the EAV model.

Exercise 3.5. Prove that any deterministic public key encryption system cannot be EAV secure.

RSA may be made secure by adding the so-called OAEP scheme, which stands for *Optimal Asymmetric Encryption Padding* [1]. This takes a message m and adds a randomized padding scheme, before running the resulting "padded message" through standard RSA as discussed above. The padding ensures that each time one encrypts the same message, a different ciphertext will occur with high probability.

3.3 Authentication and integrity

The above involved secure encryption protocols which ensure the secrecy of a message. However, in cryptography, we also care about the message's integrity. Namely, if you receive a message from someone, how do you know the message actually came from that person? How do you know it wasn't tampered with? This is where authentication methods come in.

We will first discuss the symmetric key version of authentication, known as *message authentication codes* (MACs). In the public key world, one has *digital signatures* which we will discuss, briefly, later. MACs are critical in QKD operation, as we will need an authentication method to perform QKD checks.

Let's return to the two party scenario, with our old friends Alice and Bob. Let's also assume they start off with a shared secret key k. A MAC is an algorithm that, on input a message and a secret key, will output a *tag* which can be used, later, to verify the integrity of the message. In particular, we have:

$$\texttt{MAC}(k, m) = t. \tag{3.19}$$

We also have a verification algorithm `Verify` which takes as input a secret key, a message m', and a tag t' and outputs True or False. It outputs True only if the given tag is correct for the given message. In general, we have:

$$\texttt{Verify}(k, m, \texttt{MAC}(k, m)) = True. \tag{3.20}$$

Security of a MAC should ensure that, even if the adversary knows legitimate tag/message pairs, the adversary cannot "forge" a correct tag for a message never before seen. In particular, consider the following game/experiment played between an oracle and an adversary (similar to how we defined security for encryption):

1. The oracle chooses a secret key k (simulating Alice and Bob having a secret key)
2. The adversary algorithm is allowed to ask for a tag of any messages she likes. That is, the adversary has *oracle access* to $\texttt{MAC}(k, \cdot)$. This simulates the adversary "listening" to the communication of Alice and Bob and learning valid message/tag pairs. Let $m_1, \cdots, m_p$ be the messages the adversary asks for.
3. The adversary must output a message m and a tag t such that $m \neq m_i$ for all $i = 1, \cdots, p$.

4. We say the adversary *wins* if $\texttt{Verify}(k, m, t) = True$. That is, the adversary wins if she is able to output a valid tag for a message never before seen. Note that the new message need only differ by a small amount (e.g., one bit) from the queried messages.

Note that we require the adversary to output a tag for a new message. This means that MAC protocols cannot give any security against so-called *replay attacks,* namely attacks where the adversary repeats old messages. To prevent attacks of this nature, various strategies exist, such as putting a "sequence number" on every message, ensuring it is unique. Note also that MACs do not guarantee secrecy – indeed, the security definition, above, assumes the adversary knows all messages and there is no "secrecy" game involving, say, two challenge messages. Encryption guarantees secrecy, MACs guarantee integrity. One can, in a way, combine these two notions into a CCA-secure encryption scheme (which stands for Chosen Ciphertext Attack); here messages remain secret, and it is also guaranteed that any ciphertext sent will also not be tampered with (at least, not without you knowing about it!). CCA-secure systems typically involve the careful incorporation of a MAC scheme into a CPA-secure encryption protocol.

One example of a secure MAC is actually the simple use of a PRF. Let F_k be a secure PRF (see Definition 3.1.5). Then, the following is a secure MAC:

$$\texttt{MAC}(k, m) = F_k(m). \tag{3.21}$$

Indeed, even if the adversary has multiple message tag pairs, say $m_1, t_1, \cdots, m_p, t_p$, where $t_i = F_k(m_i)$, the adversary must, at the end of the day, output a new message m and a valid tag for that message, namely $F_k(m)$. However, since F_k is a PRF, it "looks" like a random function and therefore there is no way for a computationally bounded adversary to know where $F_k(m)$ will go. Of course, that is a very high-level proof, however it gives the basic notion. Note that this MAC only works for fixed size messages (namely, for messages that are the same size as the PRF input size which, in practice, may be as small as 128 bits). However, there are common methods to extend the message size. We will not dwell too much on these here, since they are computational notions and we will not rely on them too much in this text.

In QKD security, we often need the existence of an *authenticated classical channel* connecting two parties. This channel allows Alice and Bob to send classical messages to one another in such a way that they can be sure the message received really was from the other legitimate party and that it was not tampered with. However, this channel is not secret – thus, any message they send over it will be potentially read by an adversary. Authentication of this manner can be done in an information theoretic secure manner using Carter-Wegman codes [8] which require a small pre-shared key that can only be used once. As it turns out, QKD will be able to produce a significantly larger key, so this initial pre-shared key may be "refreshed" once a QKD session ends. Unlike OTP encryption, which requires a key to be as large as the message to gain perfect secrecy (see Section 3.1.1), information theoretically secure MACs (i.e., those which do not require PPT adversaries) do not require such "large" keys. As it turns out, QKD will use "small keys" (needed for authentication) to produce much larger keys – the small key used for authentication can then be refreshed, using part of the newly produced, larger, secret key. This will be covered in more detail in later chapters.

Message authentication codes are a symmetric key primitive – they allow a secret key holder to both create and authenticate messages. Sometimes (often), however, asymmetry is needed. In particular, it is often desirable that one party can create messages, while multiple other parties can verify. The ability to verify a message should not imply the ability to create a message. For this, *digital signatures* are used. A digital signature scheme is a public key mechanism: a secret key and public key pair are first produced by some key generation algorithm. Following this, a message m can be *signed* using the secret key. This is done by some algorithm $\texttt{Sign}(sk, m) = \sigma$, where σ is the signature, linking m to the secret key holder. The public key can be used to *verify* the signature; namely an algorithm $\texttt{Verify}(pk, m, \sigma)$ is run which will return `True` or `False`. Correctness should imply that the verification algorithm returns `True` whenever the signature is correct (i.e., when the signature was actually created by the secret key holder). Security should imply that, even if the adversary has a copy of the public key and also multiple valid message/signature pairs, it should not be able to forge signatures of messages that have not been seen before. (That is, security does not guard against replay attacks, similar to MACs.)

Many digital signature schemes exist and they are used everywhere today. Most of the content in this book focuses on information theoretic constructions, and the use of MACs for authenticity, and so we will not dwell too long on digital signatures. We will, however, see them again in Chapter 10 where they will be a critical component of certain quantum public key cryptographic systems.

References

[1] Mihir Bellare, Phillip Rogaway, Optimal asymmetric encryption, in: Workshop on the Theory and Application of Cryptographic Techniques, Springer, 1994, pp. 92–111.

[2] W. Diffie, M. Hellman, New directions in cryptography, IEEE Transactions on Information Theory 22 (6) (1976) 644–654.

[3] Amir Herzberg, Applied Introduction to Cryptography and Cybersecurity, World Scientific Publishing, 2025.

[4] Jonathan Katz, Yehuda Lindell, Introduction to Modern Cryptography: Principles and Protocols, Chapman and Hall/CRC, 2007.

[5] NIST FIPS Pub, Announcing the advanced encryption standard (aes), Federal Information Processing Standards Publication 197 (2001) 1–51.

[6] Ronald L. Rivest, Adi Shamir, Leonard Adleman, A method for obtaining digital signatures and public-key cryptosystems, Communications of the ACM 21 (2) (1978) 120–126.

[7] Claude E. Shannon, Communication theory of secrecy systems, The Bell System Technical Journal 28 (4) (1949) 656–715.

[8] Mark N. Wegman, J. Lawrence Carter, New hash functions and their use in authentication and set equality, Journal of Computer and System Sciences 22 (3) (1981) 265–279.

4 Quantum key distribution (QKD) protocols

We are now, finally, in a position to discuss quantum key distribution (QKD), one of the seminal quantum technologies, and one that is commercially practical today (and has been for some time). In this chapter, we will introduce, formally, the BB84 protocol [5], while also discussing several other protocols, each chosen to introduce an interesting or unique facet of quantum secure communication. While security will not be discussed rigorously in this chapter (that will be saved for Chapter 5), we will discuss some of the intuition behind why these protocols are secure, along with discussing various trade-offs between the different protocols discussed. Finally, we will delve into some of the classical post-processing methods required to operate a QKD protocol, including error correction and privacy amplification, along with some optional classical post-processing methods, such as classical advantage distillation.

4.1 Goal of QKD and the BB84 protocol

The stated goal of a QKD protocol is simple enough, namely the establishment of a secure secret key between two parties: Alice and Bob. Unlike classical key distribution, security, here, should be guaranteed even against computationally unbounded adversaries who are bounded only by the laws of physics. As discussed previously, this is in stark contrast to the classical communication scenario whereby users can only hope to achieve security against computationally *bounded* adversaries – i.e., those where some mathematical assumption is placed upon.

4.1.1 Resource requirements

QKD protocols require, at a minimum, a *quantum channel* and a *classical authenticated channel*. The first allows for the transmission of qubits (or, as we will discuss later in Chapter 7, higher dimensional qudit systems) while the second allows for the transmission of classical messages which Alice and Bob can authenticate.

The quantum channel's properties depend, in large part, on the physical implementation, along with the *security model* under consideration. The latter is an important, and delicate, property of QKD protocols. Often, when first analyzing QKD protocols from a theoretical perspective, one assumes the quantum channel is noisy but lossless; or noisy and lossy but that Alice and Bob's devices are idealized and, in particular, Alice has access to a single-qubit source. When moving towards practice, these assumptions become unrealistic, and so one is required to analyze a security model where Alice's source is imperfect

Quantum Communication and Cryptography. https://doi.org/10.1016/B978-0-44-321532-2.00011-5

and the channel itself is both noisy and lossy. For the sake of this chapter, we will generally assume a noisy but lossless channel and idealized source and measurement devices. Stronger notions of security will be discussed in due time. Regardless, however, we almost always assume that an adversary can "splice" into the quantum channel and perform a man-in-the-middle attack.

The classical channel's properties are such that, if Alice (or Bob) sends a classical message to Bob (or Alice), the receiver of the message can verify that, first, the message actually came from Alice (or Bob) and, second, that no one tampered with the message. This can be implemented using information theoretic secure MACs, as discussed in Chapter 3.3. Note, however, that to implement such a channel, one requires a shared secret key between Alice and Bob. It seems, then, that we have a chicken and the egg problem! QKD can create a perfectly-secure secret key, but only if a perfectly-secure secret key already exists. However, unlike information theoretic secure encryption (i.e., OTP) which requires that the secret key be at least as long as the message, MACs only require a significantly shorter key than the message size. Thus, while one does require a shared secret key to "boot-strap" QKD, the actual operation of QKD will consume a small part of this key while producing a massively larger secret key as output. The (very) small portion of the secret key that was consumed by the use of the authenticated channel may be refreshed by the newly created key material. In a way, QKD can be considered a form of "key expansion" protocol: the input is a small secret key and the output is a much larger secret key. Unlike, say, PRGs (see Chapter 3), security can be guaranteed without the need for computational assumptions.

4.1.2 BB84

A typical QKD setup has Alice sending random qubit states to Bob who then measures in alternative bases. The exact encoding and decoding is protocol dependent. To give a concrete example, let's consider a "toy" BB84 protocol – this toy version is almost identical to standard BB84 but is missing some critical components, important both for security and for practicality reasons, which we will fill in later.

In "toy" BB84, Alice has a key to send to Bob, which we denote by $k^A \in \{0, 1\}^N$. To do so, she repeats the following process:

1. Let i be the current round number: $i = 1, 2, \cdots, N$.
2. Alice chooses a random bit $b_i \in \{0, 1\}$ and prepares the qubit state $|a_i\rangle = H^{b_i} |k_i^A\rangle$. Essentially, b_i is Alice's basis choice. If b_i is zero, Alice will send a Z-basis state of the form $|k_i^A\rangle$. Otherwise, if $b_i = 1$, she will send the state k_i but encoded as an X-basis state (i.e., if $k_i = 0$ she sends $|+\rangle$ otherwise she sends $|-\rangle$). The qubit is sent to Bob over the quantum channel. Note that, by H^0 we mean the identity operator.
3. Bob, on receipt of a qubit from Alice, will store the qubit in memory and inform Alice, using the classical authenticated channel, that he has received a qubit.
4. Alice, again using the authenticated channel, will tell Bob the value of b_i for this round.

5. Bob, will measure the received qubit in the correct basis indicated by b_i, leading to an outcome k_i^B. Note that, in the absence of any noise, it should hold that $k_i^B = k_i^A$. The above process repeats N times, for N set by the user.

It is not difficult to see that the toy protocol produces a correlated key between Alice and Bob, assuming the quantum channel is noiseless and that there is no additional party attacking the channel. At first glance, this toy version seems to be worthy enough as a cryptographic protocol, however there are several limitations that earn it the classification of "toy" protocol. Some are simply engineering details, for instance quantum memory is not currently feasible and, should it ever become so, it would no doubt be expensive and noisy. Thus, it does not make sense for Bob to store qubits in a memory and wait for Alice's basis information. However, some details are more critical than simply waiting for the right hardware. The protocol above, is either completely insecure, or completely unusable in that it has no noise tolerance whatever.

Indeed, consider the case of simple natural noise. Clearly it is unlikely that a perfect quantum communication channel could ever be fabricated – there is always at least some possibility of the qubit picking up some noise in transmission (which can include a $|0\rangle$ flipping to a $|1\rangle$ or a $|-\rangle$ flipping to a $|+\rangle$ with some non-zero probability). Thus, Bob's measurement result, even though he is measuring in the correct basis, will result in the wrong key-bit for some of the rounds. A cryptographic key, however, must be fully correlated – even a single bit out of place will result in bad behavior of the underlying cryptographic primitive.

However, consider a more sinister scenario, where an adversary launches a man-in-the-middle attack on the quantum channel. The attack consists of Eve intercepting every qubit sent and, with some probability p, making a Z basis measurement. How much information on the secret key k_A can she learn from this attack? On rounds where $b_i = 0$ (i.e., on rounds where Alice chooses to send Z basis states), Eve will learn k_A exactly (assuming there is no natural noise in the channel – natural noise will limit Eve's information gain, but not completely diminish it). On rounds where $b_i = 1$ (where Alice uses the Hadamard basis), Eve will learn nothing useful – however, her measurement, due to the resulting state collapse from an X basis state to Eve's observed Z basis state, will cause Bob's subsequent measurement to be totally random and uncorrelated with Alice's initial state $H|k_i\rangle$. Thus, Eve gains information from this attack on Z basis rounds, but she also creates disturbances and *noise* in the channel!

There are several important observations to be made from this toy version of BB84. First, note that the no-cloning theorem (see Section 2.5.4) forces any adversary to perform an *active* attack. That is, unlike with classical communication, Eve cannot simply copy the quantum data transmitted on the channel to attack at some future point in time (potentially decades later). Instead, Eve is forced to attack "in the moment." Second, a well-designed protocol forces Eve to betray her presence by this attack. That is, any attack which leads to information gain (even partial information gain) causes some form of disturbance in the quantum channel which the honest parties, Alice and Bob, may detect. Later, in Chapter 5, we will actually show something stronger: that, given a certain level of noise, Alice and Bob can actually upper-bound the maximal amount of information any adversary could have gained on the transmission. Given this, one may then use other means to "shrink" the transmitted key (which Eve has some partial information on), to a shorter secret key (which Eve has no, or negligible, information on). All of this will be discussed in much greater detail, and with greater rigor, later in the next chapter.

To mitigate the above shortcomings of the "toy" BB84, we need a few additions. First, to remove the need for Bob's quantum memory, Bob will, on receipt of a qubit from Alice, simply choose his own random basis to measure in, call it $c_i \in \{0, 1\}$ and immediately measure the qubit in that basis (using the Z basis if $c_i = 0$ and the X basis if $c_i = 1$). Now, his choice is independent of Alice's and so will be "wrong" half the time. Alice, therefore, must still broadcast, using the authenticated classical channel, her choice of b_i, otherwise Bob will not know if he measured correctly. Bob will tell Alice, using the same channel, whether he measured correctly or not (i.e., if $c_i = b_i$). If he did not ($c_i \neq b_i$), then both Alice and Bob will discard this round and start over on step one (making independent random choices the second time – importantly, Alice should not deterministically make the same choices on a round after discarding since Eve would know, with certainty, the basis choice as we must assume she is listening to the classical channel). This allows us to remove the memory requirements, though we do decrease the efficiency of the protocol as up to half the rounds may be discarded due to incorrect basis choices. There are ways to mitigate this concern that we discuss later. Note that, we also no longer think of QKD as sending a particular fixed key, but instead the key is built, round by round.

To remove the security concerns, we need Alice and Bob to try and detect abnormal behavior in the channel. They must sample their measurement data and determine the noise level of the quantum channel. That is, they must choose a random subset of the rounds that were not discarded and, on those rounds, disclose all relevant information – including k_A and k_B for those rounds chosen for sampling. Of course, this must be done over the authenticated channel and so Eve learns these disclosed bits – thus, they must also be thrown out afterwards and cannot be used for any secret key material.

But, sampling, of course, only lets Alice and Bob determine the noise level – it doesn't do anything about the noise itself that is still existent within the distilled key material. Thus, users must next run an error correction protocol. This process, running over the authenticated channel, necessarily leaks additional information to the adversary Eve. Indeed, if Alice (or Bob) is telling Bob (or Alice) exactly how to "fix" their key, it only makes sense that Eve will potentially be able to use this information *somehow* to gain additional information, beyond that which her attack may have given her initially. This information leakage must be taken into account in any security proof.

Now, Alice and Bob finally have key material that is perfectly correlated. But, the adversary has some information on it. Thus, one last process must be performed – *privacy amplification*. This is a process which hashes (using a two-universal hash function [9,20]) the error-corrected key material into a final, smaller, secret key. How small the final secret key needs to be depends on how much information Eve has on the original key-material – which, itself, can be upper-bounded based on the observed noise in the channel determined during the sampling stage.

In summary, the "toy" BB84 protocol may be hardened into a true QKD protocol by adding: (1) sampling; (2) error correction; and (3) privacy amplification.

4.2 Stages of a QKD protocol

From our discussion above, it is clear that at least with BB84, but in fact almost all QKD protocols, like any other communication protocol, consists of several stages of operations. Furthermore, like any quantum algorithm, quantum communication, specifically the task of QKD, also requires both quantum and classical communication and processing. In fact, taken as a whole, the actual quantum portion of a protocol may be minimal – yet it is the vital component allowing information theoretic security to be achieved.

Though there are numerous QKD protocols in existence (some of which we discuss later in this, and subsequent, chapters), in general QKD protocols operate in two distinct stages: a *Quantum Communication Stage* and a *Classical Post-processing Stage.* These stages, of course, may then be further broken up into additional sub-stages and sub-protocols.

The goal of the quantum communication stage is to distill a so-called *raw key.* This is the only part of a QKD protocol which actually utilizes quantum mechanics. Typically, Alice and Bob will stream qubits (or some other quantum resource state) to one another. Parties will also perform measurements on the quantum states. From these operations, a classical raw key is distilled which we denote r_A and r_B (Alice and Bob have their own raw keys). In the ideal case, $r_A = r_B$, however noisy quantum channels and, indeed, adversarial probing, will cause there to be discrepancies. Furthermore, an adversary may have a non-negligible amount of information on these raw keys. Thus, this key cannot be used immediately for cryptographic purposes and instead must be further processed through classical means in the second stage.

The second stage is responsible for converting the produced raw key into a final secret key. First, random sampling is performed whereby users can gain some insight into the behavior of the quantum channel. In general, they estimate the overall "noise" in the channel (though other statistics, such as mismatched measurements [3,19,15,12] can also be considered). This noise may be natural noise (which is inevitable), or due to some adversarial attack. While we discuss security of QKD in the next chapter, we point out that there is no good mechanism currently to distinguish natural from artificial (i.e., adversarial) noise and, as a consequence, all observed noise is considered adversarial, the worst-case possible for Alice and Bob. Once the channel behavior has been suitably determined (and this may require sacrificing some raw key material to also determine the noise level there), an *error correction* and a *privacy amplification* protocol are run. The first allows Alice and Bob to "fix" their respective raw keys so that $r_A = r_B$. The second, privacy amplification, is a process which hashes this error-corrected raw key into a smaller secret key upon which Eve has negligible information on (what exactly we mean by that will be more clear in the next chapter, when we discuss QKD security in detail). Returning to BB84, we may decompose it as shown in Protocol 4.1.

Before leaving this sub-section, let's discuss the several tunable parameters mentioned in Protocol 4.1. Determining values for these settings is an optimization problem that will be more clear in the next chapter. Indeed, determining the appropriate values of N along with p_Z and m (the sample size) are important to optimize both for performance and security. For example, setting $p_Z = 1$ would clearly improve the overall performance of the system, as less rounds are discarded due to incorrect basis choices. However, by doing so, Eve will get full information without causing any detectable disturbance! (See Exercise 4.1.) Indeed, recall from Chapter 3 Kerckhoffs's principle: We must always assume the adversary knows the protocol – including any public parameter choices! Thus $p_Z < 1$; but often it is not necessary to choose $p_Z = 1/2$.

Exercise 4.1. Assume that Alice and Bob set the protocol so that $p_Z = 1$. Show an attack that gives Eve full information, without causing any detectable disturbance. How is this attack "caught" if $p_Z < 1$? Repeat this question for $p_Z = 0$.

How can N affect anything? Note that QKD operates typically in "spurts." Time must be spent distilling a large block of data (namely, the system runs for N rounds in the quantum communication stage). Then this N-block of data is processed to distill a secret key of size ℓ. We will see later that the secret key size depends on many factors, including the observed noise or error rate (Q), as this will be used to provide users with an upper-bound on Eve's information. The higher the noise, the smaller ℓ will be (since Eve knows more of r^A thus it must be shrunk further to a secret key on which Eve has no information on). But, sampling can only give us a confidence interval – given an observed Q and a sample size of m, the actual noise will, with high probability, be within a range $[Q - \delta, Q + \delta]$ for some δ which vanishes as $m \to \infty$. But, m, of course, is bounded by N – thus, the larger N, the larger m can be, the smaller δ which optimizes ℓ to a point. We say "to a point" since, of course, the larger m is, the more Alice and Bob lose from their raw key material since the sample set is necessarily discarded. Clearly there is a lot of catch and play with these parameters!

4.3 Basic QKD protocols

If one were to survey the majority of QKD protocols in existence, there may be two main encoding styles. The first, like BB84 [5], encodes key material in the state itself, where a measurement produces a classical outcome which is directly translatable to a raw key bit. For instance, a "1" is encoded as either a $|1\rangle$ or a $|-\rangle$. A measurement outcome of $|+\rangle$ decodes to a classical "0." In general, to successfully distill a key-bit, Alice and Bob must both use the same basis, both for state preparation and for the subsequent measurement. While there is no standard name for this, we call it *state encoding* here.

An alternative is to encode key material in the actual basis which we call *basis encoding*. To decode the key-bit successfully, parties must use *opposite* bases! This rather counter-intuitive fact can be seen by looking at the so-called B92 [4] QKD protocol. B92 was created

Protocol 4.1 BB84: [5,14].

Public Parameters:

- N: number of rounds
- $m \leq N$: sample size
- p_Z: probability of choosing the Z basis
- $p_X = 1 - p_Z$: probability of choosing the X basis.

Quantum Communication Stage: For each *round*, $i = 1, 2, \cdots, N$, do:

1. Alice chooses a random raw key bit r_i^A and random basis $b_i^A \in \{0,1\}$, choosing $b_i^A = 0$ with probability p_Z, and creates the state $H^{b_i^A}|r_i^A\rangle$. This state is sent to Bob.
2. Bob, on receipt of a qubit will choose a basis $b_i^B \in \{0,1\}$ (independently of Alice, but choosing the Z basis with probability p_Z) and measures the state he received in basis Z if $b_i^B = 0$ or basis X if $b_i^B = 1$. The result is stored as a bit r_i^B with the understanding that a measurement of $|+\rangle$ translates to a classical 0 while a $|-\rangle$ translates to a 1.
3. Goto (1)

Alice now discloses her string of basis choices b^A. Bob sends back $b^A \oplus b^B$. Alice and Bob discard all raw key bits i, where $b_i^A \oplus b_i^B = 1$ (as that indicates Alice and Bob choose different bases for round i). This communication is all done over the classical authenticated channel. Let $N_0 \leq N$ be the size of the final raw key denoted r^A and r^B. It is expected that $N_0 \approx (p_Z^2 + p_X^2)N$.

Note: Often, only the Z basis rounds will be used for the raw key, while the X basis rounds will be used for testing the channel. We will use this variant of the protocol, in the next chapter when proving security.

Classical Post Processing Stage:

1. Alice chooses a random subset t of size $m < N_0$ of all raw key-bits that were not discarded, ensuring that a sufficient number of Z and X basis rounds are included in the subset. Alice broadcasts the following information over the authenticated classical channel:
   - t: the subset chosen.
   - r_t^A: Alice's raw key indexed by t – namely those bits of Alice's raw key that are indexed in t.
2. Bob receives this data and discloses r_t^B, his raw key string indexed by subset t. Both Alice and Bob create a new raw key r^A and r^B by discarding those bits indexed in t (since they were leaked fully to anyone snooping on the classical channel).
3. The above data allow Alice and Bob to determine the *error rate* in the quantum channel, namely $Q = w(r_t^A \oplus r_t^B)$, where $w(x)$ is the *relative Hamming weight* of x, namely the number of "ones" in x, divided by the bit-length of x. As we will see later, Alice and Bob may actually want to distinguish between the error rate in the Z basis (those bits in r_t^A where $b^A = 0$) from the X basis separately. This will be important later, but for this chapter, it is sufficient to compute the overall error rate Q – namely the relative number of 1's in the string $r_t^A \oplus r_t^B$. Let N_1 be the size of the raw key at this point (i.e., $N_1 = N_0 - m$).
4. Finally, Alice and Bob run an error correction protocol (to be discussed later in Section 4.4.1). Let $\tilde{r}$ be the error corrected raw-key (note there is no need for a superscript at this point as Alice and Bob should have the same raw-key – error correction failure will be dealt with later). Alice now chooses a random two-universal hash function $f : \{0,1\}^{N_1} \rightarrow \{0,1\}^{\ell}$ (the exact definition of this will be covered later in this chapter) and sends a description of this function to Bob (over the authenticated classical channel). Alice and Bob then compute $s = f(r)$. Here, s is the final secret key. The size of s in bits, denoted ℓ, depends on the output size of f which Alice chooses appropriately and will be the subject of the next chapter.

in 1992 by one of the same B's from BB84, namely Bennett, and it was designed to be one of the simplest QKD protocols possible, at least from a source device perspective. Consider the source of the BB84 protocol, typically Alice. If she sends one of four possible states each round, one has the BB84 protocol. Clearly, if Alice can only choose to send one possible state to Bob (i.e., her source is capable of creating only a fixed single, publicly known, quantum state), no key can possibly be distilled. (Why?) Thus, the "smallest" possible sets of states from which Alice may choose to send from, to have any hope at all of having a secure QKD protocol, is two. This was the motivation behind the B92 protocol, the quantum communication stage of which is described, in detail, in Protocol 4.2.

Protocol 4.2 B92.

Public Parameters:

- N: number of rounds
- $m \leq N$: sample size

Quantum Communication Stage: For each round $i = 1, 2, \cdots, N$, do:

1. Alice chooses a random raw key bit r_i^A for this round. If $r_i^A = 0$, Alice prepares a qubit in the state $|0\rangle$; otherwise, if $r_i^A = 1$, she prepares a qubit in the state $|-\rangle$. The qubit is sent to Bob.
2. Bob, on receipt of a qubit will choose a basis $b_i^B \in \{0, 1\}$, uniformly at random. If $b_i^B = 0$, Bob measures the qubit in the Z basis, otherwise he measures in the X basis.
3. If Bob measures in the Z basis and observes a $|1\rangle$, then he sets his raw-key bit for this round to be $r_i^B = 1$. If Bob measures in the X basis and observes a $|+\rangle$, then he sets his raw-key bit for this round to be $r_i^B = 0$. In both cases, the round is called *conclusive*; any other outcome results in an *inconclusive* round.
4. Bob informs Alice whether this round was conclusive or not. If it was inconclusive, both parties discard their data for this round. Note that parties never disclose their basis choice.
5. Goto (1)

Is the protocol correct? That is, at the end of the protocol, assuming noiseless channels, will Alice and Bob have the same raw key? Well, note that Bob calls a round *conclusive*, only if he observes a $|1\rangle$ or a $|+\rangle$. Also, note, that Alice *never* sends those two states. Thus, if Bob were to observe a $|1\rangle$, for instance, it is impossible that Alice had sent $|0\rangle$ (again, assuming no noise in the quantum channel). Thus, she must have sent a $|-\rangle$ implying that Alice's raw key-bit must have been $r_i^A = 1$, which is exactly what Bob sets his raw key bit to in this instance. Similar arguments apply for when Bob observes $|+\rangle$: Alice must have sent $|0\rangle$ since she never sends $|+\rangle$. Thus, assuming Bob finds the round conclusive, his raw key is guaranteed to match that of Alice's (assuming no error in the channel). Furthermore, note that the round can only be conclusive if Bob chooses the "wrong" basis to measure in – i.e., he chooses the opposite basis that Alice prepared in. This is in stark contrast to BB84 where Bob only gets shared key material if he measures in the same basis Alice sent in! Such an encoding scheme is an example of basis encoding – though, as you may suspect, it is not the only one.

Exercise 4.2. Building off of the discussion, above, show that the B92 protocol is correct. Namely that, if there is no channel noise, Alice and Bob will always agree on the same raw keys. Why does Bob never disclose his basis choice? Show that if he does, the protocol will no longer be secure.

4.3.1 SARG04

Let's consider another QKD protocol utilizing a basis-encoding scheme, the SARG04 protocol [1]. As is often done with the naming of early QKD protocols – though, notably, not all – this one is also named after the authors and the year of creation. Unlike B92, which was designed, really, to see if security is possible with the fewest possible states, SARG04 was designed to mitigate a particularly deadly attack against early QKD systems, the photon number splitting (PNS) attack. This attack will be discussed in Chapter 9, but imagine the following scenario: when Alice wants to send a qubit $|\psi\rangle$, sometimes, her source accidentally produces two copies of $|\psi\rangle$, namely, Alice actually sends to Bob the two-qubit state $|\psi\rangle|\psi\rangle$. This is actually a common occurrence in practical QKD sources. Practical qubit sources create single qubits with probability p_1 and two or more photons with some probability p_2 which is strictly positive (though hopefully not too large). They may also create no qubits. Unfortunately, while the distribution may be known (the p_i values), it is never clear on any particular round, whether zero, one, or more qubits left your source.

Now, imagine a round of BB84 where Alice wants to send $|+\rangle$ (i.e., she encodes a classical bit of zero on a Hadamard basis state). If her source accidentally sends $|+,+\rangle$ (two copies), Eve, who sits between Alice and Bob, intercepts this signal and captures one of the photons and sends the other to Bob. Alice and Bob are both oblivious to this interception: Alice, because she never actually knows when her source accidentally sends two photons, Bob because he cannot count photons (this is a technological limit, currently, not a physical one) and so receives a single photon as expected. Eve, who we assume is only bounded by the laws of physics and, therefore, has access to perfect quantum memory, stores the second qubit of Alice's in her memory, waiting for Alice's broadcast of b_i^A. At this point, Eve may make the correct measurement and learn, exactly r_i^A.

Thus, on any round where Alice sends two or more qubits accidentally, Eve gets full information. On any round where Alice sends one photon, Eve is limited in her information gain. A basic PNS attack can lead to Eve gaining p_2N bits of information on the raw key without creating any *detectable* noise. She is creating noise in a sense – she is taking a qubit away from the signal Alice is sending. The problem is, this noise is not detectable with current-day technology; instead we must rely on countermeasures such as SARG04, discussed below, or the decoy state protocol, discussed in Section 9.2.

Note that this does not violate the no-cloning theorem (see Section 2.5.4)! Indeed, that theorem only applies to an unknown state. However, Alice's source creates two copies of the given qubit – it (the source device) knows the qubit state since it's the one creating it. Eve, of course, is not actually cloning the qubit – she's being given a clone "for free" from Alice and is simply taking it and saving it. Note that this PNS attack can only work when the number of qubits sent on a single round is strictly greater than one thanks to the no cloning theorem. Thus, the smaller p_2, the less an issue this attack becomes (though it must still be considered!).

SARG04 was designed with this practical implementation issue in mind. In particular it uses basis encoding to ensure that, even on two photon transmissions, Eve cannot gain full information on the state. This is one of the advantages of many basis encoding QKD protocols.

Instead of presenting the entire protocol in the style of Protocols 4.1 and 4.2, we will simply give the highlights of a single round of the quantum communication stage. First, Alice chooses a random key bit k_i^A to attempt to transmit to Bob. We say "attempt" since, as may be clear from previous examples, QKD protocols where each round deterministically produces a key-bit are rare (though they do exist). Alice also chooses a basis b_i^A and sends the state $|\psi\rangle = H^{b_i^A}|k_i^A\rangle$ to Bob. On receipt of the qubit, Bob measures in a randomly chosen basis b_i^B to receive outcome $m_i^B \in \{0, 1, +, -\}$. Next, Alice, when she is certain Bob received the qubit (normally this is done at the very end of the quantum communication stage at which point Bob would accept no further qubits in his lab for processing), will pick a random state $|\phi\rangle$ in a basis opposite to the one she prepared and sent a qubit in initially. For example, if she sent the state $|+\rangle$ she may choose as her random state either $|0\rangle$ or $|1\rangle$ (each chosen uniformly at random). Alice then sends to Bob, a classical description of $|\psi\rangle$ and $|\phi\rangle$ (choosing to broadcast the computational basis state first, and then the Hadamard basis state second to avoid leaking any information to Eve). Bob will call the round conclusive only if his measurement outcome m_i^B is *not* one of the two states Alice described to him.

Exercise 4.3. Argue why the above SARG04 protocol is correct. Namely, show that, in the absence of channel noise, Alice and Bob will have the same raw key. Discuss why the protocol may be secure. How can an attack be detected?

Note the similarities to B92. Again, this is a basis encoding scheme where Bob must actually choose to measure in the opposite basis that Alice sent in to possibly get a conclusive result. Indeed, if he chose the same basis Alice used, he would always receive a measurement outcome equal to one of the two states Alice describes since it's always the case that one of the states Alice describes to Bob is the state she actually sent! The only way to get a state not in this classical message from Alice is to measure in the opposite basis (with 50% probability) and observe a state opposite to the "decoy" state (Alice's randomly chosen state in the basis opposite to that from which she sent from). In a way, this can be considered, at some level, a version of B92 where the encoding system (i.e., the choice of $|0\rangle$ for 0 and $|-\rangle$ for 1) is chosen randomly each round.

Why does this help against the PNS attack? Let's consider a round where $k_i^A = 0$ and $b_i^A = 1$. Thus, Alice sent the state $|+\rangle$ to Bob. However, her source, being imperfect, actually sent two copies, namely $|+,+\rangle$, one copy of which was captured by Eve. In BB84, Alice would eventually broadcast b_i^A to Bob, thus giving Eve full information on the transmission (as she would measure in the Hadamard basis and observe $|+\rangle$, thus decoding to a 0). However, in SARG04, Alice broadcasts the actual state sent, and a decoy state – let's say she broadcasts the classical message $M = \{``1", ``+"\}$. That is, one state which is what she sent, and one randomly chosen state in the opposite basis (computational basis in this case).

Bob, who measured in a random basis, simply tells Alice whether the round was conclusive or not. If it was not conclusive, the round is discarded so there is nothing for Eve to learn. What if the round was conclusive (and Eve now knows this)? Well, she still cannot learn, faithfully, the state that Alice sent – she can make a measurement of her own (randomly chosen) and if she gets lucky and chooses the computational basis, there is a 50% chance she will observe a $|0\rangle$ thus indicating to her, given message M, that Alice must have sent $|+\rangle$. Of course, there is a 50% chance that Eve will observe a $|1\rangle$ in this case causing Eve to be uncertain of the state Alice sent. And, of course, if Eve happens to choose the same basis Alice used, she will be uncertain again, since she would observe $|+\rangle$, a state in M. Therefore, even given the exact copy of the state Alice sent and a copy of all classical messages sent between Alice and Bob, Eve can never be absolutely certain of Alice's raw key bit all of the time in this two-photon case – a stark contrast to the 100% information leakage for two photons in BB84!

4.3.2 Additional QKD protocols

Before leaving this section, we'll briefly introduce two more QKD protocols: the so-called *three-state* BB84 protocol [12,7] and the entanglement-based E91 protocol [11]. Both are state encoding systems and both demonstrate interesting features of QKD protocol design.

First, the three state BB84 protocol is exactly as described: take BB84 and remove Alice's ability to generate one of the four qubit states. Typically, it is one of the Hadamard states. For concreteness, we'll define the protocol (or, rather, just the quantum communication stage) as repeating the following process:

1. Alice, on each round $i = 1, 2, \cdots, N$, will choose, randomly, whether this round will be a *key round* or a *test round.* If the former, Alice will choose a random k_i^A and send to Bob the qubit $|k_i^A\rangle$ (i.e., a Z basis qubit encoding the key-bit Alice is attempting to transmit to Bob). Otherwise, if this is a test round, Alice will send the state $|+\rangle$ to Bob. Unlike standard BB84 (Protocol 4.1), Alice is not allowed to send the state $|-\rangle$.
2. Bob, on receipt of a qubit, will choose a random basis to measure in (either Z or X) and measure the received qubit in that basis.
3. Alice and Bob inform one another, using the public channel, of their basis choices. If both used the Z basis (i.e., Alice chooses this round to be a key-round), then Alice will use k_i^A as her raw key bit while Bob will use his Z basis measurement outcome as his raw key bit. Otherwise, if both parties chose the X basis, this round will be used to test the noise in the channel – Bob should have received a measurement outcome of $|+\rangle$ and any other outcome is reported as "noise" in the channel.

The above three state BB84 shows an interesting feature of QKD, namely that even in state encoding systems, all basis states are not necessarily required to gain security. Keep in mind, however, that Eve's attack induces disturbances in the channel which can only be detected by making certain measurements. In the standard BB84 protocol, Alice and Bob can estimate the probability of a $|i\rangle$ flipping to a $|1-i\rangle$ and a $|\pm\rangle$ flipping to a $|\mp\rangle$. In the three-state BB84 protocol, however, Alice and Bob cannot directly observe the probability of Eve's attacking causing a $|-\rangle$ to flip to a $|+\rangle$ since the former state is never sent by Alice. This lack of information allows Eve to potentially gain more information on the raw key with the same *observable* disturbance level – in a way, Eve may "hide" a stronger attack that would normally create a lot of noise in the $|-\rangle$ to $|+\rangle$ sense – but that noise goes undetected. As it turns out, there are classical sampling countermeasures to this that hardens the protocol and makes it just as efficient, asymptotically, as BB84. These countermeasures are called *mismatched measurements* [3]. Interestingly, by looking at events such as "what's the probability that Bob observes a $|0\rangle$ given that Alice sent a $|+\rangle$?" – an event that is normally discarded since this can only occur when parties use opposite bases, i.e., *mismatched* bases – one can better bound Eve's information gain and counteract any deficiency imposed by the loss of Alice's ability to send a $|-\rangle$ [18].

Exercise 4.4. Discuss why the above three-state BB84 protocol may be secure. Show that, at least for basic measure-resend attacks where Eve measures every signal in some basis, this attack will be detected.

The last QKD protocol we'll cover, for now, is the E91 protocol [11] (actually, we'll describe the BBM92 version [6] which uses a testing stage similar to BB84). So far, all the protocols we discussed are *prepare-and-measure* protocols. Such systems involve Alice preparing a qubit and sending it to Bob who, subsequently, measures it. The preparation and measurement outcomes are distilled, somehow, to raw key material and channel noise estimations. An alternative to this mechanism are *entanglement-based* protocols, a class of protocols which E91 belongs to. In such a setup, an entangled pair is created, e.g., the standard Bell pair $\frac{1}{\sqrt{2}}|0,0\rangle + \frac{1}{\sqrt{2}}|1,1\rangle$, with one particle being sent to Alice and one particle sent to Bob. Who exactly creates the entangled pair depends on the protocol, though often we assume the worst case in that Eve, the adversary, actually creates the initial state! However, in practice, it could be an honest third-party, or even Alice who creates the pair, keeping one particle to herself and sending one particle to Bob.

Regardless of who creates the initial state, Alice and Bob each hold one part of an entangled pair. The protocol then, basically, acts like BB84, or at least the Bob-side of BB84. On receipt of their particle, Alice and Bob will each choose, independently, a basis to measure in, b_i^A and b_i^B (they may bias their choice as with standard BB84 discussed earlier). After measuring in that basis, they will broadcast their basis choices to each other over the authenticated classical channel. If their basis choice matches, they keep their measurement outcomes as their raw key bit. Otherwise, they discard the round and repeat with a new entangled pair for round $i+1$. Why this protocol is correct can be seen from the fact that $\frac{1}{\sqrt{2}}(|0,0\rangle + |1,1\rangle) = \frac{1}{\sqrt{2}}(|+,+\rangle + |-,-\rangle)$ (see Exercise 2.16). Thus, if Alice and Bob measure in the same basis, they will be guaranteed to get a correlated result.

The fact that Eve is allowed to create the quantum state is an interesting proposition, and in fact vital to many QKD security proofs. In fact, as we will discuss later, though most "practical" QKD systems are of the prepare-and-measure variety, often to prove their security, one reduces the prepare-and-measure version to an equivalent entanglement-based version. In fact, it can be shown that, mathematically, BB84 is equivalent to E91!

How can this be? It turns out that Alice sending a state in BB84 (say $|+\rangle$) is equivalent to her measuring and observing that state in E91. To see this at a high level, consider the following scenario: Let's say Alice is running the prepare-and-measure BB84 protocol and that Eve is intercepting all qubits sent and "probing" them with a unitary probe. This probe is an isometry (see Exercise 4.5) which acts on the traveling qubit and Eve's private ancilla in the following manner:

$$U|0\rangle = |0, e_0\rangle + |1, e_1\rangle$$
$$U|1\rangle = |0, e_2\rangle + |1, e_3\rangle$$

Bob will then receive the qubit system (the left-most register) and measure in either the Z or X basis. Let's consider the possible outcomes of the system before Bob measures:

$$\begin{array}{ll} |0, e_0\rangle + |1, e_1\rangle & \text{If Alice sent a } |0\rangle \\ |0, e_2\rangle + |1, e_3\rangle & \text{If Alice sent a } |1\rangle \\ \frac{1}{\sqrt{2}}|0\rangle \otimes (|e_0\rangle + |e_2\rangle) + \frac{1}{\sqrt{2}}|1\rangle \otimes (|e_1\rangle + |e_3\rangle) & \text{If Alice sent a } |+\rangle \\ \frac{1}{\sqrt{2}}|0\rangle \otimes (|e_0\rangle - |e_2\rangle) + \frac{1}{\sqrt{2}}|1\rangle \otimes (|e_1\rangle - |e_3\rangle) & \text{If Alice sent a } |-\rangle \end{array} \tag{4.1}$$

(The last two lines of the above table are easily derived by linearity of U.)

Exercise 4.5. Show that U, defined above, is an isometry. Namely, show that $\langle a|b\rangle = \langle v_a|v_b\rangle$, where $|v_a\rangle = U|a\rangle$ (and similarly for $|v_b\rangle$). Note that all isometries can be dilated to a unitary operator and, thus, is an allowed operation for Eve.

Now, let's consider E91. Assume that, instead of Alice sending qubits and Eve probing them, Eve simply creates the state:

$$|\psi\rangle = \frac{1}{\sqrt{2}}(|0,0\rangle_{AB} \otimes |e_0\rangle + |0,1\rangle_{AB} \otimes |e_1\rangle + |1,0\rangle_{AB} \otimes |e_2\rangle + |1,1\rangle_{AB} \otimes |e_3\rangle) \tag{4.2}$$

Manipulating the above, and taking advantage of the properties of tensor product (see Section 2.5), we can write the above state as:

$$|\psi\rangle = \frac{1}{\sqrt{2}}|0\rangle_A \otimes (|0, e_0\rangle_{BE} + |1, e_1\rangle_{BE}) + \frac{1}{\sqrt{2}}|1\rangle_A \otimes (|0, e_2\rangle_{BE} + |1, e_3\rangle_{BE}) \tag{4.3}$$

Now, let's run the E91 protocol – if Alice measures her system in the Z basis, the probability that she observes 0 is found to be:

$$Pr(A=0) = \frac{1}{2}(\langle e_0|e_0\rangle + \langle e_1|e_1\rangle) = \frac{1}{2}. \tag{4.4}$$

Note that the last equality follows from the fact that, since U is unitary: $1 = \langle 0|0\rangle = \langle 0|U^*U|0\rangle = \langle e_0|e_0\rangle + \langle e_1|e_1\rangle$. Similarly, one can show that $\langle e_2|e_2\rangle + \langle e_3|e_3\rangle = 1$. Of course, $Pr(A=1)$ must also be $1/2$ then. Conditioning on Alice observing $|0\rangle$ or $|1\rangle$, therefore, the BE portion of the state collapses to:

$$\begin{array}{ll} |0, e_0\rangle + |1, e_1\rangle & \text{If Alice observes } |0\rangle \\ |0, e_2\rangle + |1, e_3\rangle & \text{If Alice observes } |1\rangle \end{array} \tag{4.5}$$

Thus, if Alice chooses to measure in the Z basis, she will observe $|0\rangle$ or $|1\rangle$ equally likely and, furthermore, the BE system collapses to exactly the same state in BB84 as if she had chosen to send $|0\rangle$ or $|1\rangle$!

The same arguments can be made in the event Alice chooses to measure in the X basis, though a bit more algebra is involved. First, we take Eq. (4.3) and change basis in Alice's register to the Hadamard basis:

$$|\psi\rangle = \frac{1}{2}|+\rangle_A \otimes (|0\rangle_B \otimes (|e_0\rangle + |e_2\rangle) + |1\rangle_B \otimes (|e_1\rangle + |e_3\rangle)) \tag{4.6}$$

$$+ \frac{1}{2}|-\rangle_A \otimes (|0\rangle_B \otimes (|e_0\rangle - |e_2\rangle) + |1\rangle_B \otimes (|e_1\rangle - |e_3\rangle)) \tag{4.7}$$

Now, the probability of Alice observing $|+\rangle$ is:

$$\begin{aligned} Pr(A = \text{``}+\text{''}) &= \frac{1}{4}(\langle e_0|e_0\rangle + \langle e_2|e_2\rangle + 2Re\,\langle e_0|e_2\rangle + \langle e_1|e_1\rangle + \langle e_3|e_3\rangle + 2Re\,\langle e_1|e_3\rangle) \\ &= \frac{1}{4}(1 + 1 + 2Re(\langle e_0|e_2\rangle + \langle e_1|e_3\rangle)) \\ &= \frac{1}{2} + Re(\langle e_0|e_2\rangle + \langle e_1|e_3\rangle) \end{aligned} \tag{4.8}$$

For the above, we took advantage of the fact that $\langle x|y\rangle = (\langle y|x\rangle)^*$ (due to standard properties of inner-product as discussed in Section 2.1.2) and, also, that for any complex number z, it holds that $z + z^* = 2Re(z)$. The above is almost what we want (namely, that Alice observes $|+\rangle$ with probability $1/2$). However, note that, since U is an isometry:

$$0 = \langle 0|1\rangle = \langle 0|U^*U|1\rangle = \langle e_0|e_2\rangle + \langle e_1|e_3\rangle$$

Thus, we conclude that $Pr(A = \text{``}+\text{''}) = 1/2$ (which, naturally, forces $Pr(A = \text{``}-\text{''}) = 1/2$ also).

Finally, if Alice measures in the Hadamard basis and observes $|+\rangle$, it can now be seen that the collapsed, post-measured state, will match *exactly* that of BB84 as if she had sent a

$|+\rangle$. Similarly for observing $|-\rangle$. Thus, *any attack against BB84 can be "simulated" by Eve in the E91 protocol.* This means that if we prove security of E91, we will prove security of BB84. This is a technique we will discuss more in the next chapter; we only highlighted it here to show how close prepare-and-measure protocols can be related to entanglement-based ones, even if, at first glance, they seem very different. In fact, most QKD security proofs involve finding an equivalent entanglement-based protocol and proving security of that protocol (since, for technical reasons that will hopefully be more clear in the next chapter, proving security of entanglement-based protocols is actually often easier than prepare-and-measure ones).

Exercise 4.6. Show that if Alice observes $|-\rangle$ and Bob observes $|+\rangle$ in the entanglement-based version of the above state (Eq. (4.3)), the post measured state is identical to that of the prepare-and-measure version, if Alice had sent $|-\rangle$ and Bob observes $|+\rangle$.

4.4 Basic post processing techniques

So far, we have looked at several QKD protocols, however you will notice that when introducing a new QKD protocol, we almost exclusively only discussed the quantum communication stage of the protocol, leaving the classical post processing stage out. This is due to the "modular" nature of many QKD protocols: the quantum communication stage distills a raw key, while the classical post processing stage transforms that into a final secret key. One may use the same classical processes (to varying degrees of effectiveness) with many different quantum communication processes. For instance, the same classical stage can be used for BB84, SARG04, and B92. In fact, much current QKD research revolves around two themes: (1) designing new quantum communication stages and (2) designing faster or more effective classical strategies. Often the two research themes can run, at least in part, in parallel as any benefit in, say, (2) will most likely work for different protocols from (1). Though, of course, this is not always a hard-and-fast rule: sometimes classical strategies, or their analysis, are protocol specific. In general, there are *many* classical strategies that one can employ to boost the overall performance of a QKD protocol. Of course, here, we can only discuss some of the major topics.

4.4.1 Error correction

We will begin by discussing one of the *mandatory* QKD classical processes: error correction. This will take the noisy raw key and output a correlated key (between Alice and Bob). It may be an interactive or a non-interactive protocol, however it does require some classical communication. This classical communication is, of course, performed over the authenticated classical channel and, thus, actually leaks additional information to Eve. Thus, research efforts are constantly being performed towards (1) increasing the overall speed of error correction and (2) decreasing the amount of leakage of an error correction protocol.

Let's first consider the problem formally. Let r_A and r_B be Alice and Bob's raw keys, each of which are N bits long and where the relative error rate is no larger than some value Q, i.e., $w(r_A \oplus r_B) \leq Q$ (recall that $w(x)$ is the relative Hamming weight of the bit-string x). An error correction protocol is a two-party protocol which, when finished, consists of a transcript of classical messages sent between Alice and Bob (denoted $c \in \{0, 1\}^*$) and a new error corrected string for Alice and Bob, denoted $\tilde{r}_A$ and $\tilde{r}_B$ (which may be smaller than the original). Note it could be that $\tilde{r}_A = r_A$ and that Bob is the one who changes his string to match Alice's, or vice versa. An error correction protocol is said to be ϵ_{EC}-correct [17] if:

Definition 4.4.1. Let r_A and r_B be drawn from some distribution R_{AB}. Then an EC protocol is ϵ_{EC}*-correct on* R_{AB} if for any such r_A and r_B, it holds that:

$$Pr\left(\tilde{r}_A \neq \tilde{r}_B\right) \leq \epsilon_{EC}.$$

Furthermore, the protocol is called *fully* ϵ_{EC}*-correct* if the above holds for any probability distribution R_{AB}.

Note the probabilistic nature of the above definition. It is allowed to fail, but a fully-correct protocol will only fail with a small probability for any given r_A and r_B. Naturally, designing a good EC protocol is going to involve more than just satisfying the above definition – especially with respect to the amount of leakage an EC protocol produces. Since EC will be run on a "partially secret" key (the raw key of the QKD protocol), the more the EC protocol leaks (communicates), the more information an adversary potentially gains "for free." Thus, it is important for an EC protocol to leak as little information as possible.

4.4.1.1 The Cascade protocol

Two error correction protocols are often used in QKD: Cascade [8] and Low Density Parity Check (LDPC) codes [13]. The first was introduced by Brassard and Salvail – the first author you may recognize as one of the B's in BB84. The second is more efficient in terms of information leakage, in general, and thus more commonly used, though more complicated to understand. We'll spend some time discussing the Cascade protocol here, though only at a high level, leaving proofs of correctness to the original source.

First, the Cascade protocol [8] defines a sub-routine called `Binary` which acts on strings A and B, each of n bits, with an odd number of errors in them. That is, the number of discrepancies between A and B is odd. The protocol is an interactive one, exchanging at most $\lceil \log_2 n \rceil$ bits of information (this information exchange is considered information leakage at the end of the day). This protocol acts as follows:

1. First, Alice sends Bob the parity of the first half of her string A.
2. Bob tests the parity of the first half of his string and compares it to the parity he received from Alice. If they do not match then there must be an odd number of errors in the first half of the string.

3. The above repeats on the half of the string from which an odd number of errors occurred and, eventually, an error will be discovered.

The above procedure is basically a binary search algorithm searching for a location with an error by repeatedly splitting the string in half and seeing if the odd number of errors is in the left or in the right sub-string and recursing down that path of the search.

To illustrate, consider the following simple example where $A = 001101$ and $B = 101110$. Note that there are three errors in these strings. An application of `Binary` first has Alice transmitting the message 1 (the parity of her first half of the string, namely 001). Bob's parity on the first half of his string is 0 (the parity of 101) and so he informs Alice of this and they then repeat the procedure on the left three bits; i.e., they run `Binary` on input $A_2 = 001$ and $B_2 = 101$. Alice, sends her parity 0 on the left half (we round down in this example, so the left half of her string is simply 0). Bob then realizes that his parity, being 1, leads to the error being in the left half. Since the left half is only one bit long, he has discovered the faulty bit and can flip it. The output of the procedure are the strings $A_{fix} = 001101$ and $B_{fix} = 001110$. Of course, there are still two more errors in the string: in fact there are an even number of errors now, and so one cannot simply run `Binary` again. Instead, one must "shuffle" the remaining string and choose random subsets to look at, hoping to narrow down to subsets containing only an odd number of errors.

Of course the above leaked information to Eve. It leaks at most $O(\log n)$ bits (where n is the bit-size of the raw block being considered) which must be taken into account when considering the security of a QKD protocol – an issue we save for Chapter 5.

The actual Cascade algorithm works over multiple passes. How many, exactly, is a parameter that can be optimized by Alice and Bob: too many passes and you leak more information than needed, while too few and you may miss errors, causing the error correction protocol to fail. Again, let A and B be the raw keys that must be corrected, each of n bits (though, this time, not necessarily having only an odd number of errors). On the first pass, Alice and Bob divide their raw key into blocks of size k_1 bits (chosen by the users and something we will discuss shortly). For each block, Alice sends the parity of that block to Bob. If the parity does not match on Bob's side, then they know an odd number of errors are in that particular block, at which point they run `Binary` to correct one of those errors in that particular block. After pass one, all blocks have an even number of errors (or, of course, no errors). That concludes the first pass of Cascade.

Now, on pass two, the string is divided into blocks of size k_2, but also shuffled in some manner. In particular, a random function f_2 is chosen, which maps:

$$f_2 : \{1, 2, \cdots, n\} \to \left\{1, 2, \cdots, \left\lceil \frac{n}{k_2} \right\rceil \right\},$$

Essentially, this function clusters all n bits from the original string into n/k_2 blocks such that bit j is in block $f_2(j)$. Note, it is not required that block sizes are equal. For each of these blocks, Alice sends the parity and, if Bob's parity is different, they know an odd number of errors have occurred in that block; at that point, they run `Binary` on that block to fix one error (each execution of `Binary` leaking information as discussed, above). The above continues for multiple passes, each one choosing a random k_i, and a new function f_i. At the end, one can expect, with high probability, that all errors have been corrected, though this should be verified, as discussed below.

Exercise 4.7. Implement the Cascade protocol in a language of your choice. In particular, have your program generate random, but noisy, raw keys, where the noise level, and the number of bits, can be set by the user. Then implement a version of Cascade, keeping track of the number of bits "leaked" (or transmitted) by the `Binary` algorithm. Run your program on various noise levels, and see how the number of leaked bits increases as the noise increases.

4.4.1.2 Verifying error correction

Note that the above demonstrates how error correction is a probabilistic process which may fail (though, hopefully, the failure rate is negligible). Given this, how can one be sure that error correction actually worked? One might consider choosing a random sample of the error corrected raw key and revealing that sample to Bob over the authenticated channel. If it matches, then, assuming the sample was large enough, Alice and Bob should be certain their raw keys match. However, since this was done over the public authenticated channel, this sample must then be discarded. Since the sample must be fairly large to gain any confidence in the error correction result, and since we only need a "yes" or "no" result, this is a very wasteful method.

Instead, what is often done, is Alice and Bob send a hash of their error corrected raw keys to one another. Or, rather, Alice sends a hash of hers to Bob and Bob checks if the digest (the output of the hash function) received from Alice matches the digest he computes locally using his raw key. The type of hash, however, must be chosen carefully so as to not break information theoretic security. Commonly, a so-called *two-universal hash* function is used.

Definition 4.4.2. A family of hash functions $\mathcal{F} = \{f : \{0,1\}^n \to \{0,1\}^\ell\}$ is said to be *two-universal* [10] if for any two distinct n-bit strings $x \neq y$, it holds that:

$$Pr\left(f(x) = f(y)\right) \leq \frac{1}{2^\ell}, \tag{4.9}$$

where the probability is over the uniform choice of hash function f from $\mathcal{F}$. See [9,20,17].

How does the above help? Let r_A and r_B be Alice and Bob's raw key *after* running the error correction protocol. Alice and Bob wish to know if $r_A = r_B$ without revealing too much information on the raw key. Consider the following verification protocol:

1. Alice chooses a hash function $f \in \mathcal{F}$, uniformly at random from a two-universal family of hash functions $\mathcal{F}$ and computes $z = f(r_A)$.
2. Alice sends to Bob the value z and a description of the function f (i.e., which choice of hash function Alice used).
3. Bob computes $z' = f(r_B)$ and sends to Alice a single bit: 1 if $z' = z$ and 0 otherwise.

Now, if error correction actually did succeed, then $r_A = r_B$ and so the above test will succeed. Thus, this protocol has no "false negatives" (i.e., if error correction works, Bob will always send the correct message of 1). What is the probability of a false positive? That is, what is the probability that Bob says 1 (i.e., error correction worked) when, in fact, it failed and $r_A \neq r_B$ (where the difference may even be only a single bit). From Definition 4.4.2, it is easily seen that the probability of this occurring is no larger than $2^{-\ell}$. The size of ℓ can, of course, be set by Alice and Bob. The total information leakage of this protocol is ℓ bits. Thus, by setting $\ell = 10$ for example, leaking 10 bits only, Alice and Bob can be certain that error correction worked, except with probability at most $1/2^{10} \approx 0.098\%$. By leaking 20 bits, Alice and Bob can be sure that error correction worked, except with probability 0.000095%. Considering raw keys are usually on the order of 10^s-bits for $s \geq 4$, Alice and Bob may even choose to set $\ell = 100$, leaking 100 bits of information, but with a failure probability of less than $1/2^{100}$, a *very* small value that we won't even attempt to convert to a percentage.

4.4.2 Privacy amplification

We now discuss the second mandatory classical post-processing element, which runs after error correction: *privacy amplification*. Consider the story so far: Alice and Bob communicated, through the quantum communication stage, to distill a raw key. Eve may have some information on this due to a man-in-the-middle attack on the quantum channel, but, while we may be able to upper-bound Eve's information, we may not know exactly *what* Eve knows about the raw key so far. Next, error correction is run which leaks additional information to Eve. Finally, an error correction verification protocol may be run which leaks further information to Eve. At this point, Alice and Bob have two strings $\widetilde{r}_A$ and $\widetilde{r}_B$ of size n-bits each such that $\widetilde{r}_A = \widetilde{r}_B = r$, except with negligibly small probability. Eve, however, has, let's say m-bits of information on this string, with $m < n$. Note that, if $m = n$ then nothing can be done as Eve knows everything! Thus, *some* of r is secret, but some is not. Clearly this cannot be used as a secret key!

Exercise 4.8. Let k be a secret key shared by Alice and Bob and assume that the adversary Eve knows a single bit of this key (for simplicity, assume she knows the first bit). The remaining bits are uniform random and independent of Eve (i.e., unknown to Eve). Prove that this "secret" key cannot be used for One Time Pad (OTP) encryption while still remaining EAV-secure (this security model was defined in Chapter 3).

Now, if we knew exactly which m-bits Eve knew, we'd be fine: we simply remove those m bits from the raw key r and call that the secret key. The problem, of course, is that we don't know which bits Eve knows. Furthermore, Eve may have quantum side information on the raw key at this point through her quantum ancilla. All we know, as Alice and Bob, is that somehow, from an information theoretic perspective, Eve knows m-bits of information on our raw key. How do we create a secret key from this? The answer is privacy amplification [17].

A privacy amplification protocol is one that takes a random string x as input, upon which an adversary may have some side information on, and outputs a smaller string y which Eve has negligible information of (to be defined, more formally, in the next chapter).

In general, the protocol used for privacy amplification in most, if not all, QKD protocols involves the use of a two-universal hash function as defined earlier in Definition 4.4.2. The protocol involves the following:

1. Alice chooses a random two-universal hash function f which maps n-bit strings to ℓ-bit strings. In general, Alice should set $\ell \approx n - m$ where, recall, n is the size of the raw key, and m is the number of bits Eve "knows" including what she learned from error correction leakage (note that this has to be rigorously stated and defined and we will use information theory to do so – however, these details will be in the next Chapter).
2. Alice sends a description of f to Bob over the authenticated classical channel. Note that the hash function was chosen independently of r.
3. Alice and Bob both compute $k = f(r)$ and use k as their secret key.

Note that the function f must be sent to Bob. Thus, while it is very important to choose the hash function randomly, each time privacy amplification is used, it is not necessary to keep it secret: Eve is allowed to know the particular choice of hash function being employed this particular round.

4.4.3 Classical advantage distillation

Error correction and privacy amplification are not by any means the only classical post processing methods available. While they are mandatory, there are other additional processes that may be used, before error correction and privacy amplification are run, which have the potential to improve overall performance. We discuss one important concept here, namely *classical advantage distillation* or CAD [16,2] which is run before error correction.

CAD, is a process which takes two random strings X and Y, and creates two new, but shorter strings Z and W such that these new strings have fewer errors. That is, if there are Q errors in X and Y, there will be $Q' \leq Q$ errors in Z and W. This process involves Alice, who we assume holds X, and Bob (who holds Y), sending classical information to one another over the authenticated classical channel. Parties will take this information and distill their new strings which, though more correlated, are also potentially much shorter. CAD, therefore, when applied to QKD can improve the overall noise tolerance of the protocol, though only by shrinking the raw key size. Thus, it should be used carefully: if the noise in the channel is very high, CAD can greatly improve performance (by making it possible to distill a secret key even when, without CAD, the protocol would fail); however, if the noise is low, CAD can actually hurt performance substantially.

We discuss two main methods here from [16,2] which, for the most part, are equivalent (though, as noted in [2], for certain adversarial attacks, one is better than the other). The first involves Alice choosing a random subset of her string that "match" and sending information on this to Bob; the second involves parties sending parity information to one another. Both protocols take as input a "block parameter" $C = 1, 2, \cdots$ specified by the user. When $C = 1$ this will actually disable the CAD protocol. For higher levels of CAD, a new raw key is distilled, however the bit-size of the new key is *at most* $1/C$ the original size.

The first CAD protocol, involving choosing subsets of the raw key (produced by the quantum communication stage, but before error correction is run), operates as follows:

1. Alice chooses a random subset of C bits from her raw key such that all C bits in this subset are all zero or all one. Let τ be the subset chosen. This subset is sent to Bob (of course, the actual value of the bits of X in this subset are not disclosed).
2. Bob checks his raw key Y on this subset – if all his bits match (i.e., if $Y_\tau = 0^C$ or $Y_\tau = 1^C$), he signals to Alice to "keep" this block – otherwise he signals to Alice to "discard" this block. Note that we use 0^C to mean C bits of all zeros; similarly for 1^C.
3. If Bob said to keep the block, Alice appends to her new raw key Z, the bit 0 or 1 based on the value of the bits in X_τ (which, recall, all match); similarly, Bob appends to his raw key W the bit 0 or 1 based on the value of Y_τ.
4. Regardless of whether Bob said to keep or discard this block, the block indexed by τ is removed from X and Y. The above process repeats until $|X| = |Y| < C$. Any remaining bits in the original raw keys are discarded.

It is not difficult to see that the size of the new raw key can be no larger, and is often smaller, than $|X|/C$.

As an example, assume $X = 110000101001$ and $Y = 111000010001$. Note that there are a total of 3 errors, making for a total error rate of $3/12 = .25$. Let's trace through the execution of CAD with a block setting of 2:

1. Round 1: Alice chooses $\tau = \{1, 2\}$ (the first two bits). Bob signals "keep" as his bits both match. After this, Alice's new raw key will be $Z = 1$ and Bob's will be $W = 1$. They discard the first two bits, setting $X = 0000101001$ and $Y = 1000010001$.
2. Round 2: Alice (randomly) chooses $\tau = \{1, 2\}$ (the first two bits of the new, shorter, raw key). In this example, it works out to be the same subset, but this is not normally the case; we do so, here, for simplicity only. Bob, however signals to discard since his first two bits are 10 (Alice's were 00). Neither party adds to their new raw keys, however both parties discard the first two bits of their old raw keys, leading to $X = 00101001$ and $Y = 00010001$.
3. And so on...

Exercise 4.9. Complete the above CAD execution and write out the final raw keys. Is it unique or were there better choices Alice might have made? Use your answer to explain why it would be important for Alice to choose her subsets randomly as opposed to simply picking the "easiest" choice as we did in our example above.

Exercise 4.10. Write a simple program that simulates the CAD process. Specifically, your program should take as input two "raw keys" (of equal size) and output two new raw keys, after running the CAD process. Keep track of all transmitted data.

Let's investigate why the CAD process described above may be beneficial, given a larger amount of data (our simple example, above, did not demonstrate the effectiveness of CAD very well – however, we encourage you to try out Exercise 4.10 above to better appreciate CAD in practice). Let's assume that we have two classical strings X and Y each of size N bits. Let's further assume that the i'th bit in Y is equal to the i'th bit in X, except with probability Q (when it is flipped). That is, the error rate in the two raw keys is Q. Let's also assume bit j of each string is independent of bit $j' \neq j$ (this is not needed for CAD to be effective, but it certainly helps the analysis). In that case, what's the expected error rate $\tilde{Q}$ of the final strings Z and W if $C = 2$? Let's zoom in on a particular round of CAD: Alice sent a subset τ consisting of two indices, both of which, at least for Alice, point to two bits in X that are equal. The only way for there to be an error in the final string is if both bits Bob holds in Y are flipped (which is simply Q^2). Of course, we must also condition on Bob sending the message "keep," i.e., we must condition on both bits in Bob's string being equal in the subset τ which can only happen if both bits stay the same (probability $(1-Q)^2$) or both flip (with probability Q^2). From this analysis, we can argue that the expected error rate in the final strings Z and W is:

$$\tilde{Q} = \frac{Q^2}{Q^2 + (1-Q)^2}. \tag{4.10}$$

A graph of the above noise $\tilde{Q}$, compared to the original noise in the signal Q, can be seen in Fig. 4.1.

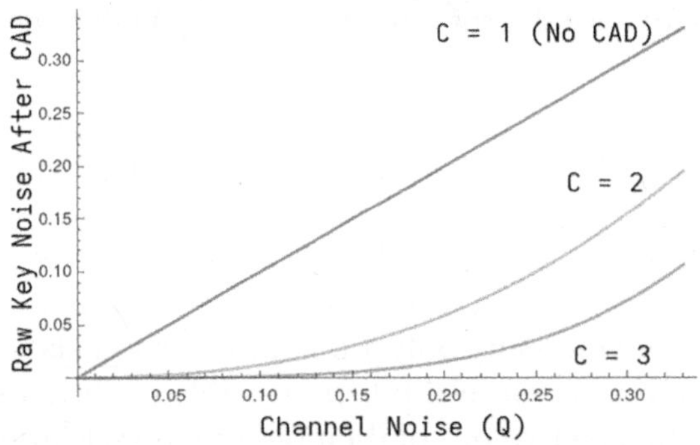

FIGURE 4.1 Showing how the CAD process can decrease the overall error rate (noise) in the raw key. Here, the x-axis represents the initial noise level (Q) in the channel while the y-axis represents the noise in the new raw key, after CAD runs. Blue line (top): $C = 1$ (equivalent to no CAD); Yellow (middle): $C = 2$ (a CAD block size of two); Green (bottom): $C = 3$.

Exercise 4.11. What is the expected error rate for arbitrary settings of C? Plot each value for different values of C to see how they compare.

The CAD protocol, above, required Alice to choose random subsets such that for every bit of her string indexed by that subset, those bits were either all zero or all one. An alternative protocol, that is perhaps simpler to implement, is described below.

The second CAD protocol operates as follows, again with a block-setting parameter $C = 1, 2, \cdots$ set by the user. First, the raw key is divided into blocks of size C bits. For each of these blocks, sequentially, the following process is run:

1. Let $X_1 \cdots X_C$ be Alice's current block of raw key bits and $Y_1 \cdots Y_C$ be Bob's. These are C-bit blocks.
2. Alice chooses a random bit r for this block. She then sends the message $M = (X_1 \oplus r, \cdots, X_C \oplus r)$ to Bob using, of course, the classical authenticated channel. That is, she sends the C bits of her raw key block, each bit XOR'd with the same randomly chosen value r (importantly, each block gets an independently chosen random r).
3. Bob takes his block of raw key bits and adds them, modulo two, to the received message. Namely, he computes $\widetilde{M} = M \oplus (Y_1 \cdots Y_C) = X_1 \oplus Y_1 \oplus r, \cdots, X_C \oplus Y_C \oplus r$. If this results in a string of all zeros or all ones, he signals to Alice to "accept" this block; otherwise he signals to Alice to "discard" this block.
4. If Bob signaled to keep this block, Alice adds to her new raw key the bit r while Bob adds to his new raw key the bit the "zero" or "one" that he found in his block, after XOR'ing with the sent message. Otherwise, no bit is added to their raw key.
5. Regardless of whether Bob signaled to keep or discard, this block of C bits is discarded from the original raw key and the above process repeats for the next C bits.

Clearly an advantage to the protocol above is that it does not require Alice to choose random subsets. The fact that it works similarly to the earlier discussed CAD protocol can be seen through an example. Let $X = 10110111$ and $Y = 10010100$. Then, a possible example execution of the above CAD protocol (for $C = 2$ as before) could be as follows:

1. Round 1: Alice chooses a random bit r, in this case let's assume it's 0. She then sends the message $M_1 = (1, 0)$ (the first two bits of her key XOR'd with zero). Bob computes $\widetilde{M}_1 = (1 \oplus 1, 0 \oplus 0) = (0, 0)$. Since this is an all zero tuple, he signals to Alice to keep this block. Alice keeps a bit 0 (the value of r) and Bob keeps a bit 0 (the value he computed). Their new raw key is then $Z = 0$ and $W = 0$. Alice and Bob discard the first two bits of their original raw keys, leading to $X = 1101$ and $Y = 0101$.
2. Round 2: Alice chooses a random bit $r = 1$ in this round. She sends the message $M_2 = (0, 0)$ (the bit string 11 XOR'd with 1). Bob computes $\widetilde{M}_2 = (0, 1)$. Since this is not an all-zero or all-one string, he signals to Alice to discard this block. Their raw keys are then $X = 01$ and $Y = 01$ while their new raw keys remain unchanged from the previous round.
3. Round 3: Alice chooses a random bit $r = 1$ and sends $M_3 = (1, 0)$. Bob computes $\widetilde{M}_3 = (1, 1)$. Since this is an all-one tuple, he signals to "keep" this string – Alice and Bob both append a 1 to their new raw keys which now consist of $Z = 11$ and $W = 11$. These two bits are then discarded from the original raw keys which, subsequently, become $X = 11$ and $Y = 00$.
4. Round 4: Alice chooses $r = 0$ and sends $M_4 = (1, 1)$. Bob computes $\widetilde{M}_4 = (1, 1)$ and signals to keep this block. Alice adds the bit 0 to her new raw key, however Bob adds

the bit 1 to his raw key. Note that an error occurred since *both* bits were flipped. At this point their original raw keys are consumed and their new raw key is $Z = 010$ and $W = 011$.

As with the first CAD protocol discussed in this section, an error can only occur if all the bits in the block under consideration are flipped. Thus, as before, the expected error rate in the new raw key will be approximately $Q^C/(Q^C + (1 - Q)^C)$ (the denominator taking into account the actual probability of accepting a block).

Exercise 4.12. Similar to Exercise 4.10, write a program that implements this second CAD protocol. Combine your program with that from Exercise 4.7 and show how the error correction leakage decreases when CAD runs, based on the noise level in your original raw keys.

References

[1] Antonio Acin, Nicolas Gisin, Valerio Scarani, Coherent-pulse implementations of quantum cryptography protocols resistant to photon-number-splitting attacks, Physical Review A 69 (1) (2004) 012309.

[2] Joonwoo Bae, Antonio Acín, Key distillation from quantum channels using two-way communication protocols, Physical Review. A, Atomic, Molecular, and Optical Physics 75 (1) (2007) 012334.

[3] Stephen M. Barnett, Bruno Huttner, Simon J.D. Phoenix, Eavesdropping strategies and rejected-data protocols in quantum cryptography, Journal of Modern Optics 40 (12) (1993) 2501–2513.

[4] Charles H. Bennett, Quantum cryptography using any two nonorthogonal states, Physical Review Letters 68 (21) (May, 1992) 3121–3124, https://doi.org/10.1103/PhysRevLett.68.3121, http://link.aps.org/doi/10.1103/PhysRevLett.68.3121.

[5] Charles H. Bennett, Gilles Brassard, Quantum cryptography: public key distribution and coin tossing, in: Proceedings of IEEE International Conference on Computers, Systems and Signal Processing, vol. 175, New York, 1984.

[6] Charles H. Bennett, Gilles Brassard, N. David Mermin, Quantum cryptography without bell's theorem, Physical Review Letters 68 (5) (1992) 557.

[7] Cyril Branciard, Nicolas Gisin, Norbert Lutkenhaus, Valerio Scarani, Zero-error attacks and detection statistics in the coherent one-way protocol for quantum cryptography, arXiv preprint, arXiv:quant-ph/0609090, 2006.

[8] Gilles Brassard, Louis Salvail, Secret-key reconciliation by public discussion, in: Workshop on the Theory and Application of Cryptographic Techniques, Springer, 1993, pp. 410–423.

[9] J. Lawrence Carter, Mark N. Wegman, Universal classes of hash functions, in: Proceedings of the Ninth Annual ACM Symposium on Theory of Computing, 1977, pp. 106–112.

[10] J. Lawrence Carter, Mark N. Wegman, Universal classes of hash functions, Journal of Computer and System Sciences 18 (2) (1979) 143–154.

[11] Artur K. Ekert, Quantum cryptography based on bell's theorem, Physical Review Letters 67 (6) (1991) 661.

[12] Chi-Hang Fred Fung, Hoi-Kwong Lo, Security proof of a three-state quantum-key-distribution protocol without rotational symmetry, Physical Review. A, Atomic, Molecular, and Optical Physics 74 (4) (2006) 042342.

[13] Robert Gallager, Low-density parity-check codes, I.R.E. Transactions on Information Theory 8 (1) (1962) 21–28.

[14] Hoi-Kwong Lo, Hoi-Fung Chau, M. Ardehali, Efficient quantum key distribution scheme and a proof of its unconditional security, Journal of Cryptology 18 (2) (2005) 133–165.

[15] Ryutaroh Matsumoto, Shun Watanabe, Key rate available from mismatched measurements in the bb84 protocol and the uncertainty principle, IEICE Transactions on Fundamentals of Electronics, Communications and Computer Science 91 (10) (2008) 2870–2873.

[16] Ueli M. Maurer, Secret key agreement by public discussion from common information, IEEE Transactions on Information Theory 39 (3) (1993) 733–742.
[17] Renato Renner, Security of quantum key distribution, International Journal of Quantum Information 6 (01) (2008) 1–127.
[18] Kiyoshi Tamaki, Marcos Curty, Go Kato, Hoi-Kwong Lo, Koji Azuma, Loss-tolerant quantum cryptography with imperfect sources, Physical Review A 90 (5) (Nov, 2014) 052314, https://doi.org/10.1103/PhysRevA.90.052314, https://link.aps.org/doi/10.1103/PhysRevA.90.052314.
[19] Shun Watanabe, Ryutaroh Matsumoto, Tomohiko Uyematsu, Tomography increases key rates of quantum-key-distribution protocols, Physical Review. A, Atomic, Molecular, and Optical Physics 78 (4) (2008) 042316.
[20] Mark N. Wegman, J. Lawrence Carter, New hash functions and their use in authentication and set equality, Journal of Computer and System Sciences 22 (3) (1981) 265–279.

5

Security of QKD

In the previous chapter, we discussed several different QKD protocols, along with their general design and layout. We are now ready to talk about the security of a QKD protocol.

5.1 Security goal and composable security

Imagine the following: two users run a communication protocol of some kind (involving the use of quantum and classical communication) leading, ultimately, to some density operator $\mathcal{K}_{ABE}$, where the A and B registers are classical and consist of ℓ-bits each while the E register is arbitrary and quantum and may consist of, for instance, intercepted communications; entangled probes; and, in general, whatever an adversary Eve could have gotten from Alice and Bob's protocol execution. The goal of this protocol that Alice and Bob performed is for them to agree on a shared secret key which is (1) fully correlated (i.e., Alice's key matches Bob's exactly); (2) uniformly distributed; and (3) totally independent of Eve. Ideally, it should hold that:

$$\mathcal{K}_{ABE} = \left(\sum_{k \in \{0,1\}^\ell} \frac{1}{2^\ell} |k\rangle \langle k|_A \otimes |k\rangle \langle k|_B \right) \otimes \rho_E, \tag{5.1}$$

where ρ_E is some density operator acting on Eve's ancilla space (we call Eve's quantum system her *ancilla*). Indeed, imagine if a protocol execution resulted in the above density operator. Then it would hold that Alice and Bob's systems are exactly as desired: uniform random, fully correlated, and completely independent of Eve's system (no matter what key k Alice and Bob have, Eve always, and only, has ρ_E so there is no information for Eve to gain on Alice's and Bob's key making it truly secret). We call this system $\mathcal{K}_{ABE}$ defined as in Eq. (5.1) the *ideal state*.

Of course, this ideal state is usually not attainable. Instead, Alice and Bob will run some protocol resulting in a density operator ρ_{ABE}, where:

$$\rho_{ABE} = \sum_{\substack{k_A \in \{0,1\}^\ell \\ k_B \in \{0,1\}^\ell}} P_{AB}(k_A, k_B) |k_A\rangle \langle k_A|_A \otimes |k_B\rangle \langle k_B|_B \otimes \rho_E^{(k_A,k_B)}. \tag{5.2}$$

Above, the probability distribution P_{AB} may not be exactly uniform; also, Eve's system may depend on the outcome of Alice and Bob's key. For instance, if they have a particular key k_A and k_B, then Eve's system will be in a state $\rho_E^{(k_A,k_B)}$. Eq. (5.2) is often called the *real state* – it describes the actual state of Alice, Bob, and the adversary Eve after running a protocol.

Quantum Communication and Cryptography. https://doi.org/10.1016/B978-0-44-321532-2.00012-7

While it is desired that $\rho_{ABE} = \mathcal{K}_{ABE}$, this will not usually be the case; however for a key distribution protocol to be secure, it should hold that the real state is not "too far" from the ideal. This, of course, can be measured using the trace distance. In particular, we want the trace distance between these two states to be ϵ-close to one another, for some small ϵ, or, formally:

$$\frac{1}{2}||\rho_{ABE} - \mathcal{K}_{ABE}|| \leq \epsilon. \tag{5.3}$$

Generally, ϵ is chosen by the user and is set to be small, such as 10^{-12} as an example. Typically the smaller the better, from a security point of view; however the smaller it is, the greater the failure probability may be, or, alternatively, the smaller ℓ may be as we will soon discover in this chapter. As with anything, there's a compromise between security and efficiency.

But, now, given a protocol and an actual execution of that protocol, leading to a mixed state ρ_{ABE}, how does one actually go about proving Eq. (5.3) holds for a given ϵ? Let's consider an ideal execution of BB84 (which was introduced in Chapter 4). By "ideal" now, we mean no noise at all in the channel and no attacker. On a particular round, Alice will choose a random key-bit and a random basis, sending a particular state $|0\rangle$, $|1\rangle$, $|+\rangle$, or $|-\rangle$ depending on this choice. Bob will choose a random basis to measure in leading to his "guess" of Alice's key. Alice and Bob will then discard all rounds where they choose opposite bases. We want to describe the density operator of this protocol.

Note that, since there is no adversary, each round really is an independent event. Let's try to describe the state of a single round, then, and the entire protocol will simply be an N-fold tensor. That is to say, if we can describe the outcome of a single round as some density operator ρ_{ABE}, then the density operator describing the entire protocol, after running it N times, is simply $\rho_{ABE}^{\otimes N}$. To model this, we'll have some scratch registers: one for Alice's basis choice, and one for Bob's basis choice, each will be two dimensional and spanned by $|``Z"\rangle$ and $|``X"\rangle$. Note we are using quotes here to signify that these are "classical" memory systems, just represented using quantum Dirac notation.

Let's slowly trace the protocol. First Alice chooses a random bit and basis choice. The density operator, now, is:

$$\begin{aligned} &\frac{1}{2}|0\rangle\langle 0|_A \otimes \left(\frac{1}{2}|``Z"\rangle\langle ``Z"| + \frac{1}{2}|``X"\rangle\langle ``X"|\right) \\ +&\frac{1}{2}|1\rangle\langle 1|_A \otimes \left(\frac{1}{2}|``Z"\rangle\langle ``Z"| + \frac{1}{2}|``X"\rangle\langle ``X"|\right) \end{aligned}$$

Note that we are assuming Alice chooses bases uniformly at random – of course biases are possible, as discussed in Chapter 4, however this is something we will return to later in this chapter.

At this point, a qubit is prepared according to the basis choice and key choice of Alice. We'll use T to denote this qubit (it is the qubit in Transit) – adding this to the above yields the mixed state:

$$\frac{1}{2}|0\rangle\langle 0|_A \otimes \left(\frac{1}{2}|\text{“}Z\text{”}\rangle\langle\text{“}Z\text{”}| \otimes |0\rangle\langle 0|_T + \frac{1}{2}|\text{“}X\text{”}\rangle\langle\text{“}X\text{”}| \otimes |+\rangle\langle +|_T\right)$$
$$+\frac{1}{2}|1\rangle\langle 1|_A \otimes \left(\frac{1}{2}|\text{“}Z\text{”}\rangle\langle\text{“}Z\text{”}| \otimes |1\rangle\langle 1|_T + \frac{1}{2}|\text{“}X\text{”}\rangle\langle\text{“}X\text{”}| \otimes |-\rangle\langle -|_T\right)$$

The transit qubit, within the T register, travels to Bob. During this transit, the qubit may normally undergo noise and potentially adversarial attack – these are issues we must deal with later. For the purpose of the current illustration, however, as we said, we're only considering noise-less and adversary-less scenarios for now. In this case, the T register arrives to B's lab undisturbed. Bob will choose his own basis, choosing randomly, and then measure in that basis. If he observes $|0\rangle$, or $|+\rangle$, he will store a "0" in a private register (to be his key-choice); otherwise, he will store a "1" in that register. This entire process leads to a final mixed state of:

$$\begin{aligned}
&\frac{1}{2}|0\rangle\langle 0|_A \otimes \Bigg[\frac{1}{4}|\text{“}Z,Z\text{”}\rangle\langle\text{“}Z,Z\text{”}| \otimes |0\rangle\langle 0|_B + \frac{1}{4}|\text{“}X,X\text{”}\rangle\langle\text{“}X,X\text{”}| \otimes |0\rangle\langle 0|_B \\
&\qquad + \frac{1}{4}|\text{“}Z,X\text{”}\rangle\langle\text{“}Z,X\text{”}| \otimes \left(\frac{1}{2}|0\rangle\langle 0|_B + \frac{1}{2}|1\rangle\langle 1|_B\right) \\
&\qquad + \frac{1}{4}|\text{“}X,Z\text{”}\rangle\langle\text{“}X,Z\text{”}| \otimes \left(\frac{1}{2}|0\rangle\langle 0|_B + \frac{1}{2}|1\rangle\langle 1|_B\right)\Bigg] \\
&+\frac{1}{2}|1\rangle\langle 1|_A \otimes \Bigg[\frac{1}{4}|\text{“}Z,Z\text{”}\rangle\langle\text{“}Z,Z\text{”}| \otimes |1\rangle\langle 1|_B + \frac{1}{4}|\text{“}X,X\text{”}\rangle\langle\text{“}X,X\text{”}| \otimes |1\rangle\langle 1|_B \\
&\qquad + \frac{1}{4}|\text{“}Z,X\text{”}\rangle\langle\text{“}Z,X\text{”}| \otimes \left(\frac{1}{2}|0\rangle\langle 0|_B + \frac{1}{2}|1\rangle\langle 1|_B\right) \\
&\qquad + \frac{1}{4}|\text{“}X,Z\text{”}\rangle\langle\text{“}X,Z\text{”}| \otimes \left(\frac{1}{2}|0\rangle\langle 0|_B + \frac{1}{2}|1\rangle\langle 1|_B\right)\Bigg]
\end{aligned} \tag{5.4}$$

Note that, whenever Alice and Bob choose different bases (i.e., when we have "Z, X" or "X, Z"), then Bob's result is random and uncorrelated; otherwise it exactly matches Alice's choice of key bit as expected.

At this point, Alice and Bob discard any round where their bases choices do not match. Thus, we can discard any of the above mentioned rounds. Conditioning on a round being *not* discarded, we find the density operator is:

$$\rho_{ABP} = \frac{1}{2}|0\rangle\langle 0|_A \otimes |0\rangle\langle 0|_B \otimes \left(\frac{1}{2}|\text{“}Z,Z\text{”}\rangle\langle\text{“}Z,Z\text{”}|_P + \frac{1}{2}|\text{“}X,X\text{”}\rangle\langle\text{“}X,X\text{”}|_P\right) \tag{5.5}$$
$$+\frac{1}{2}|1\rangle\langle 1|_A \otimes |1\rangle\langle 1|_B \otimes \left(\frac{1}{2}|\text{“}Z,Z\text{”}\rangle\langle\text{“}Z,Z\text{”}|_P + \frac{1}{2}|\text{“}X,X\text{”}\rangle\langle\text{“}X,X\text{”}|_P\right). \tag{5.6}$$

Above, we use P to denote the Public broadcast of the basis choice ("broadcasted" over the authenticated classical channel). Note that, the basis choice, even if it is publicly known, does not give any information on the key-bit of Alice and Bob in this scenario –

regardless of basis choices and their respective announcements, Alice and Bob will have a uniform random key bit. Of course, the actual final state of the entire protocol will be an N-fold tensor of the above. Thus, the actual state is $\rho_{ABP}^{\otimes N}$. How does this relate to Eq. (5.1)? Permuting subspaces, so that Alice and Bob are the left-most registers and P is the right-most, even after tensoring the state N-times, we may write the protocol state as:

$$\rho_{ABP}^{\otimes N} \cong \sum_{k\in\{0,1\}^N} \frac{1}{2^N} |k\rangle\langle k|_A \otimes |k\rangle\langle k|_B \otimes \underbrace{\left(\frac{1}{2^N} \sum_{p\in\{\text{"}ZZ\text{"},\text{"}XX\text{"}\}^N} |\text{"}p\text{"}\rangle\langle\text{"}p\text{"}|_P \right)}_{\rho_E}$$
$$= \sum_{k\in\{0,1\}^N} \frac{1}{2^N} |k\rangle\langle k|_A \otimes |k\rangle\langle k|_B \otimes \rho_E. \tag{5.7}$$

This is exactly Eq. (5.1), setting $\ell = N$. Thus, in the complete absence of noise and the total lack of any adversary, BB84 produces an ideal secret key. *Of course, in real life, there is no absence of noise* – also, we cannot discount the effect an adversary may have on the system! Thus, a QKD security proof is a bit more involved than this, however it does oftentimes involve tracing the execution of the protocol. However, one must also take into account the adversary. This leads to two immediate questions: First, how do we model noise in the channel? Second, how do we model the adversary's attack capabilities? These will be questions we answer momentarily.

Exercise 5.1. Argue why $\rho_{ABP}^{\otimes N}$ can be written as in Eq. (5.7) (after permuting subspaces appropriately).

5.2 Adversarial models and noise

In cryptography, whether classical or quantum, before one can ever prove anything, one must explicitly define the *attack model* – that is, the capabilities of the attacker. In quantum security, we must also deal with noise that may be naturally induced (no physical system is perfect after all). Due to the unique nature of quantum communication, how we deal with this noise is actually part of the attack model – this is generally unlike classical security where natural noise is usually outside of scope (though not always as in, for instance, wire-tap channels [17]). Classically, an adversary's attack usually does not create noise and, so, noise is not usually treated as part of the attack model. In the quantum setting, things are different.

5.2.1 Natural noise

First, let's consider what to do with the natural noise in a channel. Consider a basic prepare-and-measure BB84 QKD setup with Alice and Bob connected to one another through a quantum channel and a classical authenticated channel. Entanglement-based versions can be considered similarly. Now, Alice will be sending qubits to Bob; however, no physical channel is perfect and, so, there will naturally be noise in the channel (e.g., if Alice sends a $|+\rangle$ it might arrive to Bob not as $|+\rangle$ but, perhaps, as $|-\rangle$) along with loss (e.g., if Alice sends a $|1\rangle$ to Bob, it might never arrive at all). Such noise and loss naturally hurts Alice and Bob. It could also hurt any adversary (as noise will affect the adversary also). See Fig. 5.1.

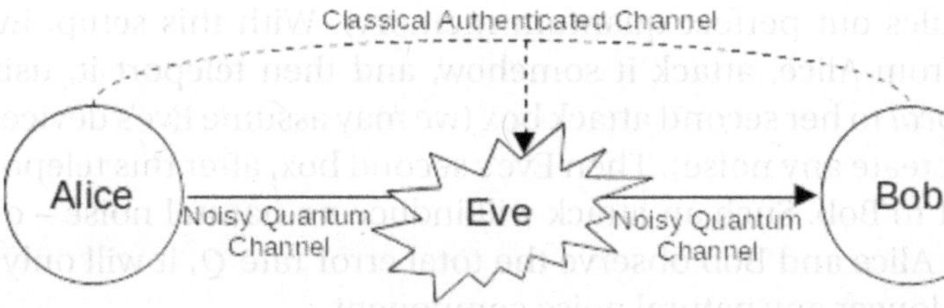

FIGURE 5.1 Showing the basic setup of a realistic QKD system with an Eavesdropping adversary (Eve). Here, the adversary probes the quantum channel, however there may be noise in this channel before and after Eve. However, we will actually assume all noise is caused by the adversary, as she may completely control the channel, as depicted in Fig. 5.2.

Now, on top of this natural noise, let's assume there's an adversary probing the channel. This adversary's attack, whatever it is, is going to introduce noise *on top of* the natural noise. The "stronger" her attack, the more information Eve will gain, but also the more noise she will create in the channel (exactly why this is true, should become clear, later in this chapter). We call this noise *adversarial noise.* Of course, Alice and Bob will run BB84 (or some other protocol) and later estimate the total noise in the channel, denoted Q (e.g., maybe 5% of Alice and Bob's raw key is incorrect). We now have a problem: on the one hand, this total noise Q is a function of both the natural noise and the adversarial noise. The higher the adversarial noise, the more information Eve has. The higher the natural noise, then *potentially* Eve has less information (in general, noise occurring in the state, before it reaches Eve, may increase her uncertainty on what Alice was trying to send to Bob). Given Q, how do we know what percentage is due to Eve?

In general, we want to make as few assumptions as possible on the adversary – so, consider the following possible scenario: Eve places her "attack box" right outside of Alice's location where the distance is so small, there is virtually no noise or loss between Alice and Eve. Eve's attack creates some noise (by altering the qubit state). Eve then passes a qubit to Bob through a fiber channel which creates natural noise and loss. Thus, in this scenario, the natural noise may not necessarily confuse Eve, though the natural noise will still be observed on top of Eve's adversarial noise. Eve's information is based on how much noise her attack created. Now, we are in a scenario where we observed Q, the total noise; the natural noise did not affect Eve, however Q is larger than the adversarial noise. How much information does Eve have in this scenario?

Let's go a step further and consider an extreme attack scenario. Here, Eve places her attack box right outside of Alice's lab and places a second attack box right outside of Bob's lab. These attack boxes are so close that there is no noise or loss between Alice and Eve's first box; nor is there noise or loss between Eve's second box and Bob. See Fig. 5.2. Let's assume that these two attack boxes share unlimited entanglement (or, at least, enough pre-shared entanglement to support her attack against Alice and Bob – Eve can always "refresh" it later). Let's also assume the worst case that Eve has perfect quantum memory to store these entangled pairs (a strong assumption, but, again, it's always safer to make assumptions that benefit an adversary, and there's no postulate of quantum mechanics that specifically rules out perfect quantum memory). With this setup, Eve can capture every qubit sent from Alice, attack it somehow, and then teleport it, using a *quantum teleportation protocol* to her second attack box (we may assume Eve's devices are so perfect that this does not create any noise). Then Eve's second box, after this teleportation, simply forwards the qubit to Bob. Such an attack will induce no natural noise – only adversarial noise. Thus, when Alice and Bob observe the total error rate Q, it will only be adversarial noise – there is no longer any natural noise component.

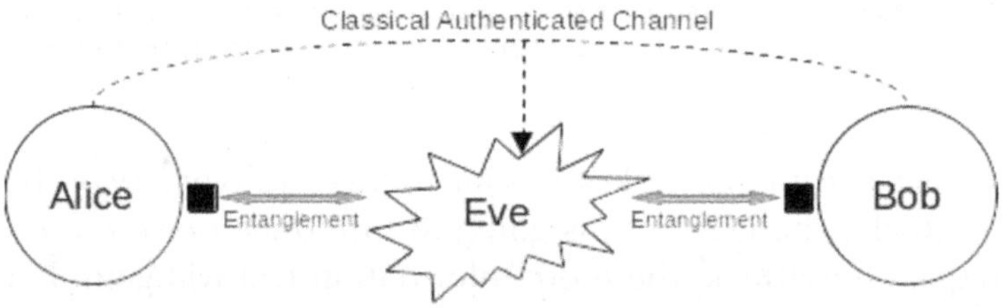

FIGURE 5.2 When proving security of QKD, we almost always assume any noise is caused by the adversary. This is because an adversary can, in theory, launch an attack depicted here, consisting of probes close to Alice and Bob which share entanglement. Quantum teleportation then would, in theory, allow Eve to create a "noiseless" channel if she wanted to. Thus, any noise, will be caused by the adversary in our security proof assumptions. In real life, this may be a strong assumption – however, it's always better to define a security model benefiting the adversary, as much as possible since, then, in real life things can only be better for the honest users.

To summarize, let Q again be the total observed noise; it is a function of the adversary's noise, denoted Q_A, and the natural noise, denoted Q_N combined. The higher the adversarial noise, Q_A, the more information Eve has. The higher the natural noise, Q_N, then potentially the less information Eve has, though that's not guaranteed. We considered three attacks above:

1. Eve attacks the fiber line in the middle (Fig. 5.1). Here, there is a large natural noise component in the fiber channel, so $Q_A < Q$. The natural noise will probably hurt Eve also. Thus her information gain for this attack will be **low**.
2. Eve places her attack box right next to Alice. Here, again, there is a large natural noise component in the fiber channel, and so $Q_A < Q$. However, the natural noise does not impede Eve's information gain on Alice's key information. For this attack, her total information gain will be **medium**.
3. Finally, Eve places an attack box right next to Alice, and a second right next to Bob (Fig. 5.2). Her boxes have sufficient entangled pairs to fully carry out her attack. In this case, there is no natural noise. Thus, all observed noise must be adversarial and

so $Q_A = Q$. Since Eve's information increases with her adversarial noise, this attack gives her the most possible information – her information gain for this attack will be **high**.

The question, now, is which of the three do we consider? The first is probably the most practical; the third clearly is the most advantageous to the adversary. Unfortunately, there is no way to distinguish between the three attack models above. Thus, keeping in line with the general rule in security that it is always better to assume the worst (or, rather, for Eve's perspective the best), the vast majority of QKD security proofs take the third approach. In this way, when Alice and Bob observe a particular noise level Q, they simply assume all noise is caused by the adversary, and none of it is natural. Assuming some of the noise is natural, can only cause Eve to have less information than she might have, had the third case been assumed.

This is clearly a "worst case" scenario for Alice and Bob. It also has the added benefit that it is easier to analyze from a security perspective as we don't need to model natural noise on top of adversarial noise – we only need to model and consider adversarial noise.

That's not to say that the first or second security models don't have merit. There have been several interesting works [7,4,5] (to list only a few), many appearing recently, that model to some degree, natural noise on top of adversarial noise. By doing so, increased efficiency and noise tolerance is possible. However, it is an assumption: one must assume Eve is not performing the third style of attack, described above. It is probably a safe physical assumption, considering the third attack above requires Eve to be right outside of both Alice and Bob's location (immediately outside to avoid any natural noise) and in general that would probably be difficult to get away with, without being detected. Beyond that, it also requires perfect quantum memory and perfect devices for teleportation. It also requires high-speed devices so that Bob is unaware of the attack from a timing-perspective. There is no law of quantum mechanics that specifically rules out such an attack, thus if we want "unconditional security" we cannot rule it out.

Exercise 5.2. Describe how the third style attack may be performed by an adversary in an entanglement-based protocol, where a source prepares signals and sends them over a fiber channel to Alice and Bob (sending qubits on two separate fiber channels). Does it still make sense to consider Eve as the source of all noise and loss?

5.2.2 Attack types

Now that we have discussed, and properly dealt with, the natural noise, let's turn to what Eve can actually do when a qubit (or qubits) arrive at her attack box. In the previous section, we motivated the notion that all noise should be considered adversarial, by considering a two-box attack where Eve teleports the qubit from the Alice-side box to the Bob-side one (Fig. 5.2). However, to simplify things (and give even more advantage to the adversary), we will not worry about the teleportation. Instead, Alice sends signals to Bob, however these signals are immediately captured by Eve who attacks them. Eve then is able to immediately, without noise or loss, send them to Bob's devices. Any noise and loss is created by Eve's attack.

Let's first consider prepare-and-measure protocols. In general, there are four main "classes" or types of attacks that are considered in QKD security. From weakest attack to strongest, they are: Intercept/Resend; Individual; Collective; and General.

Intercept/Resend attacks are the most basic type of attack that can be considered and are usually the type of attack which most people think of at first. These attacks work on each individual qubit coming out of Alice. Alice will send a qubit state to Bob, however it is immediately captured by Eve. Eve will then measure the qubit in a basis of her choice (perhaps choosing randomly for each qubit), recording the outcome as a classical bit which will be used as a "guess" as to which key-bit Alice is trying to send. Based on the outcome, Eve will prepare a new qubit state and send it to Bob (usually the qubit she sends to Bob is the post-measured state she observed). Eve attacks each qubit from Alice in an identical way. The strategy may be randomized, but the probability distribution of choices is identical for every qubit sent. That is, Eve will not adapt her attack strategy on the n'th qubit based on the outcome of her attacks on the previous $m < n$ qubits. At the end of the quantum communication stage of the protocol, Alice, Bob, and Eve all will have classical data.

Individual attacks are a slightly more generalized version of Intercept/Resend (the latter of which may be considered a particular type of individual attack). In this attack, Eve captures every qubit from Alice and attacks it in some way (e.g., through a unitary probe with a quantum ancilla, or through a more general measurement operator). She forwards a qubit to Bob who measures according to the protocol. Right after Bob measures, Eve will then immediately measure any quantum ancilla she has, leaving her with a classical memory system. Thus, as with Intercept/Resend attacks, at the end of the quantum communication stage of the protocol, Alice, Bob, and Eve all have classical data stored. Note that it is not difficult to model an Intercept/Resend attack as an Individual attack. However, it is also not difficult to see that Individual attacks give Eve potentially more attack options.

Collective attacks work by having Eve capture any qubit sent from Alice, and then probing it (attacking it) with any unitary operator which acts on the qubit state, and Eve's private quantum memory for this round (Eve has a quantum memory register for every round of the protocol). The exact unitary operator choice may be random (though the distribution of choices and the operators themselves are the same for every round of the protocol). The qubit, after passing through the unitary operator, will be sent immediately to Bob. Unlike Individual attacks, Eve will not immediately measure her ancilla and will, instead, keep it in her (assumed perfect) quantum memory. At the end of the quantum communication stage, Alice and Bob will have classical data, however Eve will have quantum data – in fact, she will have quantum data for every round of the protocol. Eve is allowed to postpone the measurement of her ancilla until any future point in time of her choice, even after she has seen a ciphertext encrypted using the derived secret key (which may be years later even)!

Finally, *general attacks*, are the strongest possible attack Eve can perform. Note that in the previous three attack types, Eve was forced to treat each qubit coming from Alice in an identical manner (possibly randomized, but with the same probability distribution of choices for each round). Eve was also forced to immediately send the qubit to Bob and to not alter her attack on the, say, n'th round based on the outcome of any previous rounds. General attacks do away with these assumptions. Eve is allowed to capture and attack multiple qubits from Alice simultaneously (possibly all qubits that Alice will eventually send). Eve is also allowed to alter her attack strategy on each round based on any input or information from previous rounds. In general, these "general" attacks model the ultimate goal of QKD security: Eve can attack in any way she likes, so long as her attack does not violate the laws of quantum mechanics. At the end of the quantum communication stage of the protocol, Alice and Bob will have classical data (of course) while Eve will have quantum and/or classical data. Note that general attacks allow Eve to create correlations between rounds which is not possible in the previous three attack models.

Exercise 5.3. Show that Intercept/Resend attacks are special types of individual attacks, which are special types of collective attacks, which are special types of general attacks. Do this by showing how, for example, any individual attack, can be modeled as a collective attack and so on. This shows that general attacks cover all the others.

For entanglement-based protocols, the story is basically the same. Though, when analyzing collective or general attacks, it is almost always assumed that the entanglement source is actually under the control of the adversary. That is, the protocol begins not with a trusted source device (e.g., Alice) creating a Bell pair (or other entangled state) and sending particles to Alice and Bob through an untrusted channel, but, instead, the protocol begins with Eve simply creating any arbitrary state, entangled with her ancilla, and then sending particles to Alice and Bob (generally through a noiseless channel – all noise is "simulated" by Eve when she creates the state – as discussed above, this is to Eve's advantage). In this regard, we alter the definitions, above, slightly to the following:

- For collective attacks, and for every round of the protocol, Eve creates an arbitrary quantum state $|\psi\rangle_{ABE}$. Thus, over the course of the protocol's N quantum communication stage rounds, Eve creates N identical copies of $|\psi\rangle_{ABE}$. The A and B registers are sent to Alice and Bob respectively, while the E register remains with Eve. Alice and Bob run the protocol as normal. Eve may postpone measurement of her ancilla to any future point in time (similar to the prepare-and-measure collective attack scenario). Note that the entire protocol state, over all N rounds, before Alice and Bob do anything, is in the tensor product state $|\psi\rangle_{ABE}^{\otimes N}$.
- For general attacks, Eve creates an arbitrary quantum state $|\psi\rangle_{ABE}$, where, now, the A and B registers consist of enough qubits for all N-rounds of the protocol. This allows Eve, for example, to create arbitrary correlations between rounds.

Letting Eve create the initial state serves several purposes. On the one hand, it's a strong assumption in favor of the adversary. On the other, it actually makes the security analysis easier. In fact, some of the best QKD security proofs for prepare and measure protocols, often involve a first step of *reducing* the protocol to an entanglement-based one (i.e., designing an entanglement based protocol that is mathematically identical to the original prepare-and-measure one). Once this is done, a security proof will assume Eve creates the initial state and Alice and Bob simply measure. Proving security in this case is generally easier than the case where Alice prepares qubit states randomly, Eve probes them unitarily, the Bob measures them randomly.

Exercise 5.4. Consider the following collective attack against an entanglement based protocol: A trusted third-party source creates the quantum state required by the protocol, $|\psi\rangle_{AB}$, sending the particles to each party. Eve then attacks with a unitary probe acting on those qubits being sent, entangling them with her private ancilla. The particles then arrive to Alice and Bob. Before Alice and Bob measure, call the state $|\phi\rangle_{ABE}$. Show that it is equivalent, mathematically, to consider the case where Eve simply creates the initial state, sending particles to Alice and Bob. I.e., show that for every possible attack against the first case (with a trusted third party source), there exists an attack in the second case (where there is no trusted third party, only Eve).

What about if Alice is the trusted source? Here, Alice creates the protocol state $|\psi\rangle_{AB}$ and sends only the B register to Bob, keeping the A register private. Eve can now only probe the B register with her unitary operator (also acting on her private ancilla). Argue that, in this case, security against the scenario where Eve creates the state will again imply security against the case where Alice creates the state. Thus, it suffices to analyze the case where Eve creates everything, initially.

5.3 Bounding the secret key length

Now that we've discussed the attack model and how we handle noise in the channel, both observed and adversarial, it's time to finally turn our attention to how we prove security, given a particular attack model. Recall Eq. (5.9) which involves the "real state" ρ_{ABE} which is the state that's output from the protocol after privacy amplification, and also $\mathcal{K}_{ABE}$ which is the "ideal state," or the state that represents a perfectly secure secret key: one that is uniform random, while also being completely independent of an adversary system. The question becomes: given a protocol which outputs ρ_{ABE}, where the A and B registers are bit-strings of size ℓ-bits, how far, in trace distance, is ρ_{ABE} from the ideal state? For this, it turns out, we can use min entropy, which we introduced, in Section 2.6.2.

Min entropy, is a vital resource in quantum cryptography, primarily for the reason that, as was shown in [8,15], it can be used to bound the difference between a quantum state and the ideal state $\mathcal{K}_{ABE}$. First, let's ignore the B system for now – assuming the protocol is *correct* (a term we will return to later in this chapter), it should hold that the A and B registers are fully correlated – i.e., they are the same random bit string, with $k_A = k_B$ always. Thus, we really care about the following distance:

$$\frac{1}{2}||\rho_{AE} - \mathcal{K}_{AE}||, \tag{5.8}$$

where we traced out the B register in both real and ideal states (they are clearly correlated in the ideal state). Then, it was shown in [8], that this trace distance may be upper-bounded by a function of the quantum min entropy. Namely, it holds that:

$$||\rho_{AE} - \mathcal{K}_{AE}|| \leq 2^{-\frac{1}{2}(H_\infty(A|E)_\sigma - \ell)}, \tag{5.9}$$

where σ is the quantum state representing the error-corrected raw key (i.e., the state before privacy amplification is run, but after all error correction information has been divulged). If one can bound the *smooth min entropy* (defined in Section 2.6) of the state before privacy amplification, namely $H_\infty^\epsilon(A|E)_\sigma$, then, one can also derive a similar bound using the following identity (also from [8]):

$$||\rho_{AE} - \mathcal{K}_{AE}|| \leq 2^{-\frac{1}{2}(H_\infty^\epsilon(A|E)_\sigma - \ell)} + 2\epsilon, \tag{5.10}$$

The general definition of security for a QKD protocol, then, is [8]:

Definition 5.3.1. A QKD protocol is said to be ϵ-*secure*, for some $\epsilon > 0$ if:

$$\frac{1}{2}||\rho_{AE} - \mathcal{K}_{AE}|| \leq \epsilon. \tag{5.11}$$

If the protocol aborts, with probability p_{abort} (in which case, Alice and Bob do not output any secret key), then it is ϵ-secure if:

$$\frac{1}{2}(1 - p_{abort})\,||\rho_{AE} - \mathcal{K}_{AE}|| \leq \epsilon. \tag{5.12}$$

Let's put this into better context. Suppose, we want to run a QKD protocol, and output a key that is ϵ-secure. Then, thanks to Eq. (5.9), so long as:

$$2^{-\frac{1}{2}(H_\infty(A|E)_\sigma - \ell)} = \epsilon, \tag{5.13}$$

our security guarantee will hold (similar equations can be found using smooth-min entropy, using Eq. (5.10)). Fixing the security parameter ϵ, the only variable above is the size of the final secret key, ℓ. Solving for this, we see that the final secret key, of size ℓ, will be ϵ secure, so long as:

$$\ell = H_\infty(A|E)_\sigma - 2\log\frac{1}{\epsilon}. \tag{5.14}$$

That is, the final size of the secret key will be, approximately, the min entropy of the system before privacy amplification is run. Looking a bit closer at this, recall that σ is the state after error correction. Error correction is a process that divulges information about Alice's and Bob's raw keys (this is a necessity, since Alice and/or Bob must send *some* information to one another, and that information must be sent "in the clear" over the authenticated channel). Later, when actually deriving a secret key length, we may compute the min entropy *before* error correction and privacy amplification, but then deduct from the final secret key size, the amount of *error correction leakage.* One may use, for instance, the chain rule of min entropy to formalize this [16,13]. These details will become clear, as we progress in this, and future, chapters.

5.3.1 Asymptotic security proof of BB84

Eq. (5.9) is general, and applies to any QKD protocol in the so-called *finite key scenario,* where the number of signals (e.g., qubits) sent between parties is finite. As the number of signals (which we also call rounds of the QKD protocol) increases, performance of the system tends to improve (for example, due to better sampling data and lower relative contribution of error terms such as the $\log \frac{1}{\epsilon}$ term in the previously referenced equation). As the number of signals approaches infinity, we attain a "best case" bound on the possible performance of the QKD protocol. Such bounds are very useful as they provide us with theoretical "speed limits" of the given QKD protocol: namely, they tell us how good they can be in the most optimal scenarios. This is called the *asymptotic scenario* of a QKD protocol. As it turns out, when analyzing this scenario, we don't need to worry about min entropy either, and can switch to the, generally easier to analyze, von Neumann entropy.

Let's define the *key-rate* of a QKD protocol to be the total secret key bits produced (denoted ℓ), divided by the total number of signals sent, which we'll denote N. Let's assume our QKD protocol required a sample size of m – these samples cannot be used to produce a secret key and, as far as the rate is concerned, can only deduct. (Of course, there is a balance when we start to consider confidence intervals, as we do in Section 5.4.) Using Eq. (5.14), the rate, then, may be written, somewhat loosely (we ignore the terms involving ϵ, here, as they will "vanish" in the asymptotic scenario), as:

$$\texttt{rate} = \frac{\ell}{N} \approx \frac{1}{N} H_{\infty}^{\epsilon}(A|E)_{\sigma} - \frac{1}{N}\lambda_{EC}, \tag{5.15}$$

where λ_{EC} is the error correction leakage, and σ is now the state of the system *before* error correction and privacy amplification (see the discussion in the previous section).

For starters, let's assume collective attacks; in this case, we can write the final state of the system after N rounds (before error correction and privacy amplification) as $\sigma = \left(\sigma^{(1)}\right)^{\otimes N}$, where $\sigma^{(1)}$ is the joint state, modeling Alice, Bob, and Eve, after a single round of the protocol that led to a raw key bit generation (i.e., where everything "went right" for parties to get a raw key bit).

Let's assume that $m = \sqrt{N}$ (i.e., $\sqrt{N}$ of the total N rounds of σ will be used for sampling and, thus, discarded from key distillation). Under this assumption, $m/N \to 0$ as $N \to \infty$. That is to say, the sample size, in the asymptotic limit, does not need to be considered as it may be made so small, relative to the number of signals, that its relative size is essentially zero in the limit. Now, taking the limit as $N \to \infty$, gives us the asymptotic rate, denoted $\texttt{rate}_{\infty}$. As it turns out, thanks to something called the *quantum asymptotic equipartition property* [12], $\frac{1}{N} H_{\infty}^{\epsilon}(A|E)_{\sigma} \to H(A|E)_{\sigma^{(1)}}$, where $H(A|E)_{\sigma^{(1)}}$ is the von Neumann entropy of the quantum system before error correction and privacy amplification, but only for a *single round* of the system. Finally, in the asymptotic setting, one can assume optimal error correction, which will imply that $\frac{1}{N}\lambda_{EC} \to H(A|B)$, where $H(A|B)$ is actually the Shannon conditional entropy of Alice and Bob's raw keys (again, only on a single round). Thus, when analyzing the asymptotic key rate, one only needs to work with von Neumann and Shannon entropy. Of course, this is assuming collective attacks – we will discuss general attacks shortly.

The above leads to an asymptotic keyrate expression of:

$$\texttt{rate}_{\infty} = H(A|E) - H(A|B), \tag{5.16}$$

which is also called the Devetak-Winter keyrate expression [3]. The above expression is actually pretty intuitive. If one considers $H(A|E)$ to be, essentially, a measure of how uncertain Eve is on Alice and, similarly, $H(A|B)$ is a measure of how uncertain Bob is on Alice, then the rate increases whenever Eve is very uncertain, and Bob is very certain which makes sense from a practical standpoint. If one can compute, or at least bound, the two entropies in the above expression, one can immediately get a bound on the asymptotic key-rate of the QKD protocol in question. This generally provides a theoretical limit on the performance of a QKD protocol (which, in practice, will not run for an infinite amount of time, but instead for a fixed number of rounds N). Now, the question of course becomes: how to bound the above entropies?

First, $H(A|B)$ is actually very straightforward to bound and, typically, is not even discussed in detail in QKD security proofs. Recall from Section 2.6.2, that $H(A|B) = H(AB) - H(B)$. Each of those expressions depends only on the probability distribution $P(A = a \text{ and } B = b)$, for $a, b \in \{0, 1\}$. Standard sampling methods (using some of the sample of size m discussed earlier), can be used to determine this distribution; furthermore, in the asymptotic scenario, one may assume that it is known exactly, without any "wiggle room" which would occur due to finite sampling imprecision (this is a luxury we will not be able to afford later, when we discuss finite key security proofs). Thus, Alice and Bob will know exactly what $H(A|B)$ is. This leaves only $H(A|E)$.

If Alice and Bob knew exactly what attack Eve did, they would be able to write out a density matrix description of the system, compute the eigenvalues of that operator, and, thus, compute $H(A|E)$ directly. Unfortunately (or, if one is actually interested in the puzzle of a QKD security proof, perhaps *fortunately*!), Alice and Bob only have limited information on Eve's attack: they only know how much noise it created (from the sampling stage of a QKD protocol). From this, they must in a way, "reverse engineer" what Eve could have done. That is, they must determine a bound on $H(A|E)$ by assuming Eve performed the most optimal attack possible which would induce the observed amount of noise. In this circumstance, the rate becomes an optimization problem in a way:

$$\texttt{rate}_{\infty} \geq \inf_E H(A|E) - H(A|B), \tag{5.17}$$

where the infimum is over all possible attacks that Eve could perform which also induce the observed amount of noise. We minimize this, as we must assume Eve performs an attack which maximizes her information gain (i.e., minimizes her uncertainty).

But, how do we determine a bound on $H(A|E)$, when we have no idea what operations Eve used to attack? This is the key point of any QKD security proof – for BB84, we will consider two alternative methods: first a direct calculation method for the prepare and measure version which gives security against collective attacks; then, second, a more powerful method using *entropic uncertainty*, which works for both the prepare and measure version, and also the entanglement based protocol, and, importantly, also can be used to prove security against arbitrary *general* attacks.

Note that these are not the only way to go about proving security in the asymptotic scenario. For instance, in the entanglement based version of BB84, an elegant approach in [8,9] is to show that the resulting state between Alice and Bob can be written as a Bell-diagonal operator, and thus Eve may hold a purification of such a system. Then, one may compute the eigenvalues directly to compute $H(A|E)$. Furthermore, this approach also can be used to prove security against general attacks. We will show the "direct calculation" method, first, as it works for many different prepare and measure variants, even though it cannot be necessarily "promoted" to a security proof against general attacks. We'll then show the entropic uncertainty version as it is a very straightforward method (and can be used for security against general attacks). However, the reader should be aware there are several other methods out there.

5.3.1.1 Directly calculating Eve's entropy

Let's trace out the BB84 protocol. However, instead of analyzing the general protocol, we will instead analyze the "biased" version [6] where only the Z basis is used for key distillation. We will still need X basis rounds to test for eavesdropping; in fact, both X and Z basis rounds will be used to test the fidelity of the channel, and derive a bound on Eve's information, but only Z basis rounds are used for key generation. The advantage to using only Z basis rounds for key distillation is that greater efficiency is possible, since one may bias the basis choice distribution in favor of Z rounds. Secondly, the security proof is somewhat simplified since we only need to bound Eve's information $H(A|E)$ on rounds where Alice and Bob use the Z basis.

When computing $H(A|E)$ in an asymptotic security proof, we may condition on those choices being made by Alice and Bob that would actually lead to a raw key bit being produced – any other outcome is discarded, and so cannot contribute to the key and, thus, should not be considered in the computation of $H(A|E)$. Those rounds are of course important, and will be used to bound the uncertainty, but we must actually bound the uncertainty in the raw key register. Thus when tracing the protocol, we will condition on events that lead to a raw key bit being produced.

Conditioning, then, on Alice sending a Z basis state, the density operator starts as:

$$\frac{1}{2}|0\rangle\langle 0|_A \otimes |0\rangle\langle 0|_T + \frac{1}{2}|1\rangle\langle 1|_A \otimes |1\rangle\langle 1|_T, \tag{5.18}$$

where the A register is Alice's private (classical) memory used to store her raw key bit and the T register is the transit qubit – i.e., the qubit that will be traveling from Alice to Bob (passing, of course, through Eve).

Now, Eve captures the qubit sent from Alice (i.e., she now has access to the T register) and attacks it. Since we are only considering collective attacks, this attack may be modeled as a unitary operator U, which acts on the T register, and Eve's private ancilla E. Furthermore, we may assume that Eve's private ancilla is cleared to some default state which Eve knows about – without loss of generality, we might as well assume Eve's ancilla starts in the state $|0\rangle_E$ (if it were some other state, she could apply an operator to map that initial state to $|0\rangle_E$; this initial operator may then be "absorbed" into her attack U, so there's no advantage to assuming anything else). Then, in this case, we may write the abstract action of U on the following basis states:

$$U\,|0,0\rangle_{TE} = |0,e_0\rangle_{TE} + |1,e_1\rangle_{TE} \tag{5.19}$$

$$U\,|1,0\rangle_{TE} = |0,e_2\rangle_{TE} + |1,e_3\rangle_{TE} \tag{5.20}$$

Since we are assuming Eve's state starts in the state $|0\rangle$, there is no need to define U on any other state – its action on other states may be completely arbitrary (though must still be unitary). The states $|e_i\rangle_E$ are arbitrary states in Eve's ancilla. Note that they are sub-normalized. They are also not necessarily orthogonal; however, unitarity of U does impose some constraints on the above states. For instance:

$$1 = \langle 0,0|0,0\rangle = \langle 0,0|U^*U|0,0\rangle = \langle 0,e_0|0,e_0\rangle + \langle 1,e_1|1,e_1\rangle = \langle e_0|e_0\rangle + \langle e_1|e_1\rangle\,. \tag{5.21}$$

Similar constraints may be derived:

$$\langle e_2|e_2\rangle + \langle e_3|e_3\rangle = 1 \tag{5.22}$$

$$\langle e_0|e_2\rangle + \langle e_1|e_3\rangle = 0.$$

Exercise 5.5. Explain how the above constraints on Eve's states follow from unitarity of her attack operator U.

Eve applies this operator to the TE registers. Continuing on from Eq. (5.18), this produces the following density operator:

$$\frac{1}{2}\,|0\rangle\,\langle 0|_A \otimes P\left(|0,e_0\rangle + |1,e_1\rangle\right) + \frac{1}{2}\,|1\rangle\,\langle 1|_A \otimes P\left(|0,e_2\rangle + |1,e_3\rangle\right), \tag{5.23}$$

where $P(|z\rangle) = |z\rangle\,\langle z|$.

The T register now travels onward to Bob, while the E register remains private, and entangled, at Eve's location. Bob, of course, will take the T register and measure it. Since we are still only concerned with events leading to a raw key bit being distilled, we will assume Bob chooses to measure in the Z basis. This causes a state collapse, including a collapse in Eve's system. Bob takes his measurement outcome and saves the result in a private register, denoted B. This will cause the state to collapse to:

$$\frac{1}{2}|0\rangle\langle 0|_A \otimes (|0\rangle\langle 0|_B \otimes |e_0\rangle\langle e_0| + |1\rangle\langle 1|_B \otimes |e_1\rangle\langle e_1|) +$$
$$\frac{1}{2}|1\rangle\langle 1|_A \otimes (|0\rangle\langle 0|_B \otimes |e_2\rangle\langle e_2| + |1\rangle\langle 1|_B \otimes |e_3\rangle\langle e_3|). \quad (5.24)$$

To see how the above is derived, refer to Exercise 2.21; specifically Eq. (2.95). For example, given $|0, e_0\rangle_{TE} + |1, e_1\rangle_{TE}$, if a measurement in the Z basis is made of the T register, the resulting state will collapse to a mixture:

$$|0\rangle\langle 0| \otimes |e_0\rangle\langle e_0| + |1\rangle\langle 1| \otimes |e_1\rangle\langle e_1|. \quad (5.25)$$

Since Eq. (5.23) is a mixture of such states, Eq. (5.24) follows from linearity.

Now, since we need to compute $H(A|E)$, we can trace out Bob, leaving us with simply:

$$\frac{1}{2}|0\rangle\langle 0|_A \otimes (|e_0\rangle\langle e_0| + |e_1\rangle\langle e_1|) + \frac{1}{2}|1\rangle\langle 1|_A (|e_2\rangle\langle e_2| + |e_3\rangle\langle e_3|). \quad (5.26)$$

But now, how do we go about computing the entropy in the above state? If we knew something about the actual vectors describing Eve's state, we could simply build the density matrix, compute the eigenvalues (using a computer program, probably), and then simply run them through the Shannon entropy function (recall, von Neumann entropy is the Shannon entropy of the eigenvalues of the system). However while we may not be able to build the vectors directly, as it turns out there is a lot of information we can learn about them, simply by looking at the observable noise in the system.

Let's first consider what the Z basis noise tells us about the system. If Alice sends a $|0\rangle$ and Bob measures in the Z basis, observing a $|1\rangle$, this is a detectable error (after sampling, of course, however in the asymptotic scenario, our samples can be as large as we like, and so we have perfect information on the probability of a $|0\rangle$ flipping to a $|1\rangle$). Let $Q_{0,1}$ denote the probability of a $|0\rangle$ being flipped to a $|1\rangle$ (after Eve's attack). Tracing this event, we have:

$$|0\rangle \to |0, e_0\rangle + |1, e_1\rangle,$$

where the right hand side of the above equation is the resulting state after Eve's attack operator U is applied. From the above, it is clear that:

$$Q_{0,1} = \langle e_1|e_1\rangle. \quad (5.27)$$

This gives us a bound on "how large" of a vector $|e_1\rangle$ can be. Of course, due to normalization constraints (see Eq. (5.21)), this also gives us $\langle e_0|e_0\rangle$. Using similar logic, we can also find:

$$\begin{aligned} Q_{0,0} &= \langle e_0|e_0\rangle \\ Q_{0,1} &= \langle e_1|e_1\rangle \\ Q_{1,0} &= \langle e_2|e_2\rangle \end{aligned} \quad (5.28)$$

$$Q_{1,1} = \langle e_3|e_3\rangle$$

The Z basis noise, above, gives us vital information on the "size" of Eve's vectors. However, by itself, it is not sufficient to get a non-trivial bound on $H(A|E)$. Indeed, consider a "CNOT" attack: here Eve just applies a CNOT operator on the traveling qubit. In this case, $|e_0\rangle = |0\rangle$ and $|e_3\rangle = |1\rangle$, while the other two "e" vectors are simply the zero-vector. Such an attack matches the above constraints, since for a CNOT attack, $Q_{0,0} = Q_{1,1} = 1$ and $Q_{0,1} = Q_{1,0} = 0$. However, it will give Eve full information; you can easily compute $H(A|E) = 0$ in this case. Thus, considering the Z basis noise is important, however not sufficient on its own (which makes sense – otherwise, the protocol would be classical in a way!).

Exercise 5.6. Show that if Alice and Bob only estimate the Z basis noise, Eve can launch an attack such that $H(A|E) = 0$. Describe your attack, and compute the resulting von Neumann entropy.

Thus, we now turn to the X basis noise where the algebra gets a little more involved. Let's look at $Q_{+,+}$, the probability that Bob observes a $|+\rangle$ in the event Alice sends a $|+\rangle$ (which should be large). Tracing this event (while taking advantage of the linearity of U) yields:

$$\begin{aligned} U|+,0\rangle_{TE} &= |0\rangle_T\left(\frac{|e_0\rangle + |e_2\rangle}{\sqrt{2}}\right) + |1\rangle_T\left(\frac{|e_1\rangle + |e_3\rangle}{\sqrt{2}}\right) \\ &= |+\rangle\left(\frac{|e_0\rangle + |e_2\rangle + |e_1\rangle + |e_3\rangle}{2}\right) + |-\rangle\left(\frac{|e_0\rangle + |e_2\rangle - |e_1\rangle - |e_3\rangle}{2}\right) \end{aligned}$$

From the above, we can now compute $Q_{+,+}$. Recalling that $\langle x|y\rangle + \langle y|x\rangle = 2Re\,\langle x|y\rangle$, we have:

$$\begin{aligned} Q_{+,+} &= \frac{1}{4}(\langle e_0|e_0\rangle + \langle e_1|e_1\rangle + \langle e_2|e_2\rangle + \langle e_3|e_3\rangle + 2Re(\langle e_0|e_1\rangle + \langle e_0|e_2\rangle + \langle e_0|e_3\rangle) \\ &\quad + 2Re(\langle e_1|e_2\rangle + \langle e_1|e_3\rangle + \langle e_2|e_3\rangle) \\ &= \frac{1}{2}(1 + Re(\langle e_0|e_1\rangle + \langle e_0|e_3\rangle + \langle e_1|e_2\rangle + \langle e_2|e_3\rangle), \end{aligned} \tag{5.29}$$

where, for the last equality, we used the fact that U is unitary; specifically, Eq. (5.22). This, by itself, does not seem overly helpful. However, by performing similar computations, we can find $Q_{-,-}$ to be:

$$Q_{-,-} = \frac{1}{2}(1 + Re[\langle e_0|e_3\rangle + \langle e_1|e_2\rangle] - Re[\langle e_0|e_1\rangle + \langle e_2|e_3\rangle]). \tag{5.30}$$

Adding the two expressions, then, yields:

$$Q_{+,+} + Q_{-,-} = 1 + Re(\langle e_0|e_3\rangle + \langle e_1|e_2\rangle). \tag{5.31}$$

Exercise 5.7. Show that Eq. (5.30) is correct.

How does this help? Well, let's consider an attack by Eve which creates no detectable noise; that is, an attack where $Q_{0,0} = Q_{1,1} = Q_{+,+} = Q_{-,-} = 1$. Using the Z basis noise tells us that $|e_1\rangle$ and $|e_2\rangle$ are both the zero vector – thus, $\langle e_1|e_2\rangle = 0$. Such a constraint collapses Eq. (5.31) to:

$$2 = Q_{+,+} + Q_{-,-} = 1 + Re\,\langle e_0|e_3\rangle \Longrightarrow Re\,\langle e_0|e_3\rangle = 1.$$

Of course, that's just the real part of the inner product; however due to the constraint that $\langle e_0|e_0\rangle \leq 1$ (and similarly for $\langle e_3|e_3\rangle$), this implies that these two vectors are actually identical. Thus, returning to the density operator description in Eq. (5.26), one can easily see that, in this case, Eve has absolutely no information on Alice's raw key – the density matrix takes the form $\rho_A \otimes \rho_E$ and, thus, $H(A|E) = 1$.

But what about *some* noise? Even in ideal scenarios where there is no adversary, there will always be some noise – which, as discussed earlier in this chapter, we must assume is adversarial noise. For this, we can turn to Theorem 2.1 discussed in Chapter 2. Applying that theorem to Eq. (5.26) will lead the following bound on the entropy (and recall, we only care about a lower-bound since more entropy is better for Alice and Bob in this scenario):

$$\begin{aligned} H(A|E) \geq &\left(\frac{\langle e_0|e_0\rangle + \langle e_3|e_3\rangle}{2}\right) \times \left(h\left[\frac{\langle e_0|e_0\rangle}{\langle e_0|e_0\rangle + \langle e_3|e_3\rangle}\right] - h[\lambda_1]\right) \\ &+ \left(\frac{\langle e_1|e_1\rangle + \langle e_2|e_2\rangle}{2}\right) \times \left(h\left[\frac{\langle e_1|e_1\rangle}{\langle e_1|e_1\rangle + \langle e_2|e_2\rangle}\right] - h[\lambda_2]\right) \\ = &\left(\frac{Q_{0,0} + Q_{1,1}}{2}\right) \times \left(h\left[\frac{Q_{0,0}}{Q_{0,0} + Q_{1,1}}\right] - h[\lambda_1]\right) \\ &+ \left(\frac{Q_{0,1} + Q_{1,0}}{2}\right) \times \left(h\left[\frac{Q_{0,1}}{Q_{0,1} + Q_{1,0}}\right] - h[\lambda_2]\right) \end{aligned} \tag{5.32}$$

where:

$$\lambda_1 = \frac{1}{2}\left(1 + \frac{\sqrt{(Q_{0,0} - Q_{1,1})^2 + 4Re\,\langle e_0|e_3\rangle}}{Q_{0,0} + Q_{1,1}}\right) \tag{5.33}$$

$$\lambda_2 = \frac{1}{2}\left(1 + \frac{\sqrt{(Q_{0,1} - Q_{1,0})^2 + 4Re\,\langle e_1|e_2\rangle}}{Q_{0,1} + Q_{1,0}}\right) \tag{5.34}$$

The only expression in the above equations that Alice and Bob cannot directly observe are $Re\,\langle e_0|e_3\rangle$ and $Re\,\langle e_1|e_2\rangle$. However, by looking at the Z and X basis noise, the following two constraints are placed on these inner products:

- $Re\,\langle e_1|e_2\rangle \in \left[-\sqrt{Q_{0,1}Q_{1,0}}, \sqrt{Q_{0,1}Q_{1,0}}\right]$ (by Cauchy-Schwarz – see Eq. (2.16))
- $Re\,\langle e_0|e_3\rangle + Re\,\langle e_1|e_2\rangle = Q_{+,+} + Q_{-,-} - 1$ (from Eq. (5.31))

This gives us everything we need to evaluate the entropy of the BB84 protocol. Indeed, we simply optimize over one parameter, namely the real part of $\langle e_1|e_2\rangle$. This is a real number, bounded within a relatively small interval. For any possible value of $Re\,\langle e_1|e_2\rangle$, this fixes what $Re\,\langle e_0|e_3\rangle$ could possibly be. Once you have that, computing the above λ_i expressions is straightforward which, subsequently, will allow you to fully evaluate Eq. (5.32). Since we must always assume the worst case, we must actually minimize $H(A|E)$ over all possible $Re\,\langle e_1|e_2\rangle$.

Exercise 5.8. Let $Q_{i,j} = Q$, for $i \neq j$, for some user-specified Q (which represents the channel error rate). Write a computer program that outputs the smallest possible $H(A|E)$, for any value of Q. Your program should minimize Eq. (5.32) over all possible real numbers $Re\,\langle e_1|e_2\rangle$ bounded by the above Cauchy-Schwarz induced constraint.

Exercise 5.9. Argue why it is sufficient to consider a unitary attack in this scenario. For instance, Eve could perform a measurement, or apply different unitary operators with some fixed probability distribution. Argue why this, and other such *collective* attacks, can all be modeled as Eve applying a single unitary U to the TE registers, while the E register starts in the state $|0\rangle$. Your argument may be high-level.

We still need to bound $H(A|B)$; however before we do this, let's look at an alternative proof method to bound $H(A|E)$.

5.3.1.2 Bounding Eve's entropy using entropic uncertainty

The above method was a "direct calculation" style approach: we trace out the density matrix representing an actual raw key bit (i.e., we trace the protocol conditioning on events leading up to an actual raw key being distilled). Then, we apply Theorem 2.1 from Section 2.6. Finally, to evaluate, we look at what the noise in the channel tells us about the inner products of the state in question. Sometimes, we can get exact values (e.g., as in $\langle e_0|e_0\rangle$); other times we can only get bounds (e.g., we used Cauchy-Schwarz to bound $Re\,\langle e_1|e_2\rangle$). This is a useful method for handling prepare and measure protocols under collective attack scenarios, however it is admittedly laborious. That method also only works for security against collective attacks.

An alternative, and very powerful, proof technique utilizes entropic uncertainty, discussed in Section 2.6.2.2; however, now, we're going to actually apply it to our QKD protocol. This technique is powerful in the sense that it provides an almost immediate bound on the entropy, though it is most useful for entanglement based protocols.

Recall that BB84 is actually mathematically equivalent to the E91 protocol (as discussed in Chapter 4). Thus, any analysis on the E91 protocol, which is entanglement-based, will imply a result for the prepare-and-measure BB84. In E91, Eve starts the day off by creating a quantum state and sending part to Alice and part to Bob. Alice and Bob subsequently measure their system without any further interaction from Eve. Let $|\psi\rangle_{ABE}$ be the (pure) state that Eve creates initially before Alice and Bob measure. We may assume the state is pure, otherwise we can purify it and give the purification to Eve. Since, as before, we are considering collective attacks, Eve actually creates N copies of this state (with N going to infinity as we are also in the asymptotic setting). Then, as discussed in Section 2.6.2.2, it holds that:

$$H(A_Z|E) + H(A_X|B) \geq 1, \tag{5.35}$$

where, the A_Z and A_X random variables, above, are induced by measuring the Alice system of $|\psi\rangle$ in the corresponding basis. Note the right-hand size of the bound is one due to our particular measurement bases.

Now, since measurements cannot decrease uncertainty, if Bob measures his system in the X basis, leading to random variable B_X, it holds that $H(A_X|B_X) \geq H(A_X|B)$ and, so:

$$H(A_Z|E) + H(A_X|B_X) \geq 1 \Longrightarrow H(A_Z|E) \geq 1 - H(A_X|B_X). \tag{5.36}$$

$H(A_Z|E)$ is exactly the entropy we want to bound in order to determine the asymptotic key rate (Eq. (5.16)). By entropic uncertainty, we only need to know what $H(A_X|B_X)$ is. However, this is exactly based on the X-basis noise, in particular the probability values $Q_{+,-}$ and $Q_{-,+}$. From this alone, we're finished! We end up with:

$$\texttt{rate}_\infty \geq 1 - H(A_X|B_X) - H(A_Z|B_Z), \tag{5.37}$$

which is a function only of observable values, namely based on the distribution of Alice and Bob's various measurements. Note that, we're using the A_Z and B_Z registers for the last entropy to measure the information leaked during error correction (see Eq. (5.16)).

Exercise 5.10. Let $Q_{+,-} = Q_{-,+} = Q$. Show algebraically that:

$$H(A_X|B_X) = h(Q).$$

Exercise 5.11. Using Exercise 5.8, show numerically, that both methods produce the same bound on $H(A_Z|E)$, in this instance.

5.3.1.3 Putting it all together: the key-rate

The above sections showed you how to bound Eve's uncertainty. However, you'll notice that we're still missing $H(A_Z|B_Z)$, namely Bob's uncertainty on Alice's raw key bit. However, this is simply based on the probability values $Q_{i,j}$, for $i, j \in \{0, 1\}$. If we let $Q_{0,1} = Q_{1,0} = Q_{+,-} = Q_{-,+} = Q$, it can be shown (see also Exercise 5.10), that $H(A_Z|B_Z) = h(Q)$. Thus, the key-rate of BB84 (and/or E91), using the entropic uncertainty method (Eq. (5.37)), simplifies to:

$$\texttt{rate}_\infty \geq 1 - H(A_X|B_X) - H(A_Z|B_Z) = 1 - 2h(Q). \tag{5.38}$$

Using the direct computation method, the equation does not simplify as nicely (and is, in fact, an optimization problem over one real parameter to find $H(A_Z|E)$) – however the final $H(A_Z|B_Z)$ term is exactly the same. And, in fact, if you write out a computer program to evaluate the minimum $H(A_Z|E)$ value (see Exercise 5.8), you can actually check that program will agree *exactly* with the above $1 - 2h(Q)$. The resulting key-rate is also shown in Fig. 5.3.

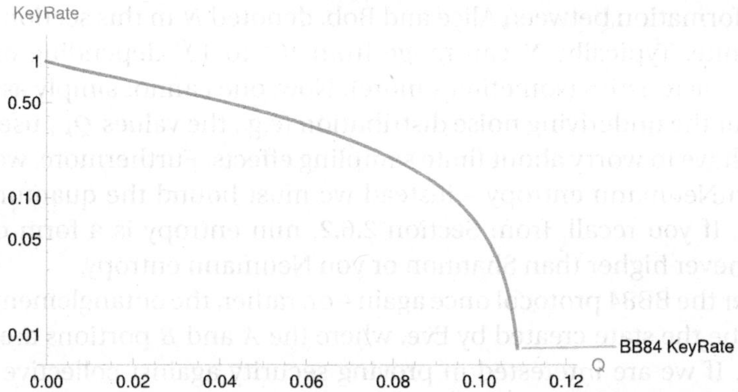

FIGURE 5.3 Showing the key-rate of BB84, namely Eq. (5.38) as the noise increases. We note that the key-rate goes to zero when Q approaches 11%.

Deriving a key-rate expression is the main objective in a QKD security proof. Once one derives bounds on the two important entropy functions, Eq. (5.16) guarantees that the resulting key will be ϵ-secure (though, to be really formal, in the asymptotic scenario, we take $\epsilon \to 0$).

This also gives us the protocols *noise tolerance*. Indeed, if one has a bound on the rate, as a function of the noise (as we do above), then by setting the equation to zero and solving for the noise parameter, we know exactly how much noise the protocol can withstand before the rate becomes zero (i.e., before Alice and Bob simply have to abort, since their resulting key is zero-bits long). For example, if you set $1 - 2h(Q) = 0$ and solve for Q (just do this numerically using your favorite graphing calculator), one gets $Q = 11\%$. This, or any smaller, value of Q will produce a positive key-rate.

Note that, the above proof of security was against collective attacks. However, when working with entanglement based protocols, where Eve creates an i.i.d. signal state $|\psi\rangle_{ABE}^{\otimes N}$, in general, such security proofs can be promoted to security against *general attacks* using post-selection techniques [2]. This does require that the protocol is permutation invariant – however, this can typically be enforced by having Alice and Bob randomly permute their systems. Note that while post-selection techniques can be applied to the entanglement based methods and entropic uncertainty results, it's not clear if this applies to the prepare-and-measure "direct calculation" method, discussed earlier, where Eve probes with a given unitary operator. This is another advantage to the entropic uncertainty proof methodology, assuming one can derive an equivalent entanglement based protocol.

5.4 Finite key security

Asymptotic analyses are great for understanding the theoretical limits of a QKD protocol. They are very often a "first step" in a security proof. However, to be useful in practice, a *finite key security* proof is required. Here, the number of signals (or rounds) used to transmit key information between Alice and Bob, denoted N in this section, is not infinite, but, instead, finite. Typically, N can range from 10^4 to 10^9 depending on the protocol and channel characteristics (sometimes more). Now, one cannot simply assume we know everything about the underlying noise distribution (e.g., the values $Q_{i,j}$ used previously) – instead we will have to worry about finite sampling effects. Furthermore, we can no longer directly use von Neumann entropy – instead we must bound the quantum min entropy (see Eq. (5.10)). If you recall, from Section 2.6.2, min entropy is a form of "worst-case" entropy and is never higher than Shannon or von Neumann entropy.

Let's consider the BB84 protocol once again – or, rather, the entanglement based version and let $|\psi\rangle_{ABE}$ be the state created by Eve, where the A and B portions are N qubits each (for a finite N). If we are interested in proving security against collective attacks, it will hold that $|\psi\rangle_{ABE}$ has a tensor-product form. Namely, we can write this state as $|\psi\rangle_{ABE} = (|\psi_0\rangle_{ABE})^{\otimes N}$. Thus, after running the QKD protocol, the final mixed state modeling Alice and Bob's raw keys and Eve's quantum ancilla will be of the form $(\rho_{ABE})^{\otimes n}$ (before error correction and privacy amplification). Note that n represents the size of the distilled raw key and that $n < N$ (since at least some rounds will be used for error testing).

Computing a bound on the min entropy of a product state, such as this, can be done by first bounding the von Neumann entropy of a single round $H(A|E)_\rho$ (which we did in the previous section!) and then using the so-called *asymptotic equipartition property* (AEP) [12] to derive a bound on the (smooth) min entropy of the form:

$$H_\infty^\epsilon(A|E) \geq nH(A|E)_\rho - O\left(\sqrt{n}\right). \tag{5.39}$$

(The exact error term will depend on the dimension of the system.) Of course, one must take into account finite sampling error, also, when using this technique. This method allows one to use the techniques discussed earlier, to bound von Neumann entropy, to derive finite key results. The penalty, though, is in the error term which one must subtract from the final key length (the $O(\sqrt{n})$ term).

However, if we are considering general attacks, then the state does not have any form of i.i.d. structure – that is, we no longer assume that all rounds are identical. We do not assume anything about the state of Eve's private ancilla, other than it is finite dimensional. Two common approaches for handing this scenario are to use AEP as above and then apply post selection techniques [2] to promote the security analysis to general attacks. However, doing so introduces additional error terms which shrink the final key length (that is, using such methods produce a sub-optimal finite key result). An alternative method is to compute a bound on the min entropy directly, without the use of AEP. For this, we will discuss two methods: entropic uncertainty (Section 5.4.1), and quantum sampling (Section 5.4.2).

5.4.1 Using Entropic Uncertainty

Let $|\psi\rangle_{ABE}$ be the state Eve prepares initially, where the A and B systems are N qubits each. We do not assume any structure to this state (e.g., we are not assuming collective attacks).

From this state, Alice and Bob will sample (test) m of the qubits, by measuring them in the X basis to determine the noise in the channel (call it q). Since this is a finite sample, there will be sampling error to account for. If $q > Q$ for some user-specified *noise threshold* Q, parties will abort the protocol.

The remaining $n = N - m$ signals will be measured in the Z basis to derive the secret key (following error correction and privacy amplification). The question, then, is what is the size of the secret key ℓ and what is the key-rate ℓ/N. Note that the key-rate will no longer be one, even in the noiseless case, since $m > 0$.

One method to prove security and determine the size of the final secret key will be to use an entropic uncertainty relation from [14] which bounds the *smooth* min entropy based on the overlap of the two measurements performed, along with the quantum max entropy, denoted H_{max}^{ϵ} (see Section 2.6.2). This method, which proves security against arbitrary, general, attacks is discussed in detail in [13] and the interested reader is encouraged to look closely at that proof. The main highlights are that Eve's uncertainty, now measured through quantum min entropy, can be bounded based on the correlation between Alice and Bob's X basis measurements. By taking into account sampling errors from these measurements, one can find very tight finite key length bounds for BB84 of the form:

$$\ell = n(1 - 2h(Q + \mu)) - \log\frac{2}{\epsilon}, \tag{5.40}$$

where ϵ is related to the failure probability and also the security of the key (see Eq. (5.3)), while μ is a function of the sampling error, namely:

$$\mu = \sqrt{\frac{(n+m)(m+1)}{nm^2}\ln\frac{2}{\epsilon}}. \tag{5.41}$$

A graph of the resulting keyrate (namely ℓ/N) for various noise levels as N increases, is shown in Fig. 5.4. Note that the overall key-rate improves as N increases; however larger N means more signals must be sent before a key can actually be derived. Thus, in practice, there is a balance that users must choose: larger N will be more efficient in the long run, but you may be waiting longer for a key – also, error correction and privacy amplification will take more time to process larger data sizes too. Note that, in Fig. 5.4, we optimize over m, the sample size. The optimal choice of sample "rate" (namely, the value of r which gives the highest key-rate, if we set $m = rN$) is shown in Fig. 5.5. We see that, as the number of signals increases, the needed sample rate $r = m/N$, approaches zero, as expected. In particular, Alice and Bob can choose a sample size that maximizes ℓ.

5.4.2 Alternative security proof technique: Quantum Sampling

We'll end this chapter by discussing an alternative finite key approach, which is based on a sampling framework introduced by Bouman and Fehr in [1]. This method also works against arbitrary, general attacks as with most typical entropic uncertainty proofs. We'll first go over the basics of their sampling result; then we'll show how it can be used for QKD security proofs.

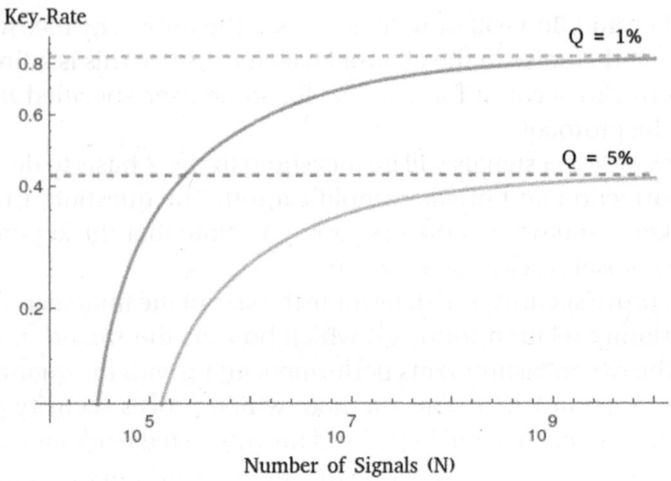

FIGURE 5.4 Evaluating the BB84 finite key rate ℓ/N, where ℓ is defined in Eq. (5.40) (Solid lines). We test two noise levels, $Q=1\%$ (Blue line – upper line) and $Q=5\%$ (Yellow line – lower line) and we set m, the sample size, to be $m=rN$, for an optimally chosen r. We also compare with asymptotic results $1-2h(Q)$ (Dashed lines). We see that, as the number of signals N increases, the finite key rate approaches the asymptotic bound, as expected. Note that, if the number of signals is "too low" then no key can be established. As the noise increases, the "minimal number" of signals needed, also increases, as we see here.

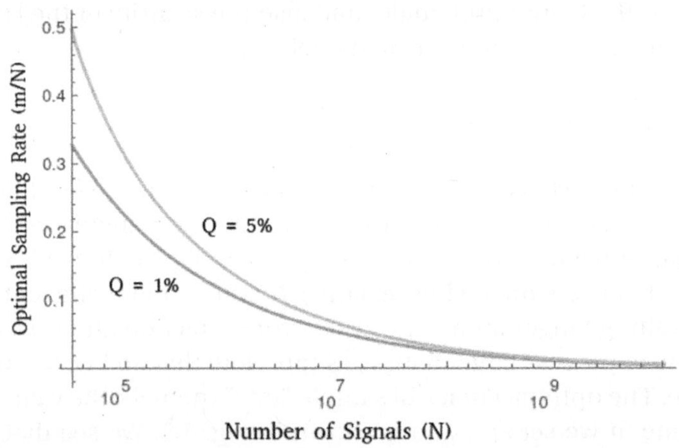

FIGURE 5.5 Showing the optimal sampling rate $r=m/N$ which maximizes the key-rate (Eq. (5.40)). Note that, since the sampling rate is a parameter users can choose, they can choose this so as to maximize their final secret key rate. Here, we see that as the number of signals increases, the sampling rate decreases (making the system more efficient, asymptotically). We also see that, when the noise is higher, more rounds must be devoted to sampling.

Consider the following game: Let's say someone chooses a classical bit-string $q \in \{0,1\}^N$ and keeps it hidden (though you know the size N). You are allowed to choose a random subset $t \subset \{1, \cdots, N\}$ and send it to the "referee" party. That party will tell you exactly what q_t is – that is, what the substring of q is, indexed by t. From this, you must guess at some property of q_{-t}; namely some property of the unobserved portion of q (note the notation: we use q_{-t} to mean the substring of q indexed by the *complement* of t). You can assume the referee who chooses q and interacts with you is always honest.

Let's be a little more concrete and say that the property of the string we're interested in is the relative Hamming weight. Namely, given any bit-string $x \in \{0,1\}^n$, the relative Hamming weight, denoted $w(x)$, is defined to be:

$$w(x) = \frac{|\{i : x_i = 1\}|}{n} = \frac{\text{Number of one's in } x}{\text{Size of } x}. \tag{5.42}$$

Thus, for this particular property, we observe $w(q_t)$, the relative Hamming weight of the substring of q indexed by our subset t; and we want to use this to guess at the relative Hamming weight of the unobserved portion of q, namely $w(q_{-t})$. If we pick our subset t randomly, and it's "pretty large" then we should expect the sampled/observed q_t to behave similarly to the unobserved portion.

This is a type of *classical sampling strategy*. Such a strategy consists of three important items: A probability distribution over subset choices P_T, which dictates how to choose subsets; a *guess function* $g : \{0,1\}^* \to \mathbb{R}$; and a *target function* $r : \{0,1\}^* \to \mathbb{R}$. The strategy works by first picking a subset t, according to the given distribution, observing q_t and computing $g(q_t)$. The output of $g(q_t)$ is used to guess at the value of the unobserved $r(q_{-t})$. A "good" sampling strategy should produce a guess that is δ-close to the target, for some user specified δ, except, perhaps, with some small failure probability (which, itself, depends on δ).

Let's formalize this a bit. Fix some $\delta > 0$ and some sampling strategy. Now, fix some subset t such that $P_T(t) > 0$. Then, we may define a set of *good words* for this specific subset t and sampling strategy, to be:

$$\mathcal{G}^t = \{q \in \{0,1\}^N \ : \ |g(q_t) - r(q_{-t})| \leq \delta\}. \tag{5.43}$$

Note that this means if we have a word $q \in \mathcal{G}^t$, and it just so happens that subset t is chosen by the sampling strategy, it will hold that the guess function, evaluated on the observed portion of the string will be δ-close to the target value evaluated on the unobserved portion. Namely, it will hold that $g(q_t)$ and $r(q_{-t})$ will be δ-close to one another.

But, of course, it is not guaranteed that t will be chosen. Thus, for *any* word q, we must consider the worst case failure probability of the strategy. In particular, the failure probability of the sampling strategy is defined to be:

$$\epsilon_\delta^{cl} = \max_{q \in \{0,1\}^N} Pr\left(q \notin \mathcal{G}^t\right), \tag{5.44}$$

where the probability above is over all subset choices t, chosen according to P_T. The above essentially says that, for any word q, the probability that the guess will be far from the target will be no more than ϵ_δ^{cl}. Note that the "cl" superscript is used to emphasize that this is a *classical* failure rate and sampling strategy.

Of course, this is all in the classical setting – however given such a sampling strategy, it may be promoted to a quantum one in a fairly natural way. First: fix an orthonormal qubit basis – let's say $|0\rangle$ and $|1\rangle$ (though the choice may be arbitrary actually). Let $|\psi\rangle_{AE}$ be some quantum state where the A register consists of N qubits and the E system is arbitrary (and may be entangled arbitrarily with A). The sampling strategy will first choose a subset t, then, it will measure those qubits indexed by t (in the A register), using the given orthonormal basis. This creates a classical measurement outcome $i \in \{0,1\}^{|t|}$, and also a collapsed, post-measured state $|\phi\rangle_{A'E}$ (which may depend on i), where the A' register consists, now, of $N-|t|$ qubits. The main result of Bouman and Fehr in [1] was to say that this post measured state behaves almost like a superposition of good words. Formally, the state should collapse to something that (almost) looks like:

$$|\phi\rangle_{A'E} = \sum_{j\in J_i} |j\rangle\,|E^t_{j|i}\rangle \tag{5.45}$$

where

$$J_i = \{j \in \{0,1\}^{N-|t|} : |g(i) - r(j)| \le \delta\}. \tag{5.46}$$

Consider the following subspace:

$$\mathcal{G}^t_Q = \text{span}\left\{|q\rangle : q \in \mathcal{G}^t\right\}, \tag{5.47}$$

that is, the space of all states which are superpositions (linear combinations) of "good words" $|q\rangle$, with respect to the given fixed basis (which in this case is just the Z basis though it may be arbitrary as mentioned above). Let $|\phi^t\rangle_{QE} \in \mathcal{G}^t_Q \otimes \mathcal{H}_E$. Such a state is called a *good state* due to the following reason: assume subset t is chosen by the sampling strategy, and that a measurement is made resulting in outcome i. Then it is guaranteed, the post-measured state collapses to one of the form shown in Eq. (5.45).

The main result of Bouman and Fehr in [1], besides defining the above notions, was to prove that any state behaves, on average, as a mixture of good states. Formally:

Theorem 5.1. *Let $|\psi\rangle_{AE}$ be a quantum state, where the A register consists of N qudits (d-dimensional systems). Then, for any sampling strategy, there exists a collection of good states $|\phi^t\rangle_{AE}$, where each $|\phi^t\rangle_{AE} \in$ span$\{|q\rangle : q \in \mathcal{G}^t\} \otimes \mathcal{H}_E$, and:*

$$\frac{1}{2}\left\|\sum_t P_T(t)\,|t\rangle\langle t| \otimes |\psi\rangle\langle\psi|_{AE} - \sum_t P_T(t)\,|t\rangle\langle t| \otimes |\phi^t\rangle\langle\phi^t|_{AE}\right\| \le \sqrt{\epsilon^{cl}_\delta}, \tag{5.48}$$

where ϵ^{cl}_δ is the failure probability of the given sampling strategy (see Eq. (5.44)).

Essentially, the above says that, on average over all subset choices, the given input state will behave like a mixture of good states – i.e., a mixture of states where sampling *always* works. Note that span$\{|q\rangle : q \in \mathcal{G}^t\}$ is defined with respect to an arbitrary basis – we used the Z basis, but it could have been the X basis (e.g., span$\{|q\rangle^X = H^{\otimes N}|q\rangle : q \in \mathcal{G}^t\}$) or some other basis. Also note that the dimension of the A system is arbitrary and we can even, as we do later, sub-divide this system into two parties: one for Alice and one for Bob.

5.4.2.1 Application to BB84 security

Now that the sampling framework of Bouman and Fehr has been introduced, we show how it can be used to provide an alternative proof of security for BB84, using methods from [18]. As before, we will actually prove security of E91, the entanglement based protocol.

Let $|\psi\rangle_{ABE}$ be the state that Eve creates. Now, we have two parties Alice and Bob, whereas the sampling framework above was discussed with respect to one party. However, we can consider AB as one party for the sake of sampling, so long as we are careful to keep track of what public information Alice and Bob reveal to each other, and that we don't do any joint measurement that Alice and Bob could not do privately (since they are, in reality, separate parties, located geographically apart). To use the sampling framework, we first need to settle on an appropriate sampling strategy. However, the protocol gives us this.

Consider E91: Alice and Bob will choose some subset t and measure their qubits in the X basis. (In practice, this subset is the result of their random X basis measurement decisions on each round.) They will then check the parity of their outcomes – this should be the all zero string. Namely, Alice and Bob agree on a subset t (which for the purposes of this section we will assume Alice picks the subset and sends it to Bob over the public authenticated channel). Alice and Bob measure their qubits in the X basis, resulting in strings q_A and q_B. If $|t| = m$, then these are m-bit strings (thinking of $|+\rangle$ as a 0 and $|-\rangle$ as a 1). They both broadcast the outcomes of their measurements and compute $q = q_A \oplus q_B$. If Eve is honest (and there is no noise), Alice and Bob should have fully correlated outcomes, namely $q = 0 \cdots 0$. Alice and Bob compute $w(q)$. If they were to measure the remainder of their qubits in the X basis, it should hold that the Hamming weight of that outcome (or, rather, the parity of those outcomes), should be very close to $w(q)$, with high probability.

Let's try to formalize this classically – i.e., define a classical sampling strategy. Our "word" is actually a tuple: $(a, b) \in \{0,1\}^N \times \{0,1\}^N$. A subset t is chosen, uniformly at random, from all subsets of $\{1, 2, \cdots, N\}$ of size $m < N/2$. Then, the "guess function" is $w(a_t \oplus b_t)$ while the target function is $w(a_{-t} \oplus b_{-t})$. This induces a set of good words:

$$\mathcal{G}^t = \{(a,b) \in \{0,1\}^N \times \{0,1\}^N \ : \ |w(a_t \oplus b_t) - w(a_{-t} \oplus b_{-t})| \le \delta\}. \tag{5.49}$$

This adequately models the above scenario: we want to measure some of Alice and Bob's qubits and ensure that the remainder will behave similarly had we measured all of it. Of course, we now need the failure probability of this strategy.

For this, we first review a simple strategy discussed and analyzed in [1]. Here, the words are N-bit strings and the subsets are of size m and are chosen uniformly at random. The guess and target functions are simply the relative Hamming weights. This induces a set of good words:

$$\mathcal{G}^t_{HW} = \{q \in \{0,1\}^N \ : \ |w(q_t) - w(q_{-t})| \le \delta\}. \tag{5.50}$$

Then, it was proven in [1] using Serfling's bound [10], that the failure probability of this "Hamming Weight" strategy, is:

$$\epsilon_\delta^{cl} \le 2\exp\left(\frac{-\delta^2 mN}{N+2}\right). \tag{5.51}$$

Given this, we can prove the following:

Lemma 5.1. *(From [18]): Consider the two-party sampling strategy with good-words defined in Eq. (5.49). Then the failure probability of this strategy is upper-bounded by:*

$$\epsilon_\delta^{cl} = \max_{(a,b)} Pr\left((a,b) \notin \mathcal{G}^t\right) \le 2\exp\left(\frac{-\delta^2 mN}{N+2}\right). \tag{5.52}$$

Proof. We only sketch the proof here and leave the full proof as an exercise. Consider a word $(a,b) \in \{0,1\}^N \times \{0,1\}^N$. Define $q = a \oplus b$. Then $(a,b) \notin \mathcal{G}^t$ implies $q \notin \mathcal{G}^t_{HW}$. The result follows then from Eq. (5.51). □

Exercise 5.12. Complete the proof of Lemma 5.1.

Now, let's return to the E91 protocol. Assume Eve creates some large initial state $|\psi\rangle_{ABE}$, where the A and B systems are N qubits each. Let ϵ be given by the users. Solving the square root of Eq. (5.52) for δ leads us with:

$$\delta = \sqrt{\frac{N+2}{mN}\ln\frac{2}{\epsilon^2}}. \tag{5.53}$$

Choosing δ as above will ensure that Theorem 5.1 leads us to a simple ϵ difference in the trace distance (where $\epsilon > 0$ will be given by the user).

Using the above sampling strategy for two parties, and Theorem 5.1, we know that there exists a collection of ideal states $|\phi^t\rangle_{ABE}$, where:

$$|\phi^t\rangle_{ABE} \in \text{span}\{|a,b\rangle^X \ : \ (a,b) \in \mathcal{G}^t\} \otimes \mathcal{H}_E,$$

and where

$$\frac{1}{2}\left\|\sum_t P_T(t)\,|t\rangle\langle t| \otimes |\psi\rangle\langle\psi|_{ABE} - \sum_t P_T(t)\,|t\rangle\langle t| \otimes |\phi^t\rangle\langle\phi^t|_{ABE}\right\| \le \epsilon.$$

(Again, this is assuming we set δ as above.) Above, we use $|a,b\rangle^X$ to mean that state in the X basis. Namely $|0\rangle^X = |+\rangle$ and $|1\rangle^X = |-\rangle$. So, for example, if $a = 101$ and $b = 001$, then $|a,b\rangle^X = |-+-\rangle_A |++-\rangle_B$.

Now, instead of analyzing the "real state" $\rho_{TABE} = \sum_t P_T(t)\,|t\rangle\langle t| \otimes |\psi\rangle\langle\psi|_{ABE}$, let's analyze the "ideal state" $\sigma_{TABE} = \sum_t P_T(t)\,|t\rangle\langle t| \otimes |\phi^t\rangle\langle\phi^t|_{ABE}$. A subset is chosen: this is equivalent to measuring the T register, observing some t with probability $P_T(t)$, and causing the entire ideal state to collapse to $|\phi^t\rangle\langle\phi^t|_{ABE}$. Next, a measurement is performed on the AB registers in the X basis, for those qubits indexed by t resulting in outcome q_A and q_B. Let $q = q_A \oplus q_B$. Since $|\phi^t\rangle$ is a good word, it is guaranteed that the post-measured state (for those $n = N - m$ qubits that were not yet measured) must be of the form:

$$|\phi_q^t\rangle_{ABE} = \sum_{(a,b)\in J_q} |a,b\rangle^X \otimes |E^t_{a,b|q}\rangle, \tag{5.54}$$

where:

$$J_q = \{(a,b) \in \{0,1\}^{N-m} \times \{0,1\}^{N-m} \ : \ |w(q) - w(a \oplus b)| \leq \delta\}. \tag{5.55}$$

Now, we need to compute $H_\infty(A|E)$, after measuring the above in the Z basis (note that the above state is represented in the X basis – this is important). Let's trace out Bob first. This results in a density operator which may be written in the form:

$$\sigma_{AE} = \sum_{b\in\{0,1\}^n} Pr(b)\, \underbrace{P\left(\sum_{a\in J_{q,b}} |a\rangle^X \otimes |E^t_{a|b,q}\rangle\right)}_{\sigma^{(b)}_{AE}}, \tag{5.56}$$

where:

$$J_{q,b} = \{a \in \{0,1\}^n \ : \ |w(a \oplus b) - w(q)| \leq \delta\}, \tag{5.57}$$

where, recall from earlier in this chapter, we use $P(|z\rangle) = |z\rangle\langle z|$. Note that the Eve vectors are sub-normalized.

Exercise 5.13. Explain how tracing out B from Eq. (5.54) results in a density operator which may be written in the form shown in Eq. (5.56).

Let's zoom in on one of the density operators conditioning on a particular b, namely: $\sigma^{(b)} = P(\sum_{a\in J_{q,b}} |a\rangle^X |E^t_{a|b,q}\rangle)$. This is a superposition with a "small" number of terms (assuming the noise, $w(q)$ is small). Eq. (2.127) from Section 2.6.2 can be used in this case! Namely, using Eq. (2.127), we have:

$$H_\infty(A|E)_{\sigma^{(b)}} \geq H_\infty(A|E)_{mix} - \log_2 |J_{q,b}|, \tag{5.58}$$

where $H_\infty(A|E)_{mix}$ is the min entropy resulting from a Z basis measurement of the A register of the following mixed state:

$$\sum_{a\in J_{q,b}} |a\rangle\langle a|^X_A \otimes |E^t_{a|b,q}\rangle\langle E^t_{a|b,q}|. \tag{5.59}$$

Note this is the "mixed" form of $\sigma^{(b)}$. We need two things: first, we need to determine $H_\infty(A|E)_{mix}$, and second we need a bound on $|J_{q,b}|$. Let's tackle the first item. We will need to measure the above mixed state (which has the A system written in the X basis), in the Z basis and save the result. Recall that mixed states, unlike superposition, are really classical mixtures. Thus, the above state represents a system that really is $|a\rangle^X \otimes |E^t_{a|b,q}\rangle$ with some probability $p(a) = \langle E^t_{a|b,q}|E^t_{a|b,q}\rangle$. If the state really is a particular "$|a\rangle^X$", then when we measure this in the Z basis, we'll get some outcome $|z\rangle$, where $z \in \{0,1\}^n$, with some probability $p(z|a) = \frac{1}{2^n}$ (since the probability of measuring an X basis state in the Z basis yields a uniform random outcome). Applying this logic to the entire mixed state (Eq. (5.59)) yields:

$$\sum_{z\in\{0,1\}^n} \frac{1}{2^n}|z\rangle\langle z|_Z \otimes \left(\sum_{a\in J_{q,b}} |E^t_{a|b,q}\rangle\langle E^t_{a|b,q}|_E\right) \tag{5.60}$$

From this, we see the Z basis outcome is actually independent of the E system! Thus, it can be readily shown (see Eq. (2.120) and the discussion afterwards) that $H_\infty(A|E)_{mix} = n$. That is, there are n bits of uncertainty in the measured A system, which makes sense since E, in this case, is completely separable from A, and the A register is a uniform random n-bit string.

Now that we know what the entropy of the mixed state is, we need to bound the size of the superposition set $|J_{q,b}|$. For this, we have:

$$\begin{aligned}|J_{q,b}| &= |\{a \in \{0,1\}^n : |w(a\oplus b) - w(q)| \le \delta\}|\\ &= |\{x \in \{0,1\}^n : |w(x) - w(q)| \le \delta\}|\\ &\le |\{x \in \{0,1\}^n : w(x) \le w(q)+\delta\}| \le 2^{nh(w(q)+\delta)}.\end{aligned} \tag{5.61}$$

The last inequality follows from a well known bound on the volume of a Hamming ball. Thus, combining all of the above with Eq. (5.58) yields:

$$H_\infty(A|E)_{\sigma^{(b)}} \ge n(1 - h(w(q)+\delta)). \tag{5.62}$$

Finally, Eq. (2.124) (also from Section 2.6.2), allows us to conclude that:

$$H_\infty(A|E)_\sigma \ge \min_b H_\infty(A|E)_{\sigma^{(b)}} \ge n(1 - h(w(q)+\delta)). \tag{5.63}$$

Of course, this was just the ideal state but the ideal state is "close" in trace distance to the real state. A lemma from [11], stated below, allows us to directly use this ideal state analysis to bound the secret key size of the real state.

Lemma 5.2. *(From [11], though simplified slightly for our application here): Assume $|\psi\rangle_{ABE}$ is the initial state created by Eve and let $|\phi^t\rangle_{ABE}$ be the constructed ideal states using Theorem 5.1. Then, if it holds that $H_\infty(A|E)_\sigma \ge \gamma$, for the ideal state σ (before error correction and privacy amplification, and also assuming Alice and Bob do not abort), it holds that the real QKD protocol (running on $|\psi\rangle_{ABE}$) will be ϵ_{PA}-secure, according to Definition 5.3.1, where:*

$$\epsilon_{PA} = 2^{-\frac{1}{2}((\gamma-\lambda_{EC})-\ell)} + 4\sqrt{\epsilon^{cl}_\delta}, \tag{5.64}$$

where λ_{EC} is the error correction leakage.

Note that we made some simplifications in the presentation of the above lemma; one also has to deal with scenarios where parties abort, and define a set of "abort conditions." Full details can be found in [11], though the above version is sufficient to complete our discussion of the proof, here.

Assuming Alice and Bob abort the protocol if their observed noise $w(q)$ is greater than some given threshold Q, then, using the above Lemma and our ideal state analysis earlier (namely, Eq. (5.63)), we derive a final secret key length of ℓ-bits, where:

$$\ell = n(1 - h(Q + \delta)) - \lambda_{EC} - 2\log\frac{1}{\epsilon} \tag{5.65}$$

and we will have a $\epsilon_{PA} = 5\sqrt{\epsilon_\delta^{cl}} = 5\epsilon$ -secure key (recall, our choice of δ ensures that $\sqrt{\epsilon_\delta^{cl}} = \epsilon$, for user-given ϵ). Full details can be found in [18,11]).

In general, when evaluating your expressions theoretically, one can set $\lambda_{EC} = f \cdot n \cdot h(Q + \delta)$, where f is some constant, depending on the error correction protocol used (asymptotically, $f \to 1$ while in practice $f \approx 1.1$ [13]).

This agrees, asymptotically, with previous proof methods described above (the δ from Eq. (5.53) and the μ from Eq. (5.41) differ slightly, however both will converge to $1 - 2h(Q)$ in the asymptotic limit). For more details on the proof itself, including all the technical parts we glazed over above, the reader is referred to [18,11].

So why go into all this detail for an alternative proof? Well, for one thing, alternative proofs are important, as different methods may be more applicable in other scenarios. Indeed, this sampling based approach can sometimes tackle protocols that alternative methods cannot easily handle. We leave it to the reader whether this method is something to be appreciated or not.

This section introduced three methods to prove finite key security of the BB84 protocol. These methods can be broadly applied to other QKD protocols. While some of the details we only covered at a high-level, the information, here, should provide the reader with a sufficient background knowledge to begin to understand the finer details of other finite key security proofs for alternative protocols. We'll also return to these methods later in this book, and expand on them for other protocols.

References

[1] Niek J. Bouman, Serge Fehr, Sampling in a quantum population, and applications, in: Annual Cryptology Conference, Springer, 2010, pp. 724–741.

[2] Matthias Christandl, Robert König, Renato Renner, Postselection technique for quantum channels with applications to quantum cryptography, Physical Review Letters 102 (2) (2009) 020504.

[3] Igor Devetak, Andreas Winter, Distillation of secret key and entanglement from quantum states, Proceedings of the Royal Society A. Mathematical, Physical and Engineering Sciences 461 (2053) (2005) 207–235.

[4] Michael Graifer, Yuval Kochman, Ofer Shayevitz, Quantum key distribution with state replacement, in: 2023 59th Annual Allerton Conference on Communication, Control, and Computing (Allerton), IEEE, 2023, pp. 1–8.

[5] Adrian Harkness, Walter O. Krawec, Bing Wang, Security of partially corrupted quantum repeater networks, Quantum Science and Technology 10 (1) (2024) 015005.
[6] Hoi-Kwong Lo, Hoi Fung Chau, Mohammed Ardehali, Efficient quantum key distribution scheme and a proof of its unconditional security, Journal of Cryptology 18 (2) (2005) 133–165.
[7] Markus Mertz, Hermann Kampermann, Zahra Shadman, Dagmar Bruß, Quantum key distribution with finite resources: taking advantage of quantum noise, Physical Review. A, Atomic, Molecular, and Optical Physics 87 (4) (2013) 042312.
[8] Renato Renner, Security of quantum key distribution, International Journal of Quantum Information 6 (01) (2008) 1–127.
[9] Renato Renner, Nicolas Gisin, Barbara Kraus, Information-theoretic security proof for quantum-key-distribution protocols, Physical Review. A, Atomic, Molecular, and Optical Physics 72 (1) (2005) 012332.
[10] Robert J. Serfling, Probability inequalities for the sum in sampling without replacement, The Annals of Statistics (1974) 39–48.
[11] Trevor N. Thomas, Walter O. Krawec, New key rate bound for high-dimensional bb84 with multiple basis measurements, in: 2025 IEEE International Conference on Quantum Computing and Engineering (QCE), vol. 1, IEEE, 2025, pp. 926–936.
[12] Marco Tomamichel, Roger Colbeck, Renato Renner, A fully quantum asymptotic equipartition property, IEEE Transactions on Information Theory 55 (12) (2009) 5840–5847.
[13] Marco Tomamichel, Charles Ci Wen Lim, Nicolas Gisin, Renato Renner, Tight finite-key analysis for quantum cryptography, Nature Communications 3 (1) (2012) 634.
[14] Marco Tomamichel, Renato Renner, Uncertainty relation for smooth entropies, Physical Review Letters 106 (11) (2011) 110506.
[15] Marco Tomamichel, Christian Schaffner, Adam Smith, Renato Renner, Leftover hashing against quantum side information, IEEE Transactions on Information Theory 57 (8) (2011) 5524–5535.
[16] Alexander Vitanov, Frederic Dupuis, Marco Tomamichel, Renato Renner, Chain rules for smooth min- and max-entropies, IEEE Transactions on Information Theory 59 (5) (2013) 2603–2612.
[17] Aaron D. Wyner, The wire-tap channel, The Bell System Technical Journal 54 (8) (1975) 1355–1387.
[18] Keegan Yao, Walter O. Krawec, Jiadong Zhu, Quantum sampling for finite key rates in high dimensional quantum cryptography, IEEE Transactions on Information Theory 68 (5) (2022) 3144–3163.

6

Alternative quantum key distribution protocols

So far, we've discussed several of the most commonly used point-to-point QKD systems, along with their security. This includes the BB84 protocol, and also the E91, B92, and SARG04 protocols. In this chapter, we delve into some alternative QKD protocols and their designs, discussing, in turn, their advantages and disadvantages both from a theoretical perspective, along with a practical one. We'll begin by discussing so-called *two-way* QKD protocols which rely on two-way quantum communication. This will lead us to an alternative model known as *semi-quantum cryptography*. We'll also discuss a fascinating, and relatively new, discovery in QKD research known as *twin field* protocols.

6.1 Two-way QKD

We begin by discussing some general two-way QKD protocols. So far, all the protocols we looked at involved one-way quantum communication: Either Alice sent qubits to Bob (or vice versa), or a third party sent qubits to Alice and Bob. However, nowhere did qubits "return" to the sender. Two-way QKD changes this scenario by allowing the initial recipient of the qubit to return a quantum signal to the sender. Such a communication model opens up a lot of possibility including direct communication systems [4,30] which allow a party to deterministically encode a message on a qubit (as opposed to letting the random measurement outcomes dictate the "message/key"). Some physical implementations actually hold benefits when using two-way communication [31,3] since polarization drift (essentially, noise) may be compensated for when the photon returns; and, indeed, two-way QKD is experimentally feasible [25,14]. Finally, it opens the door to many interesting theoretical questions, important in studying the foundations of quantum cryptography.

We'll begin by discussing the so-called *Ping Pong* protocol introduced in [4]. First, we actually discuss a "toy" version of the protocol which is insecure. Later, we show the full Ping Pong protocol and prove its security.

To discuss and analyze the protocol, we'll use a different notation for the four Bell states. Let $x, y \in \{0, 1\}$, then we define the Bell state $|\phi_x^y\rangle$ to be:

$$|\phi_x^y\rangle = \frac{1}{\sqrt{2}}(|0, x\rangle + (-1)^y |1, \bar{x}\rangle), \tag{6.1}$$

where $\bar{x} = 1 - x$.

Quantum Communication and Cryptography. https://doi.org/10.1016/B978-0-44-321532-2.00013-9

The Ping Pong protocol proceeds as follows. First, Alice begins by creating a Bell state $|\phi_0^0\rangle = \frac{1}{\sqrt{2}}(|0,0\rangle_{AT} + |1,1\rangle_{AT})$. She keeps one qubit (the A qubit) private and sends the second qubit (the T qubit where "T" here stands for Transit qubit) to Bob. Bob chooses a random bit to send to Alice $k \in \{0,1\}$. If $k = 0$, Bob applies the identity operator to the received qubit; otherwise, if $k = 1$, he applies the Pauli X gate σ_X which maps $|x\rangle$ to $|\bar{x}\rangle$. The qubit returns to Alice who, subsequently, performs a Bell measurement on it which allows her to determine Bob's key choice. Indeed, note that Bob's σ_X operation will transform the Bell state $|\phi_0^0\rangle$ to $|\phi_1^0\rangle$. Thus, if Alice observes $|\phi_0^0\rangle$ she knows Bob must have encoded the key-bit 0; otherwise, if she observes $|\phi_1^0\rangle$ she knows Bob's key-bit is 1. Of course, this all assumes there is no noise, natural or adversarial.

Exercise 6.1. Expand the above protocol to allow Bob to encode two message/key bits per communication round.

Of course there are many issues with the above protocol in terms of security. In fact, there are no security checks at all in the above toy version. Bob, on receipt of a qubit, needs to know, with high probability, that the qubit he holds came from Alice and not from an adversary Eve. Indeed, the protocol above is open to a simple attack: Eve captures Alice's qubit and holds on to it, while preparing her own Bell pair, sending the second particle to Bob. When Bob replies, Eve measures her Bell pair in the Bell basis to determine Bob's key. Eve then performs the correct encoding operation on the qubit from Alice that she saved. This is returned to Alice who will measure and also learn Bob's key. Eve in this way gets an exact copy of the key and Alice and Bob are unaware of Eve's interference. There is also the technical point of deciding what to do if Alice observes, say $|\phi_1^0\rangle$ (i.e., there is a bit-flip error in the channel).

The first problem, dealing with the security of the channel, can be dealt with by adding a sampling procedure. The second problem can be dealt with in a variety of ways – we'll just assume any measurement of $|\phi_i^k\rangle$ will translate to Alice's guess of Bob's key as k, despite the value of i. Thus, observing $|\phi_1^0\rangle$, for example, will translate to Alice guessing Bob's key bit is 0. The full, and secure, protocol is described below in Protocol 6.1.

There are, by now, several proofs of security for the Ping Pong protocol. One interesting approach was developed in [3] which first reduces the two-way protocol into a standard one-way entanglement based protocol, however this requires an extension to the protocol (an extension you can develop in the above Exercise 6.1). For the "vanilla" protocol, there are other techniques – we will consider two: First we will prove security assuming Eve is restricted to collective attacks and then compute the von Neumann entropy directly. Second, we will prove general security using a sampling-based approach in the finite key setting (requiring us to bound the quantum min entropy).

As discussed in the previous section, collective attacks are those where the adversary performs the same operation each round of the protocol. In general, such attacks often imply security against arbitrary, coherent, attacks using de Finetti style arguments or post selection methods [12,17]; however whether this is the case for all protocols operating on two-way channels remains an open problem. Thus, here, we do not claim that post

Protocol 6.1 Ping Pong Protocol [4].

Public Parameters:

- N: number of rounds
- p_{test}: probability of a Test round

Quantum Communication Stage: For $i = 1, 2, \cdots, N$, do:

1. Alice creates a Bell state $|\phi_0^0\rangle_{AT}$. She keeps the A particle private while sending the T particle to Bob through an insecure quantum channel.
2. Bob, on receipt of a qubit from Alice will perform a Test or an Encoding as follows:
   - With probability p_{test}, Bob measures his particle in the X basis, saving the result as $q_B \in \{0, 1\}$ with a 0 outcome meaning an observation of $|+\rangle$. In this case, he informs Alice over the authenticated channel, that this was a "test round." Alice measures her particle in the X basis in this case and reports the outcome q_A. It should be that these values are correlated (i.e., that $q_A \oplus q_B = 0$ while any non-zero result will be considered noise).
   - Otherwise, with probability $1 - p_{test}$, this will be an Encoding round. In this case, Bob chooses a random raw key bit $r_B \in \{0, 1\}$. If $r_B = 0$, he performs the Identity operation on his qubit; otherwise he performs a Pauli X operation. The qubit is returned to Alice.
3. Alice, on receipt of a qubit (on Encoding rounds), will measure in the Bell basis. If she observes outcome $|\phi_k^i\rangle$, for any $i = 0, 1$, she will set her raw key bit for this round to be $r_A = k$.
4. Goto (1)

Classical Post Processing Stage: Alice chooses a random subset t of size $m < N_0$ of all raw key-bits. Here N_0 is the size of the raw key (the number of Encoding rounds). Alice broadcasts the following information over the authenticated classical channel:

- t: the subset chosen.
- r_t^A: Alice's raw key indexed by t – namely those bits of Alice's raw key that are indexed in t.

Bob receives this data and discloses r_t^B, his raw key string indexed by subset t. Both Alice and Bob create a new raw key r^A and r^B by discarding those bits indexed in t (since they were leaked fully to anyone snooping on the classical channel). This is used to determine the error rate in the raw key data needed for error correction. If the error rate in the raw key, or in the Test rounds, is too high, parties abort.

Alice and Bob now run an error correction protocol and privacy amplification protocol to distill their final secret key.

selection methods work on the below two-way protocols and, instead, focus on collective attack security.

We will need a bound on the von Neumann entropy. For this, we must first trace the protocol's execution and write out a density operator description of the joint quantum state, held by Alice, Bob, and Eve, in the event a raw key is established. A collective attack in this protocol case is actually modeled using two unitary operators, as opposed to the single operator used in the BB84 collective attack proof. Alice begins the protocol by preparing the Bell state $|\phi_0^0\rangle_{AT} = \frac{1}{\sqrt{2}}(|00\rangle + |11\rangle) = \frac{1}{\sqrt{2}}(|{+}{+}\rangle + |{-}{-}\rangle)$, sending the T register to Bob. Eve, of course, captures the qubit and probes it using her first attack operator U_1 which acts on the T register and Eve's private ancilla which, we assume without loss of generality, is cleared to some $|0\rangle$ state. We may write the action of this unitary operator as follows:

$$U_1 |t, 0\rangle = |+, e_{+,t}\rangle + |-, e_{-,t}\rangle, \tag{6.2}$$

for $t \in \{+, -\}$. Note that the $|e_{b,t}\rangle$ states are sub-normalized and not necessarily orthogonal. Note also that, unlike the BB84 security proof we derived in Chapter 5, we are defining U_1's action with respect to the Hadamard basis. The reason for this is that Alice and Bob's first test will be in that basis – we could have defined the action in the computational Z basis however the proof of security would be more complex. Many times the hardest part of a problem in quantum computing and cryptography is simply finding the right basis to work in; choose the wrong basis and the proof is magnified in complexity; choose the right basis and the proof becomes much simpler.

Unitarity of U_1 does impose certain constraints on these states, however, namely that for each $t \in \{+, -\}$:

$$\langle e_{+,t}|e_{+,t}\rangle + \langle e_{-,t}|e_{-,t}\rangle = 1 \tag{6.3}$$

$$\langle e_{+,t}|e_{+,1-t}\rangle + \langle e_{-,t}|e_{-,1-t}\rangle = 0 \tag{6.4}$$

where, above, we abused notation slightly and by $1 - t$ we mean the opposite symbol t could be (i.e., $1 - t =$ " $-$ " if $t =$ " $+$ "); this notation makes sense if one considers a $+$ as a zero and a $-$ as a one.

Exercise 6.2. Explain why the constraints listed in Eqs. (6.3) and (6.4) are true.

Eve applies this operator, causing the joint state to evolve to:

$$\frac{1}{\sqrt{2}}|+\rangle_A \otimes \left(|+, e_{+,+}\rangle_{TE} + |-, e_{-,+}\rangle_{TE}\right) + \frac{1}{\sqrt{2}}|-\rangle_A \otimes \left(|+, e_{+,-}\rangle_{TE} + |-, e_{-,-}\rangle_{TE}\right) \tag{6.5}$$

The T qubit is now sent to Bob. At this point, as with most QKD security proofs, it is always a good idea to see what information Alice and Bob can learn at this stage. If Bob chooses this round to be a test round, both parties will measure their particles in the Hadamard basis. Let's see what information this yields on Eve's vectors. If we denote by $p_{a,b}$, for $a, b \in \{+, -\}$ to be the probability that Alice observes a a and Bob observes a b after their measurements, it is not difficult to see that

$$p_{a,b} = \frac{1}{2} \langle e_{b,a} | e_{b,a} \rangle . \tag{6.6}$$

Note that in the ideal, noiseless case, it should hold that $p_{+,+} = p_{-,-} = \frac{1}{2}$ and, of course, $p_{+,-} = p_{-,+} = 0$. Thus this noiseless scenario, in particular, would require $|e_{+,-}\rangle$ and $|e_{-,+}\rangle$ to both be zero vectors. This would make the security proof very simple with only two adversarial states to consider.

Of course there will be noise and, so, $p_{+,-}, p_{-,+} > 0$; however they should be "small" and, as we show, if that's the case Eve cannot get full information on the raw key bit in the Encoding case (Eve may get partial information, and so privacy amplification will be required as always). We continue the analysis by encoding the key state information onto the transit qubit (now in Bob's lab). If Bob's key-bit choice for this particular round is a zero, he will do nothing to the state. Otherwise, he will perform a Pauli X operation. Note that $X|+\rangle = |+\rangle$ and $X|-\rangle = -|-\rangle$. This entire process can be modeled as a mixed state where we will use the B register, below, to store Bob's choice of key bit (which is uniformly distributed):

$$\begin{aligned}
&\frac{1}{2}|0\rangle\langle 0|_B \otimes P\left(\frac{|{+}{+}, e_{+,+}\rangle_{ATE} + |{+}{-}, e_{-,+}\rangle_{ATE} + |{-}{+}, e_{+,-}\rangle_{ATE} + |{-}{-}, e_{-,-}\rangle_{ATE}}{\sqrt{2}}\right) \\
&+\frac{1}{2}|1\rangle\langle 1|_B \otimes P\left(\frac{|{+}{+}, e_{+,+}\rangle_{ATE} - |{+}{-}, e_{-,+}\rangle_{ATE} + |{-}{+}, e_{+,-}\rangle_{ATE} - |{-}{-}, e_{-,-}\rangle_{ATE}}{\sqrt{2}}\right)
\end{aligned} \tag{6.7}$$

where $P(|z\rangle) = |z\rangle\langle z|$.

We actually have enough information now to compute a lower-bound on the von Neumann entropy. Normally, Eve would attack again with a second unitary operator U_2, acting on both the T register and the E register (which, itself, was already acted on in the initial attack). The T qubit would then be returned to Alice and Eve would be left only with the E register from which to attempt to extract information on. However, it is not difficult to see that if we compute a bound on the entropy if Eve holds both the T and the E register at this point in time (and does not yet apply U_2), this will give us a lower-bound on the actual entropy in the case where Eve is not allowed to hold on to the T qubit indefinitely. Why this is the case is left as an exercise below.

Exercise 6.3. Argue (and prove!) why bounding the von Neumann entropy when Eve holds the T and E registers (but not the A register) in Eq. (6.7) but does not attack yet with U_2 gives us a lower-bound on the von Neumann entropy of the actual protocol where Eve must attack with U_2 and then forward the T qubit to Alice (thus, losing access to it). You should use in your proof the fact that for any quantum system ρ_{XYZ}:

$$H(X|Y)_\rho \geq H(X|YZ)_\rho . \tag{6.8}$$

(The above is also true for quantum min entropy.)

Let's trace out the A register in Eq. (6.7), needed to compute the entropy of the B register (which holds the actual raw key-bit) and Eve's information. Note that we are using so-called *reverse reconciliation*. The resulting state is found to be:

$$\rho_{BTE} = \frac{1}{2}\left|0\right\rangle\left\langle 0\right|_B \otimes \left(\frac{1}{2}P(\left|+,e_{+,+}\right\rangle + \left|-,e_{-,+}\right\rangle) + \frac{1}{2}P(\left|+,e_{+,-}\right\rangle + \left|-,e_{-,-}\right\rangle)\right)$$
$$+\frac{1}{2}\left|1\right\rangle\left\langle 1\right|_B \otimes \left(\frac{1}{2}P(\left|+,e_{+,+}\right\rangle - \left|-,e_{-,+}\right\rangle) + \frac{1}{2}P(\left|+,e_{+,-}\right\rangle - \left|-,e_{-,-}\right\rangle)\right) \quad (6.9)$$

Why would this be secure? As a sanity check, let's return to considering the ideal, noiseless case, where $\langle e_{t,1-t}|e_{t,1-t}\rangle = 0$. (Always a good idea when proving security) In this noiseless case, the above state simplifies to:

$$\frac{1}{2}\left|0\right\rangle\left\langle 0\right|_B \otimes \left(\frac{1}{2}P(\left|+,e_{+,+}\right\rangle) + \frac{1}{2}P(\left|-,e_{-,-}\right\rangle)\right)$$
$$+\frac{1}{2}\left|1\right\rangle\left\langle 1\right|_B \otimes \left(\frac{1}{2}P(\left|+,e_{+,+}\right\rangle) + \frac{1}{2}P(-\left|-,e_{-,-}\right\rangle)\right)$$

Since $P(-\left|z\right\rangle) = (-1)^2\left|z\right\rangle\left\langle z\right| = P(\left|z\right\rangle)$, the above becomes simply:

$$\frac{1}{2}\left|0\right\rangle\left\langle 0\right|_B \otimes \left(\frac{1}{2}P(\left|+,e_{+,+}\right\rangle) + \frac{1}{2}P(\left|-,e_{-,-}\right\rangle)\right)$$
$$+\frac{1}{2}\left|1\right\rangle\left\langle 1\right|_B \otimes \left(\frac{1}{2}P(\left|+,e_{+,+}\right\rangle) + \frac{1}{2}P(\left|-,e_{-,-}\right\rangle)\right)$$
$$= \left(\frac{1}{2}\left|0\right\rangle\left\langle 0\right|_B + \frac{1}{2}\left|1\right\rangle\left\langle 1\right|_B\right) \otimes \left(\frac{1}{2}P(\left|+,e_{+,+}\right\rangle) + \frac{1}{2}P(\left|-,e_{-,-}\right\rangle)\right)$$
$$= \frac{I_B}{2} \otimes \rho_E.$$

That is, in the noiseless case, Bob's register is actually completely independent of Eve's system and, furthermore, it is uniform random. Thus the von Neumann entropy in the noiseless case is maximal, i.e., $H(A|E) = 1$.

Of course, that was the ideal noiseless case. While useful to consider when working on a proof of security to determine if you're on the right path or not, it is not terribly interesting. Instead, we want to know what $H(A|E)$ is in the noisy case. Note that, von Neumann entropy is continuous [38], so it can't jump "too much" as noise is introduced. But, we need to make that claim rigorous.

Returning to the actual state ρ_{BTE} in Eq. (6.9), let's use Theorem 2.1 from Section 2.6. To use the theorem, let's define the following states:

$$\left|E_0\right\rangle = \frac{1}{\sqrt{2}}(\left|+,e_{+,+}\right\rangle + \left|-,e_{-,+}\right\rangle)$$
$$\left|E_1\right\rangle = \frac{1}{\sqrt{2}}(\left|+,e_{+,-}\right\rangle + \left|-,e_{-,-}\right\rangle)$$

$$|F_0\rangle = \frac{1}{\sqrt{2}}(|+, e_{+,+}\rangle - |-, e_{-,+}\rangle)$$

$$|F_1\rangle = \frac{1}{\sqrt{2}}(|+, e_{+,-}\rangle - |-, e_{-,-}\rangle)$$

From this, we have:

$$\rho_{BTE} = \frac{1}{2}|0\rangle\langle 0|_B \otimes (|E_0\rangle\langle E_0| + |E_1\rangle\langle E_1|) + \frac{1}{2}|1\rangle\langle 1|_B \otimes (|F_0\rangle\langle F_0| + |F_1\rangle\langle F_1|) \tag{6.10}$$

as required by the theorem and, so:

$$\begin{aligned} H(B|TE)_\rho \geq &\frac{1}{2}(\langle E_0|E_0\rangle + \langle F_0|F_0\rangle)\left(h\left(\frac{\langle E_0|E_0\rangle}{\langle E_0|E_0\rangle + \langle F_0|F_0\rangle}\right) - h(\lambda_0)\right) \\ &+ \frac{1}{2}(\langle E_1|E_1\rangle + \langle F_1|F_1\rangle)\left(h\left(\frac{\langle E_1|E_1\rangle}{\langle E_1|E_1\rangle + \langle F_1|F_1\rangle}\right) - h(\lambda_1)\right), \end{aligned} \tag{6.11}$$

where:

$$\lambda_j = \frac{1}{2}\left(1 + \frac{\sqrt{(\langle E_j|E_j\rangle - \langle F_j|F_j\rangle)^2 + 4Re^2\langle E_j|F_j\rangle}}{\langle E_j|E_j\rangle + \langle F_j|F_j\rangle}\right). \tag{6.12}$$

Our goal is to show how all of the inner products, appearing in the above expressions, can be bounded using only observable quantities.

First, consider the inner products $\langle E_j|E_j\rangle$ and $\langle F_j|F_j\rangle$. From our discussion earlier and our definition of $p_{a,b}$, the following are easily derived:

$$\langle E_0|E_0\rangle = \frac{1}{2}\langle +, e_{+,+}|+, e_{+,+}\rangle + \frac{1}{2}\langle -, e_{-,+}|-, e_{-,+}\rangle = p_{+,+} + p_{-,+}$$

$$\langle E_1|E_1\rangle = \frac{1}{2}\langle +, e_{+,-}|+, e_{+,-}\rangle + \frac{1}{2}\langle -, e_{-,-}|-, e_{-,-}\rangle = p_{+,-} + p_{-,-}$$

$$\langle F_0|F_0\rangle = \frac{1}{2}\langle +, e_{+,+}|+, e_{+,+}\rangle + \frac{1}{2}\langle -, e_{-,+}|-, e_{-,+}\rangle = p_{+,+} + p_{-,+}$$

$$\langle F_1|F_1\rangle = \frac{1}{2}\langle +, e_{+,-}|+, e_{+,-}\rangle + \frac{1}{2}\langle -, e_{-,-}|-, e_{-,-}\rangle = p_{+,-} + p_{-,-}$$

Clearly, these inner products are observable quantities.

Next, consider $\langle E_j|F_j\rangle$ which are determined to be:

$$\langle E_0|F_0\rangle = \frac{1}{2}(\langle e_{+,+}|e_{+,+}\rangle - \langle e_{-,+}|e_{-,+}\rangle) = p_{+,+} - p_{-,+}$$

$$\langle E_1|F_1\rangle = \frac{1}{2}(\langle e_{+,-}|e_{+,-}\rangle - \langle e_{-,-}|e_{-,-}\rangle) = p_{+,-} - p_{-,-}.$$

Note that, if the noise is "small" (in particular if $p_{-,+}$ and $p_{+,-}$ are small), the above inner products are either large positive, or large negative. Looking closely at Eq. (6.11), this will lead to a large conditional entropy and, thus, a large key-rate. But let's evaluate this using our usual depolarization channel. We'll assume that the probability of error in a measurement is symmetric, and so $p_{+,-} = p_{-,+}$, and can be parameterized by a variable $Q \in [0, 1]$. We set $p_{+,-} = p_{-,+} = \frac{1}{2}Q$, and thus $p_{+,+} = p_{-,-} = \frac{1}{2}(1 - Q)$, and our entropy expression, Eq. (6.11) simplifies nicely to:

$$H(B|E)_\rho \geq H(B|TE)_\rho \geq 1 - h(1 - Q) = 1 - h(Q) \tag{6.13}$$

Of course, for a key-rate expression, we also require the error correction leakage, namely $H(B|A)$ (again, we are using reverse reconciliation, so the key-rate expression will be $H(B|E) - H(B|A)$). However, this will be $H(B|A) = h(Q_{bit})$, where Q_{bit} is the probability of Alice's raw key bit being different from Bob's. Exercise 6.5 asks you to work out Q_{bit} assuming the reverse channel is a depolarization channel.

Exercise 6.4. Show how Eq. (6.13) is derived from Eq. (6.11) and the assumption of a depolarization channel with parameter Q.

Exercise 6.5. Derive Q_{bit}, the probability that Alice's raw key bit is different from Bob's, assuming the forward and reverse channels are modeled as independent depolarization channels, with parameter Q. Trace out the protocol, in this event to determine the expected bit error rate. Then, complete the key-rate expression, using Eq. (6.13), and plot the result. Compare to BB84.

6.1.1 Finite key proof of security

The previous section proved security in the asymptotic case. As discussed in Chapter 5, this is useful as a theoretical upper-bound on a QKD protocol's performance and is often the first-step in proving security of a new protocol. However, to be useful in practice, a finite-key analysis is required, where the protocol is only run N times (i.e., N rounds). This requires us to deal with finite sampling effects, and also forces us to bound the quantum min entropy, introduced in Section 2.6.2, a generally more difficult quantity to bound. For this, we will use an approach from [22], though in this case restricted to the two dimensional case ([22] dealt with a higher-dimensional variant of the protocol). The proof uses the quantum sampling mechanism of Bouman and Fehr [6] which we discussed, and used for BB84, in Section 5.4.2 – the reader may want to review that chapter before proceeding with this section.

For the proof, we must first set up the classical sampling strategy required. Looking at the protocol, the main test is when Alice and Bob test the fidelity of the entangled pair sent through the first channel. Let's assume the worst case: in the first stage of the protocol, it's not Alice creating entangled pairs and sending a single particle to Bob, instead Eve creates the particles and sends one to Alice and one to Bob. Let's go a step further and, as with BB84's security proof, we will assume that Eve can create particles for all N rounds of the protocol simultaneously and send them to Alice and Bob. This will allow Eve to create arbitrary correlations in the rounds and also allow her to become arbitrarily entangled with

Alice and Bob's system. Clearly, in real life, this would be challenging for Eve to do and so security in real life can only be better.

Now, once particles are distributed, Alice and Bob will test a subset of the received rounds, measuring those Test rounds in the X basis and broadcasting the measurement result. Ideally, their X basis measurements should match; any mismatches will be counted as errors in the signal. Our sampling strategy, then, is to first choose a random subset $t \subset \{1, 2, \cdots, N\}$, of size m, and second measure those qubits indexed by t in the X basis, resulting in outcomes q_A (for Alice) and q_B (for Bob) to determine $w(q_A \oplus q_B)$, namely the relative number of errors in the sample (recall, $w(x)$ is the relative Hamming weight of x, or the relative number of ones in the bit-string). It should be expected that, should Alice and Bob have measured the remaining $n = N - m$ signals in the X basis, the relative error would be roughly the same. Of course, Alice and Bob will not be measuring the remaining signals in the X basis, but this gives us an idea as to what our classical sampling strategy should do. In fact, it is the exact same strategy analyzed in Section 5.4.2.1.

In particular, we have $P_T(t)$ will choose a subset t of size m uniformly at random. A classical word to be sampled will be from the alphabet $\{0,1\}^N \times \{0,1\}^N$. The process will observe $w(q_t^A \times q_t^B)$ where $w(\cdot)$ is the relative Hamming weight function. Thus, the strategy will observe those bits indexed by t in both the A and B portions of the string and compute their relative error rate. The target function will be the Hamming weight of the unobserved portion. This all induces the set of good words (again, see Section 5.4.2 for details on the sampling framework used here), for some subset t and given $\delta > 0$ to be:

$$\mathcal{G}^t = \left\{(a,b) \in \{0,1\}^N \times \{0,1\}^N : |w(a_t \oplus b_t) - w(a_{-t} \oplus b_{-t})| \leq \delta\right\}. \tag{6.14}$$

Lemma 5.1 bounds the error probability of this sampling strategy by:

$$\epsilon_\delta^{cl} = \max_{q^A \times q^B} Pr\left(q^A \times q^B \notin \mathcal{G}^t\right) \leq 2\exp\left(\frac{-\delta^2 mN}{N+2}\right). \tag{6.15}$$

Now, let's return to the protocol. Eve, as mentioned, is allowed to create the initial state, sending N particles to Alice and N to Bob, where N is the number of rounds specified by the users; Eve keeps a private ancilla that is arbitrarily entangled with the $2N$ qubits. Call this state $|\psi\rangle_{ABE}$ (it is pure, without loss of power to Eve). By the quantum sampling theorem (see Theorem 5.1 from Section 5.4.2, and the text surrounding it), using the above sampling strategy with respect to the X basis, there exists a collection of ideal states $|\phi^t\rangle$, one for each possible subset t of size m, such that:

$$|\phi^t\rangle \in \text{span}\left\{|a\rangle^X |b\rangle^X \ : \ (a,b) \in \mathcal{G}^t\right\} \otimes \mathcal{H}_E, \tag{6.16}$$

where we use $|a\rangle^X$ to mean the string a written in the X basis (e.g., $|010\rangle^X = |+,-,+\rangle$). The sampling theorem also guarantees that:

$$\frac{1}{2}\left\| \sum_t P_T(t) |t\rangle\langle t| \otimes |\psi\rangle\langle\psi| - \sum_t P_T(t)|t\rangle\langle t| \otimes |\phi^t\rangle\langle\phi^t| \right\| \leq \sqrt{\epsilon_\delta^{cl}}. \tag{6.17}$$

Given some $\epsilon > 0$, if we set δ to be:

$$\delta = \sqrt{\frac{N+2}{mN} \ln \frac{2}{\epsilon^2}}. \tag{6.18}$$

then $\sqrt{\epsilon_\delta^{cl}} = \epsilon$. Since the two states (the real and ideal states) are ϵ-close, we can analyze the ideal state and then promote that analysis to the real-state case.

Consider the ideal state – assume sample t is chosen, causing the state to collapse to a particular $|\phi^t\rangle$. Next, Alice and Bob measure their qubits indexed by t in the X basis, getting result $q_A, q_B \in \{0,1\}^m$. Let $q = w(q_A \oplus q_B)$. From Eq. (6.16), the state necessarily collapses to one of the form (after tracing out those qubits which have been measured):

$$|\phi_q^t\rangle = \sum_{(a,r)\in J_q} \alpha_{a,r} |a,r\rangle_{AT}^X \otimes |E_{a,r}\rangle, \tag{6.19}$$

where:

$$J_q = \{(a,r) \in \{0,1\}^n \times \{0,1\}^n \; : \; |w(a \oplus r) - q| \le \delta\}. \tag{6.20}$$

Note we are using r, here, to index the "remaining" qubits in the transit register, which will travel back to Eve after Bob applies his operation.

By manipulating, or more specifically, permuting subspaces (a nice trick in quantum cryptography when analyzing multi-party states – just be careful when permuting subspaces to do it to all kets in the superposition), we can write the above state in the following, easier to analyze yet equivalent (up to permutation) form:

$$|\phi_q^t\rangle \cong \sum_{a\in\{0,1\}^n} \beta_a |a\rangle_A^X \otimes \sum_{r\in J_{q,a}} \beta_{r|a} |r\rangle_B^X \otimes |E_{a,t}\rangle, \tag{6.21}$$

where:

$$J_{q,a} = \{r \in \{0,1\}^n \; : \; |w(a \oplus r) - q| \le \delta\}. \tag{6.22}$$

Note that, so far, our arguments are similar to those in Section 5.4.2 for proving the security of BB84 using the sampling method. Now, however, they are about to diverge as the key is not going to be based on a measurement result, but instead, Bob is actually going to manipulate the state, encoding his key choice onto it, and then send the qubit back. We create a new register to hold Bob's key choice which is simply a completely mixed state: $\sum_{k\in\{0,1\}^n} \frac{1}{2^n} |k\rangle\langle k|_B$. Bob encodes his key choice by applying a Pauli X gate to the i'th qubit of the current T register (after the above sampling is performed), if $k_i = 1$. Applied to an n-qubit state his key encoding maps $|r\rangle_T = |r_1 \cdots r_n\rangle_T \mapsto (-1)^{k_1r_1+k_2r_2+\cdots+k_nr_n} |r_1 \cdots r_n\rangle = (-1)^{k\cdot r} |r\rangle$, where by $k \cdot r$ we mean the bit-wise dot-product. This entire process creates a joint state:

$$\sigma_{BATE}^{t,q} = \frac{1}{2^n} \sum_{k\in\{0,1\}^n} |k\rangle\langle k|_B \otimes P\left(\sum_{a\in\{0,1\}^n} \beta_a |a\rangle_A^X \otimes \sum_{r\in J_{q,a}} (-1)^{r\cdot k} \beta_{r|a} |r\rangle^X \otimes |E_{a,r}\rangle\right). \tag{6.23}$$

We now bound $H_\infty(B|ET)_{\sigma^{t,q}}$. Tracing out A yields:

$$\sigma_{BTE}^{t,q} = \frac{1}{2^n}\sum_{k\in\{0,1\}^n} |k\rangle\langle k|_B \otimes \sum_a |\beta_a|^2 P\left(\sum_{r\in J_{q,a}} (-1)^{r\cdot k}\beta_{r|a}|r\rangle^X \otimes |E_{a,r}\rangle\right)$$

$$= \sum_a |\beta_a|^2 \left(\underbrace{\frac{1}{2^n}\sum_{k\in\{0,1\}^n} |k\rangle\langle k|_B \otimes P\left(\sum_{r\in J_{q,a}} (-1)^{r\cdot k}\beta_{r|a}|r\rangle^X \otimes |E_{a,r}\rangle\right)}_{\sigma_{BTE}^{t,q,a}}\right) \tag{6.24}$$

From Eq. (2.124), we have:

$$H_\infty(B|ET)_{\sigma^{t,q}} \geq \min_a H_\infty(B|ET)_{\sigma^{t,q,a}}. \tag{6.25}$$

Bounding the entropy in that state, conditioned on a particular a, may be done using a result similar to Eq. (2.127) (we say similar, since we have an additional B system, here, however the same end-result may be proven [22]). Since $J_{q,a}$ is a "small" subset of all n-bit strings, the entropy will not differ too greatly from an equivalent mixed state which we can define as:

$$\frac{1}{2^n}\sum_{k\in\{0,1\}^n} |k\rangle\langle k|_B \otimes \sum_{r\in J_{q,a}} |(-1)^{r\cdot k}|^2 |\beta_{r|a}|^2 |r\rangle\langle r|^X \otimes |E_{a,r}\rangle\langle E_{a,r}|$$
$$= \frac{1}{2^n}\sum_{k\in\{0,1\}^n} |k\rangle\langle k|_B \otimes \sum_{r\in J_{q,a}} |\beta_{r|a}|^2 |r\rangle\langle r|^X \otimes |E_{a,r}\rangle\langle E_{a,r}|$$

Note that the above has no dependence on k at all! Thus the min entropy of the mixed state will be n; the difference in the entropy of the mixed state versus the actual pure state will be no more than $\log|J_{q,a}|$. A bound on this set was derived in Section 5.4.2.1 (see Eq. (5.61)), and so we conclude:

$$H_\infty(B|ET)_{\sigma^{t,q,a}} \geq n - \log|J_{q,a}| \geq n(1 - h(q+\delta)). \tag{6.26}$$

Completing the proof can be done, now, similar to the BB84 proof using quantum sampling, discussed in Section 5.4.2. Note that $h(q) = h(1-q)$, and so this finite key result approaches the asymptotic bound found in Eq. (6.13). Of course, to derive a full key-rate bound, we also require the error correction leakage. However, Exercise 6.5 may be used for this to determine the bit error rate of the raw key.

Exercise 6.6. Above, we analyzed the min entropy of the ideal state. Using the same methodology from Section 5.4.2, complete the proof and derive a bound on the final secret key size ℓ of this protocol.

6.2 Semi-quantum cryptography

Semi-quantum cryptography was first introduced in 2007 by Boyer, Kenigsberg, and Mor [8] as a way to study the question "how quantum" does a protocol need to be to gain an advantage over classical communication. In particular, it is a mathematical fact that, if Alice and Bob use classical communication only, then key distribution is impossible unless computational assumptions are made on the power of the adversary. As we've seen throughout this book, however, key distribution is possible if Alice and Bob utilize quantum communication (in addition to classical communication). Can anything exist in the "gap" between classical and quantum? Clearly such a question is interesting, though it's also very hard to exactly define what it means to exist in the gap; one proposed framework (though by no means the only meaningful one in this case) is the semi-quantum model.

In the semi-quantum model, at least one party is restricted to being so-called semi-quantum or "classical." This party has access to a quantum channel, but can only operate on it in a restricted manner. In particular, the party can only:

- **Measure**: The party can perform a measurement in a fixed, publicly known, basis, typically the computational basis. Measuring in any other basis or manner is not allowed.
- **Send**: The party can prepare and send a qubit but only using the same basis that measurements are allowed in (again, typically the computational basis).
- **Reflect**: The party can disconnect from the quantum channel and simply reflect any quantum signal received, back to the original sender.
- **Permute**: Given two (or more) quantum signals, the semi-quantum party can permute them.

Why is this a reasonable model? Consider the case if *both* parties, Alice and Bob, were restricted in the above sense and there was no third-party help. Then the protocol, whatever it is, would be mathematically equivalent to a classical one. Indeed, if Alice and Bob were only sending and measuring in a single fixed basis (e.g., the Z basis, though the exact choice is irrelevant as long as it's the same for both measuring and sending), then the resulting system is classical. Thus, calling a party restricted to the above a "classical" party makes some sense. An alternative wording is this party is performing *classical operations on quantum data* [24].

How can such a protocol function? We discuss two protocols here, first the original SQKD protocol [8] and, second, a so-called *mediated* SQKD protocol introduced in [18], where both parties are "classical" but they utilize the services of a quantum server (which may actually be adversarial). Both protocols offer interesting insights into QKD protocol design that may be useful outside the semi-quantum model. Also, interestingly, the study of semi-quantum cryptography results in fascinating insights into the development and employment of classical post-processing methods that benefit key-rates – such insights are useful in general QKD work also.

At a high level, the first SQKD protocol developed [8] uses the two way channel in the following way: The "forward" channel (connecting quantum Alice to "classical" Bob) is used to transmit the key – a transmission done in the computational Z basis. The joint forward and reverse channels (the reverse channel connecting Bob to Alice) are used to police adversarial activity. So long as Eve never knows when a reflection event happens (used to test the fidelity of the channel), the protocol is secure. The protocol, in detail, can be found in Protocol 6.2. See also Fig. 6.1.

Protocol 6.2 Semi-Quantum Key Distribution [8].

Public Parameters:

- N: number of rounds
- p_{ref}: probability of Bob choosing Reflect.
- p_Z: probability of Alice choosing the Z basis.

Quantum Communication Stage: For $i = 1, 2, \cdots, N$, do:

1. Alice prepares, randomly, one of the four BB84 states $|0\rangle$, $|1\rangle$, $|+\rangle$, or $|-\rangle$ remembering the choice (she chooses a Z basis state with probability p_Z). If she happens to choose a Z basis state, she'll record the actual state (i.e., 0 or 1) in a classical bit k_A. She sends the qubit to Bob.
2. Bob, on receipt of a qubit from Alice will choose either to Reflect, or Measure-and-Resend:
   - With probability p_{ref}, Bob reflects the particle. This gives Bob no information on the quantum state and, also, does not disturb it. The qubit returns to Alice.
   - Otherwise, with probability $1 - p_{ref}$ Bob will first measure the qubit in the Z basis, recording the result in a private register as k_B. He then resends his measurement result as a Z basis state. In particular, if he observes $|r\rangle$, he will send $|r\rangle$ for $r = 0$ or 1.
3. Alice, on receipt of a qubit, will measure in the same basis she initially sent in and record the result.
4. Goto (1)

Sampling:

- Alice and Bob will choose a random subset of rounds where (1) Alice sent in the Z basis and Bob Reflected; (2) Alice sent in the X basis and Bob Reflected; (3) Alice sent in the Z basis and Bob measured
- It should hold that on (1) and (2), Alice should measure the same state she initially sent; on type (3) rounds, Alice and Bob should observe the same values.

Classical Post-processing:

- Alice and Bob now run an error correction protocol and privacy amplification protocol to distill their final secret key from their raw key bits k_A and k_B.

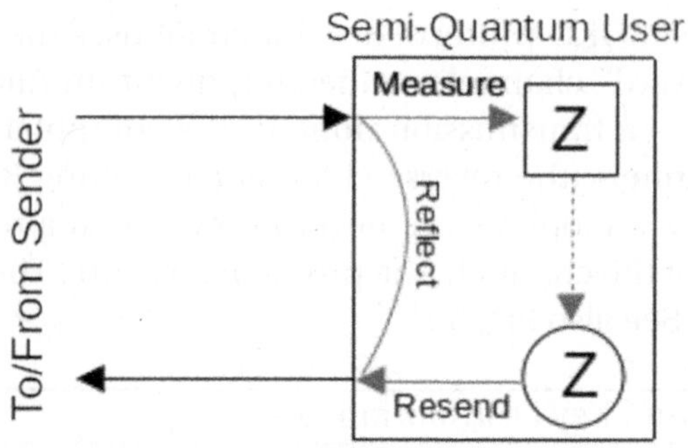

FIGURE 6.1 Showing a schematic of the "semi-quantum" or "classical" user's capabilities. A sender can input a quantum state which can either be "Measured" or "Reflected." If the former, the measurement can only be done in the Z basis; the classical result can be sent to a Z basis source device to prepare a new qubit in the same state. If the latter (Reflected), then the quantum state simply passes, undisturbed, through the user's device and back to the sender.

Exercise 6.7. Trace out the execution of the protocol to show it is correct (i.e., that Alice and Bob will agree on the same secret key in the absence of any noise).

Due to the complex nature of the channel, allowing qubits to travel from Alice to Bob, then back to Alice again, in addition to Bob's inability to measure in all bases or translate qubits to alternative bases, makes the security proof challenging. The limited information that party's can gain on the adversary's action, in addition to the increased attack surface of a two-way channel, make many "traditional" methods of proof ineffective. Some proof methods exist, however. We'll discuss a basic collective attack security proof now following [19].

As with BB84's collective attack security, and the Ping Pong protocol above, the first step is describing the density operator of system modeling Alice, Bob, and Eve conditioning on their using this particular round for key distillation. Then, one needs to look at how statistics can help to gather information on Eve's attack and information gain.

To begin, Alice will send a random Z basis state (only a Z basis state as we must condition on a key being distilled which can only happen in the case when Alice sends computational states at the start of the round). Eve will probe with some isometry (which may be dilated to a unitary and, therefore, is something Eve can do according to the laws of quantum mechanics) U_1 which sends $|0\rangle_T \mapsto |0, e_0\rangle + |1, e_1\rangle$ and $|1\rangle_T \mapsto |0, e_2\rangle + |1, e_3\rangle$. Bob will then measure in the computational basis, saving the result in a private register, and resending the qubit. So far, the state is modeled by the density operator:

$$\frac{1}{2}|0\rangle\langle 0|_A \otimes (|0,0,e_0\rangle\langle 0,0,e_0|_{BTE} + |1,1,e_1\rangle\langle 1,1,e_1|_{BTE})$$
$$+\frac{1}{2}|1\rangle\langle 1|_A \otimes (|0,0,e_2\rangle\langle 0,0,e_2|_{BTE} + |1,1,e_3\rangle\langle 1,1,e_3|_{BTE})$$

At this point the travel qubit (in the T register) returns to Eve who is allowed to probe the system again using a second unitary U_2 acting on the T and E registers. We may write the action of U_2 abstractly, as follows:

$$U_2 |t, e_i\rangle_{TE} = |0, e^0_{t,i}\rangle + |1, e^1_{t,i}\rangle . \tag{6.27}$$

Following the application of this operator, the T register returns to Alice who subsequently discards it (i.e., we trace it out – this is mathematically equivalent to measuring and then discarding). The final density operator, therefore, becomes:

$$\begin{aligned}\rho =& \frac{1}{2} |0,0\rangle\langle 0,0|_{AB} \otimes \left(|e^0_{0,0}\rangle\langle e^0_{0,0}| + |e^1_{0,0}\rangle\langle e^1_{0,0}|\right) + \frac{1}{2} |0,1\rangle\langle 0,1| \otimes \left(|e^0_{1,1}\rangle\langle e^0_{1,1}| + |e^1_{1,1}\rangle\langle e^1_{1,1}|\right)\\ &+\frac{1}{2} |1,1\rangle\langle 1,1|_{AB} \otimes \left(|e^0_{1,3}\rangle\langle e^0_{1,3}| + |e^1_{1,3}\rangle\langle e^1_{1,3}|\right) + \frac{1}{2} |1,0\rangle\langle 1,0| \otimes \left(|e^0_{0,2}\rangle\langle e^0_{0,2}| + |e^1_{0,2}\rangle\langle e^1_{0,2}|\right)\end{aligned} \tag{6.28}$$

As with other collective attacks, we can use Theorem 2.1 from Chapter 2 to bound $H(A|E)_\rho$. For this, we will need information on $\langle e^j_{t,i}|e^j_{t,i}\rangle$ and also other inner products such as $\langle e^0_{0,0}|e^1_{1,3}\rangle$. Of course, whatever bounds we derive for these inner products must be functions of observable statistics – and here is where Bob's inability to make certain measurements becomes especially challenging for the proof writer.

First let's tackle the "easy" ones – namely $\langle e^j_{t,i}|e^j_{t,i}\rangle$. Noting that $\langle e_0|e_0\rangle = Pr(B=0|A_1 = 0)$ (and similarly for other vectors $|e_i\rangle$ from the first attack operator), it can be shown that:

$$\begin{aligned}\langle e^j_{0,0}|e^j_{0,0}\rangle &= Pr(B=0 \wedge A_2 = j|A_1 = 0)\\ \langle e^j_{1,1}|e^j_{1,1}\rangle &= Pr(B=1 \wedge A_2 = j|A_1 = 0)\\ \langle e^j_{0,2}|e^j_{0,2}\rangle &= Pr(B=0 \wedge A_2 = j|A_1 = 1)\\ \langle e^j_{1,3}|e^j_{1,3}\rangle &= Pr(B=1 \wedge A_2 = j|A_1 = 1)\end{aligned}$$

Above, A_1 is the random variable storing A's initial choice of Z basis bit to send, B is Bob's Z basis measurement outcome, and A_2 is Alice's second Z basis measurement result. Clearly all of these are directly observable in test cases of the SQKD protocol. Furthermore, all but $\langle e^0_{0,0}|e^0_{0,0}\rangle$ and $\langle e^1_{1,3}|e^1_{1,3}\rangle$ should be "small" if the Z basis noise in the channel is small. But, for sake of clarity, let's put some parameters to these. If we denote by Q to be the probability of a Z basis bit flip in either direction of the channel, then:

$$\begin{aligned}\langle e^0_{0,0}|e^0_{0,0}\rangle &= (1-Q)^2 & \langle e^1_{0,0}|e^1_{0,0}\rangle &= Q(1-Q)\\ \langle e^0_{1,1}|e^0_{1,1}\rangle &= Q^2 & \langle e^1_{1,1}|e^1_{1,1}\rangle &= Q(1-Q)\\ \langle e^0_{0,2}|e^0_{0,2}\rangle &= Q(1-Q) & \langle e^1_{0,2}|e^1_{0,2}\rangle &= Q^2\\ \langle e^0_{1,3}|e^0_{1,3}\rangle &= Q(1-Q) & \langle e^1_{1,3}|e^1_{1,3}\rangle &= (1-Q)^2\end{aligned} \tag{6.29}$$

Of course, this is just one parameterization. In general, Alice and Bob will simply observe actual values – but since we must simulate the protocol's behavior, we need to put something reasonable to these values; assuming a depolarization source such as this is a reasonable setup. One can also look into other natural parameterizations such as differing noise in each direction of the channel. This is expanded on in the following exercise.

Exercise 6.8. Let Q_F be the probability of a Z basis bit flip in the forward channel (connecting Alice to Bob) and Q_R the bit flip error in the reverse. Re-derive the above expressions in Eq. (6.29) for this assumption.

We'll compute the key-rate assuming reverse reconciliation, namely $H(B|E) - H(B|A)$. Tracing out Alice in Eq. (6.28), yields:

$$\begin{aligned}\rho_{BE} = &\frac{1}{2} |0\rangle\langle 0|_B \otimes \left(|e^0_{0,0}\rangle\langle e^0_{0,0}| + |e^0_{0,2}\rangle\langle e^0_{0,2}| + |e^1_{0,0}\rangle\langle e^1_{0,0}| + |e^1_{0,2}\rangle\langle e^1_{0,2}| \right) \\ &+ \frac{1}{2} |1\rangle\langle 1|_B \otimes \left(|e^1_{1,3}\rangle\langle e^1_{1,3}| + |e^1_{1,1}\rangle\langle e^1_{1,1}| + |e^0_{1,3}\rangle\langle e^0_{1,3}| + |e^0_{1,1}\rangle\langle e^0_{1,1}| \right)\end{aligned} \tag{6.30}$$

At this point, one can compute a bound using Theorem 2.1 from Chapter 2. In particular, taking Eq. (6.30), and applying that theorem to it, we can derive the following bound:

$$\begin{aligned}H(B|E) \geq &(1-Q)^2(1-h(\lambda_1)) \\ &+ Q(1-Q)(1-h(\lambda_2)) \\ &+ Q(1-Q)(1-h(\lambda_2)) \\ &+ Q^2(1-h(\lambda_4)),\end{aligned} \tag{6.31}$$

where:

$$\begin{aligned}\lambda_1 &= \frac{1}{2}\left(1 + \frac{|Re \langle e^0_{0,0}|e^1_{1,3}\rangle|}{(1-Q)^2}\right) \\ \lambda_2 &= \frac{1}{2}\left(1 + \frac{|Re \langle e^0_{0,2}|e^1_{1,1}\rangle|}{Q(1-Q)}\right) \\ \lambda_3 &= \frac{1}{2}\left(1 + \frac{|Re \langle e^1_{0,0}|e^0_{1,3}\rangle|}{Q(1-Q)}\right) \\ \lambda_4 &= \frac{1}{2}\left(1 + \frac{|Re \langle e^1_{0,2}|e^0_{1,1}\rangle|}{Q^2}\right)\end{aligned} \tag{6.32}$$

Exercise 6.9. Use Theorem 2.1 from Chapter 2, applied to Eq. (6.30), and show how Eq. (6.31) is derived.

We now need bounds on the inner products of those vectors appearing in the λ_i expressions above. There are two approaches to this. We will show the "easy" approach, here which, though it works, will produce a sub-optimal bound on the overall keyrate (indeed, the noise tolerance will be low with this approach). An alternative approach, shown in [20], is to utilize mismatched measurements (themselves, originally introduced in [2] for "fully quantum" QKD protocols). In the semi-quantum case, this involves looking at events such as: Alice sending an X basis state, Bob measuring and observing a $|0\rangle$, and then Alice observing a $|-\rangle$ when she measures the returning system. Normally these events are discarded, but by looking at their distribution (the outcome should be nearly uniform), one can get a much tighter bound on Eve's uncertainty and, thus, derive a much more optimistic key-rate equation (which matches BB84's expression for depolarization channels), however the algebra is more involved. For the mismatched measurement case, see [20] for details on the derivation.

For the "regular" case (without mismatched measurements), we'll first look at Q_X, which is the probability that Alice observes the wrong X basis state in the event she initially sent a Hadamard state, and Bob chose to Reflect. For instance, Alice observing $|-\rangle$ if she initially sent $|+\rangle$ and Bob Reflected. For this, we need to model the joint channel when Bob reflects. This is given by the unitary $V = U_2U_1$. By tracing its execution on the state $|+\rangle$ and $|-\rangle$ and then computing the probability that Alice observes the opposite state she initially sent, we can derive:

$$Q_X = 1 - Re\left(\langle g_0|g_3\rangle + \langle g_1|g_2\rangle\right), \tag{6.33}$$

where:

$$\begin{aligned}
|g_0\rangle &= |e^0_{0,0}\rangle + |e^0_{1,1}\rangle \\
|g_1\rangle &= |e^1_{0,0}\rangle + |e^1_{1,1}\rangle \\
|g_2\rangle &= |e^0_{0,2}\rangle + |e^0_{1,3}\rangle \\
|g_3\rangle &= |e^1_{0,2}\rangle + |e^1_{1,3}\rangle
\end{aligned}$$

Exercise 6.10. Let $V = U_2U_1$ and trace the execution of $V|+\rangle$. Compute the probability of observing a $|-\rangle$. Next, trace the execution of $V|-\rangle$ and compute the probability of observing $|+\rangle$. Add the two together and show that you arrive at Eq. (6.33). You may need to take advantage of unitarity of V to cancel some terms out.

Now, we need to bound Eq. (6.31). Since we are not computing the optimal bound (which requires mismatched measurements, as discussed earlier), let's compute a "simple" bound which evaluates to one, whenever the noise is zero, and smoothly decreases as the noise increases. In this case, we can find a non-optimistic bound on $H(B|E)$ by taking Eq. (6.31) and noting that each term in the summation is non-negative. Thus:

$$H(B|E) \geq (1-Q)^2(1-h(\lambda_1)), \tag{6.34}$$

where λ_1 remains the same as before, and is based on $\langle e^0_{0,0}|e^1_{1,3}\rangle$. This term can be found by looking at $\langle g_0|g_3\rangle$. In particular, from Eq. (6.33), we have:

$$\begin{aligned}
&Re\,\langle g_0|g_3\rangle = 1 - Q_X - Re\left(\langle g_1|g_2\rangle\right) \\
&\Longrightarrow Re\,\langle e^0_{0,0}|e^1_{1,3}\rangle = 1 - Q_X - Re\left(\langle g_1|g_2\rangle - \langle e^0_{0,0}|e^1_{0,2}\rangle - \langle e^0_{1,1}|e^1_{0,2}\rangle - \langle e^0_{1,1}|e^1_{1,3}\rangle\right)
\end{aligned} \tag{6.35}$$

This seems like a mess! However, we can actually now use the Cauchy Schwarz inequality to finish the proof. Indeed, all inner-products on the right hand side of the above equation are zero, whenever the noise is zero. This implies that, in the no-noise case, $Re\langle e^0_{0,0}|e^1_{1,3}\rangle = 1-0-0=1$ and, thus, $\lambda_1 = 1$ and so Eq. (6.34) will also be one. As the noise increases, the innerproducts increase, but only proportionally to Q. For example, using Eq. (6.29), and the Cauchy-Schwarz inequality, we can find:

$$|Re\langle e^0_{1,1}|e^1_{0,2}\rangle| \leq \sqrt{Q^2 \times Q^2} = Q^2 \tag{6.36}$$

$$|Re\langle e^0_{0,0}|e^1_{0,2}\rangle| \leq \sqrt{(1-Q)^2 \times Q^2} = Q(1-Q). \tag{6.37}$$

Others appearing in Eq. (6.35) can be found similarly. For $\langle g_1|g_2\rangle$, Cauchy-Schwarz may be used also by expanding each in terms of $|e^k_{i,j}\rangle$ vectors. See Exercise 6.11. In particular, we can derive the following expression:

$$Re\langle e^0_{0,0}|e^1_{1,3}\rangle \geq 1 - Q_X - 6Q(1-Q) - Q^2 \tag{6.38}$$

This is sufficient to determine a bound on $H(B|E)$ using Eq. (6.34). Bounding $H(B|A)$ is trivial (it is simply $H(B|A) = h(Q)$ in our channel model). This allows us to finish the key-rate proof, against collective attacks, for this SQKD protocol. A plot of the resulting equation is shown in Fig. 6.2. Note that, in this two-way channel, we set $Q_X = 2Q(1-Q)$ when evaluating (where, again, Q is the depolarizing noise in one of the channels).

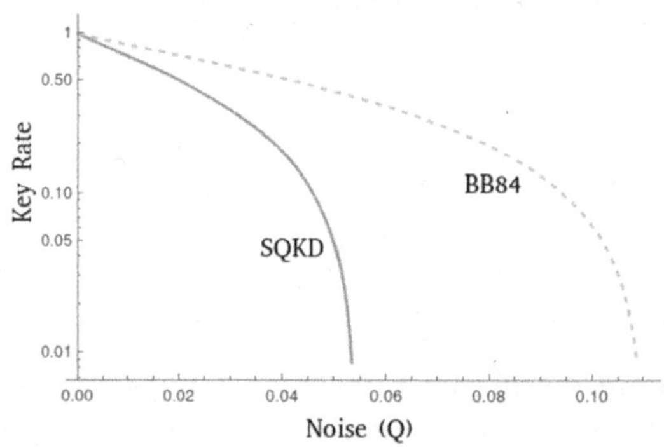

FIGURE 6.2 Evaluating the key-rate of the SQKD protocol and comparing to standard BB84.

Note it is not nearly as efficient or noise tolerant as BB84. This would normally be expected, since the semi-quantum protocol is a "restricted" protocol in the sense that one party does not have full quantum capabilities. However, as shown in [20], mismatched measurements can be used to improve the result, for depolarization channels, against collective attacks. In particular, mismatched measurements allow one to derive exact expressions for the inner-products, above, instead of using "worst-case" bounds via Cauchy-Schwarz. Doing so, results in a key-rate of the form $1 - h(2Q(1-Q)) - h(Q)$ (see [20]). While still not as efficient as BB84's rate of $1 - 2h(Q)$, this difference is mostly due to the fact that X basis rounds have to travel down both channels, thus picking up a higher error rate, depending on the channel model.

Exercise 6.11. Show that all inner-products, appearing on the right-hand-side of Eq. (6.35), are zero when the noise is zero. Write out bounds for all inner-products appearing in that equation, and show that you arrive at Eq. (6.38).

6.2.1 Alternative SQKD protocols

From a practical perspective, SQKD protocols clearly have shortcomings in that two-way quantum communication is required, along with the ability for the "classical" user to choose, deterministically, whether to reflect or measure/resend (by deterministically, we mean generally this user must be able to actually choose which action to take, as opposed to using, for example, passive optics to make the choice). Such optical circuitry is generally slow and lossy. Furthermore, if the "classical" party has such optics, often they also have the ability to perform measurements in alternative bases and, thus, might as well perform BB84! Nonetheless, SQKD protocols are very interesting from a theoretical perspective as they help us study what requirements are needed to achieve unconditional security for key distribution. To that end, it is worth looking at some alternative SQKD protocols, introduced after the original discussed above, which place fewer resource requirements on either one, or both, users.

There are two main areas to limit quantum resource requirements: in state preparations and in state measurements. For the first, perhaps the simplest protocol is the so-called Single State SQKD protocol introduced in [39]. Here, the quantum user Alice only needs to be able to prepare and send a single state $|+\rangle$ (unlike the original SQKD protocol where the quantum users need the ability to prepare $|0\rangle$, $|1\rangle$, $|+\rangle$, and $|-\rangle$). The quantum communication stage, at a very high level, of this protocol acts as follows:

1. Alice prepares and sends the state $|+\rangle$ to Bob (the classical user)
2. Bob chooses randomly to Measure and Resend or to Reflect.
3. Alice measures in either the Z or X basis.

The raw key is distilled from rounds when Bob chooses to Measure and Resend and Alice chooses to measure in the Z basis (they should have a correlated bit under such circumstances). Rounds where Bob reflects and Alice measures in the X basis, along with rounds where Alice measures in the Z basis and Bob Measures and Resend, can be used to detect the presence of an adversary (and, ultimately, bound $H(A|E)$ as was done before). Of course, error correction, and privacy amplification are required in the usual way, along with sampling. We will not go into all these details since, at this point, it should be clear how they are added to turn the above three high-level steps into a real protocol. The reader is encouraged to work out these details.

Next, we look at a protocol which requires fewer measurements and only three states introduced in [23]:

1. Alice prepares $|0\rangle$, $|1\rangle$, or $|+\rangle$, choosing randomly.
2. Bob chooses to Measure and Resend or Reflect.
3. Alice measures in the X basis.

Above, a key is distilled whenever Alice chooses to send a Z basis state and Bob chooses to Measure and Resend. Other rounds can be used to determine the presence of an adversary system. This includes rounds where Alice sends Z basis states and Bob Reflects (so called *mismatched* rounds). The above protocol actually introduces an interesting novelty: the need for collecting information on these *mismatched measurements.*

Mismatched measurements are a concept introduced in [2] which involved the collection of measurement data on rounds which, ordinarily, would be discarded. For example, with BB84, one normally discards rounds where Alice and Bob send/measure in different bases. As it turns out, however, collecting information on these rounds (for instance, ensuring that these measurements produce a 50/50 outcome), can be vital to security, or significantly improve performance [2,37,20]. This particular SQKD protocol actually requires them [23].

To see why, consider the following, fairly simple, attack Eve might perform: In the forward channel, Eve simply ignores the qubit. In the reverse channel (when the qubit returns from Bob to Alice), Eve attacks with unitary operator U which acts as follows:

$$U\,|+,0\rangle_{TE} = |+,0\rangle_{TE}$$
$$U\,|-,0\rangle_{TE} = |+,1\rangle_{TE}\,.$$

Exercise 6.12. Show that the above operator, U, is unitary.

First, note that this attack will create no noise in the channel, with noise being the probability of an incorrect measurement result (this attack does create "noise" in the sense that it will create an imbalance in mismatched measurements, however, as we soon see). If Alice sends $|+\rangle$, the state returning to her will be $|+\rangle$ and so no X basis measurement will report an error. On the other hand, this attack gives Eve *full information* on the raw key. Indeed, if Alice sends $|0\rangle$ and Bob Measures and Resends, the state returning to Alice will be $|0\rangle_T$. Eve probes with her unitary operator, causing the joint state to evolve to:

$$U\,|00\rangle_{TE} = U\left(\frac{1}{\sqrt{2}}(|+\rangle_T + |-\rangle_T)\,|0\rangle_E\right) = |+,+\rangle_{TE}\,. \tag{6.39}$$

Whereas, if Alice sends a $|1\rangle_T$ initially and Bob Measures and Resends the state evolves to:

$$U\,|10\rangle_{TE} = U\left(\frac{1}{\sqrt{2}}(|+\rangle_T - |-\rangle_T)\,|0\rangle_E\right) = |+,-\rangle_{TE}\,. \tag{6.40}$$

Note that if Eve were to make a simple X basis measurement of her ancilla, she will know exactly what Alice and Bob's raw key is. On the other hand, this attack will create no Z basis noise in the forward channel and no X basis noise in the reverse (the only noise that Alice and Bob can detect in this case). However, this attack does create a lot of "noise" during mismatched rounds: rounds where Alice sends a Z basis state and Bob Reflects. Ordinarily, such a round should cause Alice's subsequent X basis measurement to be uniform random: either she will observe $|+\rangle$ or $|-\rangle$. However, Eve's attack will cause Alice to always observe $|+\rangle$. Similarly for rounds where Alice sends $|1\rangle$. By looking at the noise in the channel, along with the distance of the distribution of mismatched rounds from uniform, Alice and Bob are able to bound $H(A|E)$. Details of the proof can be found in [23].

The last alternative SQKD protocol we'll look at in this chapter is a so-called *mediated* SQKD protocol originally introduced in [18], with several other protocols being developed since then [28,40,26,29,11,21,16]. Here, there are two "classical" (or restricted) users who wish to agree on a secret key. Of course, by themselves, they cannot achieve information theoretic security (as the restrictions on these users forces them to only communicate using Z basis states – thus, if the protocol only involved these two users, it would be mathematically equivalent to a classical protocol). To achieve security, these users will rely on the services of a "fully quantum" server (or mediator). However, this server need not be trusted and, in fact, may be the adversary. See Fig. 6.3.

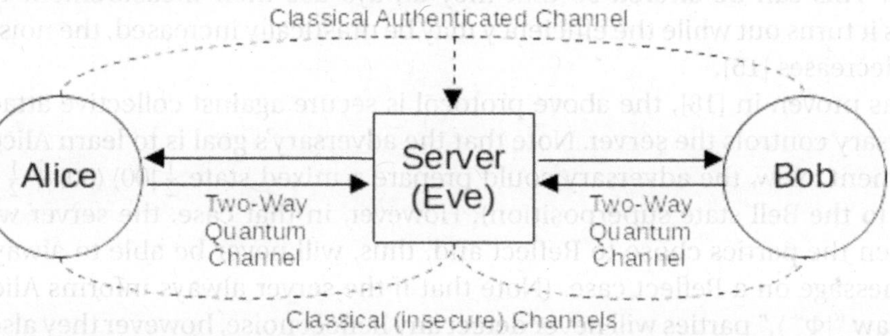

FIGURE 6.3 Showing a general mediated SQKD setup. Here two semi-quantum/"classical" users (Alice and Bob) are connected to a fully quantum server which is untrusted (and may be the adversary Eve). Connecting the users to the server is a two-way quantum channel allowing qubits to be sent from the server, to a user, then back to the server. Alice and Bob share an authenticated classical channel (top) allowing them to communicate (Eve and the server may listen on this channel). A regular classical channel connects the server to the two users – this channel may not be authenticated. Security can be proven even if the adversary controls the server and the lower, insecure, classical channel.

The original protocol from [18] operated as follows (we'll describe the protocol assuming an honest and trusted server; security, however, can be proven even if the server does not follow the protocol):

1. The server prepares a Bell state $|\Phi^+\rangle = \frac{1}{\sqrt{2}}(|00\rangle + |11\rangle)$, sending one particle to Alice and the other to Bob.
2. Alice and Bob, who are the limited "classical users" choose, independently of one another, to Measure and Resend, or to Reflect their particle.
3. Two particles return to the server who performs a Bell measurement on the two systems, reporting the outcome to Alice and Bob.

4. Alice and Bob reveal to one another, using the public authenticated classical channel, their choice to Measure and Resend, or to Reflect.
5. If both parties chose to Measure and Resend, and if the Server stated it observed outcome $|\Phi^-\rangle = \frac{1}{\sqrt{2}}(|00\rangle - |11\rangle)$, then Alice and Bob will keep their measurement results to use as their respective raw keys (which should be correlated in the absence of noise and if the server is honest). Otherwise, parties discard this round.
6. If both parties chose to Reflect, the server should have stated it observed outcome $|\Phi^+\rangle$. Any other observation is counted as an error.
7. Alice and Bob will also determine the error rate in the rounds where they both measure (which should result in correlated outcomes). That is, they compare their Z basis measurement outcomes on rounds they both chose to measure and resend. This is done by sampling a random subset of such rounds; the remaining, unsampled portions, are used as their raw keys.
8. Error correction and Privacy Amplification are run on their remaining raw key.

Note that parties only use a round for their raw key if the server observed one particular Bell state. This can be altered so that they always use their measurement results [15], though as it turns out while the efficiency may be drastically increased, the noise tolerance actually decreases [15].

Now, as proven in [18], the above protocol is secure against collective attacks, even if the adversary controls the server. Note that the adversary's goal is to learn Alice and Bob's measurement. Now, the adversary could prepare a mixed state $\frac{1}{2}|00\rangle\langle 00| + \frac{1}{2}|11\rangle\langle 11|$ (as opposed to the Bell state superposition). However, in that case, the server would never know when the parties chose to Reflect and, thus, will never be able to always send the correct message on a Reflect case. (Note that if the server always informs Alice and Bob that she saw "$|\Phi^+\rangle$," parties will never detect any Reflect noise, however they also will never have a key, since they only use rounds where they both measure and where the server says she saw "$|\Phi^-\rangle$.") Similarly, if the server tries to measure in the Z basis before the Bell basis on return of the qubits, the same problem will occur and this attack will be detected. If both parties measured, they are technically transmitting their raw key bit back to the server; however the server must perform a Bell measurement (or something like it) in order to not "get caught" on the Reflect cases. This Bell measurement projects the computational states returned from the parties in the Measure-Resend case, into a Bell state, which causes the original Z-basis information to be lost!

Of course, the above is not a rigorous proof of security. Instead, the security proof in [18] follows a similar method as discussed earlier, bounding $H(A|E)$ using a result similar to Theorem 2.1. In [15], the proof method was improved by designing an equivalent *entanglement based version* of the protocol and analyzing that (though, again, using Theorem 2.1).

We leave this section with a comment on the experimental feasibility of semi-quantum protocols. In general, the Measure and Resend operation introduces several practical attack possibilities an adversary may perform. Since measurements involve single photon detectors (see Chapter 9), they are destructive. Thus, the "Resend" part of this operation involves creating a new photon and sending it back to the original party. However, Eve may "tag" photons and realize that a new photon was created, without creating any detectable noise [35,9]. This attack allows Eve to distinguish between rounds where the classical party measures and resends, or reflects, thus breaking the security of any SQKD protocol. However, it is possible to build a SQKD protocol that is also secure against such attacks, though at the cost of decreased efficiency. This is done through so-called Mirror Protocols [7] and involves *time bin encoding* [5] of qubits. In this setup, when a party chooses to Measure and Resend, they never actually have to create a new photon – instead they perform a partial measurement of a particular time bin. If it's not observed, then the qubit collapses to the other time bin and returns to the original sender. This is a way to implement the Measure/Resend operation without actually creating a new photon (thus securing against these so-called photon tagging attacks, along with several other practical attacks). We refer the reader to [7] for more details. An alternative approach was presented in [33] using a Mach-Zehnder interferometer setup where, again, parties did not have to create a new photon when they chose to Measure and Resend.

6.3 Twin field QKD

An interesting, and fairly recent (in the grand scheme of things), development in QKD research has been the design of so-called *twin field* QKD protocols (TF-QKD) [32,13]. One of the main limitations to QKD operation is distance: as the distance increases, the probability of successful transmission of your qubit decreases (exponentially so if communicating over fiber). As it turns out, there is a strict upper-bound on the number of secret bits one can extract from a lossy channel. Let η be the probability of a successful qubit transmission over your channel (i.e., it is the transmittance of the channel). Then, for any quantum protocol, even if you assume free classical communication, there is a limit to how many secret key bits you can distill without the help of a quantum repeater. This is known as the PLOB bound [34] after the four authors: Pirandola, Laurenza, Ottaviani, and Banchi.

Since the first TF protocol was introduced in [32], numerous other protocols have been proposed (see [1] for a recent survey), along with experimental implementations [27,10]. Distances in experiments as of writing this can even exceed 800 km [36]. In this chapter, we'll just talk about one of them, introduced in [13]. The protocol consists of Alice and Bob preparing systems and sending photons to a central node C. This node must perform a measurement and report on the outcome.

A single round of the TF-QKD protocol introduced in [13] consists of the following operations:

1. Alice creates a state of the form:

$$|\psi_a\rangle = \sqrt{p}\,|0, vac\rangle_{Aa} + \sqrt{1-p}\,|1, 1\rangle_{Aa}, \tag{6.41}$$

where the A register is Alice's private memory (a single qubit system) and the a system consists of either a vacuum state $|vac\rangle_a$ or a single photon $|1\rangle_a$. The value of p is set by Alice and Bob and is public knowledge. Similarly, Bob creates a state of the form:

$$|\psi_b\rangle = \sqrt{p}\,|1, vac\rangle_{Bb} + \sqrt{1-p}\,|0, 1\rangle_{Bb}\,. \tag{6.42}$$

The A and B registers are kept private while the a and b photons are sent to the central node C.

2. The central node routes incoming photons through a 50 : 50 beam splitter with two detectors labeled 0 and 1. The outcome of the detector that receives a photon is reported to Alice and Bob. If no photon is detected, or if two photons are detected (one on each detector – note the central node does not typically have the ability to count the number of photons hitting a single detector), these outcomes are also reported.
3. If the server reported a vacuum event or a double click event, this round is discarded. If the server reported an outcome of 1 (i.e., it hit the detector labeled 1), Bob applies a Pauli Z gate to his private ancilla.
4. If the server reported a single photon event on either detector 0 or 1, they will then measure their private ancillas in either the X basis (for testing) or the Z basis (for key generation). Essentially, parties run the E91 protocol (see Section 4.3.2) using their private AB registers.

Why does this work at all? Let's trace the protocol, assuming no noise (i.e., to show correctness of the protocol). First, the joint state of the system is $|\psi_a\rangle \otimes |\psi_b\rangle$ which, by permuting subspaces, may be written as:

$$\begin{aligned}|\psi_a\rangle \otimes |\psi_b\rangle \cong\; & p\,|0, 1, vac, vac\rangle_{ABab} + (1-p)\,|1, 0, 1, 1\rangle_{ABab} \\ & + \sqrt{p(1-p)}\left(|0, 0, vac, 1\rangle_{ABab} + |1, 1, 1, vac\rangle_{ABab}\right).\end{aligned} \tag{6.43}$$

Note we use the $\cong$ symbol to indicate that the systems are equal up to a permutation of subspaces. As in previous cases, this permutation is done to make the algebra easier – so long as one is consistent to permute the subspaces in all kets of a superposition, it is safe to do so.

Now, the ab register travels to the server which routes the photons through a beam splitter. This may be modeled as a unitary operation:

$$\begin{aligned} BS\,|1, vac\rangle_{ab} &= \frac{1}{\sqrt{2}}(|D_0\rangle - |D_1\rangle) \\ BS\,|vac, 1\rangle_{ab} &= \frac{1}{\sqrt{2}}(|D_0\rangle + |D_1\rangle) \end{aligned} \tag{6.44}$$

This is a Hadamard transform. Of course, given as input $|vac, vac\rangle$, the output will be $|vac, vac\rangle$; if the input is $|1, 1\rangle$, then we will call the output $|\phi_2\rangle$ which will be arbitrary.

This beamsplitter is a unitary operation and, so, acts coherently on the state. This BS operation is applied to the ab system only yielding:

$$\begin{aligned}
I_{AB} \otimes BS\,|\psi_a, \psi_b\rangle &= p\,|01\rangle_{AB}\,|vac, vac\rangle + (1-p)\,|10\rangle_{AB} \otimes |\phi_2\rangle \\
&\quad + \sqrt{p(1-p)}\left(|00\rangle_{AB}\left(\frac{|D_0\rangle + |D_1\rangle}{\sqrt{2}}\right) + |11\rangle_{AB}\left(\frac{|D_0\rangle - |D_1\rangle}{\sqrt{2}}\right)\right) \\
&= p\,|01\rangle_{AB}\,|vac, vac\rangle + (1-p)\,|10\rangle_{AB} \otimes |\phi_2\rangle \\
&\quad + \sqrt{p(1-p)}(|\Phi^+\rangle\,|D_0\rangle + |\Phi^-\rangle\,|D_1\rangle),
\end{aligned}$$

where, $|\Phi^\pm\rangle = \frac{1}{\sqrt{2}}(|00\rangle \pm |11\rangle)$ are two (of the four) Bell states. Ignoring, for now, the multi-photon case $|\phi_2\rangle$, if the server "sees" a photon at Detector 0 (i.e., measures and observes $|D_0\rangle$), then the state will collapse to $|\Phi^+\rangle$; otherwise it will collapse to $|\Phi^-\rangle$ (which will be transformed to $|\Phi^+\rangle$ by Bob's Z gate). Either state may be used to distill a random key bit using the E91 protocol. Thus, even though Alice and Bob started off separate, the server's operation actually causes their systems to become entangled. Furthermore, a key-bit is established when one photon survives the journey to the center node, thus allowing TF-QKD protocols to beat the PLOB bound. Normally, in a "standard" QKD system involving a center node, both qubits must survive the journey for a key-bit to be established.

Of course, this disregarded the double photon term – however, if we set p to be close to 1 (so that Alice and Bob are creating superposition states which are heavily weighted in favor of $|vac\rangle$), then $1 - p < \sqrt{p(1-p)}$ so these double photon events won't occur as often as the "good" single photon events.

References

[1] Syed M. Arslan, Syed Shahmir, Noureldin Mohammad, Saif Al-Kuwari, Muataz Alhussein, Twin-field quantum key distribution: protocols, security, and open problems, arXiv preprint, arXiv:2510.26320, 2025.

[2] Stephen M. Barnett, Bruno Huttner, Simon J.D. Phoenix, Eavesdropping strategies and rejected-data protocols in quantum cryptography, Journal of Modern Optics 40 (12) (1993) 2501–2513.

[3] Normand J. Beaudry, Marco Lucamarini, Stefano Mancini, Renato Renner, Security of two-way quantum key distribution, Physical Review. A, Atomic, Molecular, and Optical Physics 88 (6) (2013) 062302.

[4] Kim Boström, Timo Felbinger, Deterministic secure direct communication using entanglement, Physical Review Letters 89 (18) (2002) 187902.

[5] William Boucher, Thierry Debuisschert, Experimental implementation of time-coding quantum key distribution, Physical Review. A, Atomic, Molecular, and Optical Physics 72 (6) (2005) 062325.

[6] Niek J. Bouman, Serge Fehr, Sampling in a quantum population, and applications, in: Annual Cryptology Conference, Springer, 2010, pp. 724–741.

[7] Michel Boyer, Matty Katz, Rotem Liss, Tal Mor, Experimentally feasible protocol for semiquantum key distribution, Physical Review A 96 (6) (2017) 062335.

[8] Michel Boyer, Dan Kenigsberg, Tal Mor, Quantum key distribution with classical Bob, Physical Review Letters 99 (140501) (2007).

[9] Michel Boyer, Dan Kenigsberg, Tal Mor, Boyer, Kenigsberg, and Mor reply, Physical Review Letters 102 (9) (2009) 098902.

[10] Jiu-Peng Chen, Chi Zhang, Yang Liu, Cong Jiang, Wei-Jun Zhang, Zhi-Yong Han, Shi-Zhao Ma, Xiao-Long Hu, Yu-Huai Li, Hui Liu, et al., Twin-field quantum key distribution over a 511 km optical fibre linking two distant metropolitan areas, Nature Photonics 15 (8) (2021) 570–575.
[11] Lingli Chen, Qin Li, Chengdong Liu, Yu Peng, Fang Yu, Efficient mediated semi-quantum key distribution, Physica A. Statistical Mechanics and Its Applications 582 (2021) 126265.
[12] Matthias Christandl, Robert König, Renato Renner, Postselection technique for quantum channels with applications to quantum cryptography, Physical Review Letters 102 (2) (2009) 020504.
[13] Marcos Curty, Koji Azuma, Hoi-Kwong Lo, Simple security proof of twin-field type quantum key distribution protocol, npj Quantum Information 5 (1) (2019) 64.
[14] Pavel Gurevich, Meir Orenstein, Tal Mor, Experimental quantum key distribution with classical Alice, Technion-Israel Institute of Technology, Faculty of Computer Science, 2012.
[15] Julia Guskind, Walter O. Krawec, Mediated semi-quantum key distribution with improved efficiency, Quantum Science and Technology 7 (3) (2022) 035019.
[16] Tzonelih Hwang, Chia-Wei Tsai, et al., Mediated semi-quantum key distribution in randomization-based environment, arXiv preprint, arXiv:2010.04441, 2020.
[17] Robert König, Renato Renner, A de Finetti representation for finite symmetric quantum states, Journal of Mathematical Physics 46 (12) (2005).
[18] Walter O. Krawec, Mediated semiquantum key distribution, Physical Review A 91 (3) (2015) 032323.
[19] Walter O. Krawec, Security proof of a semi-quantum key distribution protocol, in: 2015 IEEE International Symposium on Information Theory (ISIT), IEEE, 2015, pp. 686–690.
[20] Walter O. Krawec, Quantum key distribution with mismatched measurements over arbitrary channels, Quantum Information & Computation 17 (3&4) (2017) 209–241.
[21] Walter O. Krawec, Multi-mediated semi-quantum key distribution, in: 2019 IEEE Globecom Workshops (GC wkshps), IEEE, 2019, pp. 1–6.
[22] Walter O. Krawec, Security of a high dimensional two-way quantum key distribution protocol, Advanced Quantum Technologies 5 (10) (2022) 2200024.
[23] Walter O. Krawec, Eric P. Geiss, Semi-quantum key distribution with limited measurement capabilities, in: 2018 International Symposium on Information Theory and Its Applications (ISITA), IEEE, 2018, pp. 462–466.
[24] Walter Oliver Krawec, Rotem Liss, Tal Mor, Security proof against collective attacks for an experimentally feasible semiquantum key distribution protocol, IEEE Transactions on Quantum Engineering 4 (2023) 1–16.
[25] Rupesh Kumar, Marco Lucamarini, Giovanni Di Giuseppe, Riccardo Natali, Giorgio Mancini, Paolo Tombesi, Two-way quantum key distribution at telecommunication wavelength, Physical Review. A, Atomic, Molecular, and Optical Physics 77 (2) (2008) 022304.
[26] Po-Hua Lin, Chia-Wei Tsai, Tzonelih Hwang, Mediated semi-quantum key distribution using single photons, Annalen der Physik 531 (8) (2019) 1800347.
[27] Hui Liu, Cong Jiang, Hao-Tao Zhu, Mi Zou, Zong-Wen Yu, Xiao-Long Hu, Hai Xu, Shizhao Ma, Zhiyong Han, Jiu-Peng Chen, et al., Field test of twin-field quantum key distribution through sending-or-not-sending over 428 km, Physical Review Letters 126 (25) (2021) 250502.
[28] Zhi-Rou Liu, Tzonelih Hwang, Mediated semi-quantum key distribution without invoking quantum measurement, Annalen der Physik 530 (4) (2018) 1700206.
[29] Yu-Chin Lu, Chia-Wei Tsai, Tzonelih Hwang, Collective attack and improvement on "mediated semi-quantum key distribution using single photons", Annalen der Physik 532 (9) (2020) 1900493.
[30] Marco Lucamarini, Stefano Mancini, Secure deterministic communication without entanglement, Physical Review Letters 94 (14) (2005) 140501.
[31] Marco Lucamarini, Stefano Mancini, Quantum key distribution using a two-way quantum channel, Theoretical Computer Science 560 (2014) 46–61.
[32] Marco Lucamarini, Zhiliang L. Yuan, James F. Dynes, Andrew J. Shields, Overcoming the rate–distance limit of quantum key distribution without quantum repeaters, Nature 557 (7705) (2018) 400–403.
[33] Francesco Massa, Preeti Yadav, Amir Moqanaki, Walter O. Krawec, Paulo Mateus, Nikola Paunković, André Souto, Philip Walther, Experimental semi-quantum key distribution with classical users, Quantum 6 (2022) 819.
[34] Stefano Pirandola, Riccardo Laurenza, Carlo Ottaviani, Leonardo Banchi, Fundamental limits of repeaterless quantum communications, Nature Communications 8 (1) (2017) 15043.

[35] Yong-gang Tan, Hua Lu, Qing-yu Cai, Comment on "quantum key distribution with classical Bob", Physical Review Letters 102 (9) (2009) 098901.
[36] Shuang Wang, Zhen-Qiang Yin, De-Yong He, Wei Chen, Rui-Qiang Wang, Peng Ye, Yao Zhou, Guan-Jie Fan-Yuan, Fang-Xiang Wang, Wei Chen, et al., Twin-field quantum key distribution over 830-km fibre, Nature Photonics 16 (2) (2022) 154–161.
[37] Shun Watanabe, Ryutaroh Matsumoto, Tomohiko Uyematsu, Tomography increases key rates of quantum-key-distribution protocols, Physical Review. A, Atomic, Molecular, and Optical Physics 78 (4) (2008) 042316.
[38] Andreas Winter, Tight uniform continuity bounds for quantum entropies: conditional entropy, relative entropy distance and energy constraints, Communications in Mathematical Physics 347 (1) (2016) 291–313.
[39] Xiangfu Zou, Daowen Qiu, Lvzhou Li, Lihua Wu, Lvjun Li, Semiquantum-key distribution using less than four quantum states, Physical Review. A, Atomic, Molecular, and Optical Physics 79 (5) (2009) 052312.
[40] Xiangfu Zou, Zhenbang Rong, Nan-Run Zhou, Three attacks on the mediated semi-quantum key distribution without invoking quantum measurement, Annalen der Physik 532 (8) (2020) 2000251.

7

High-dimensional quantum cryptography

Ever since the development of the BB84 protocol [2] in 1984, QKD research has expanded rapidly. One of the great themes to this area of research is in improving its *efficiency* (in terms of key-rate) and *noise tolerance.* These are the two main metrics by which QKD protocols are often judged. One potential avenue to improve both of these is in the use of *high-dimensional* (HD) quantum states. Namely, moving beyond qubits, which exist in a two-dimensional Hilbert space, to *qudits* which exist in a d-dimensional Hilbert space. Protocols which harness these high dimensional states have been shown to exhibit numerous theoretical benefits in these two key metrics, over qubit based systems. Efficiency is often improved due to the increased dimension of the system as, now, a single round can lead to multiple key-bits at once. This means more information can be carried per round/signal. Improved noise tolerance is rather striking and very interesting, and often depends on the channel noise type (with HD-QKD protocols being highly resilient to depolarizing noise).

Surprisingly, HD-QKD systems also have several uniquely interesting properties and benefits. For instance, the so-called *Round Robin* QKD protocol [15] (discussed below in Section 7.2) allows users to bound Eve's uncertainty based only on the dimension d of the quantum system and not necessarily on the noise in the channel (as with all other known qubit-based QKD protocols). HD-QKD systems are also experimentally feasible with today's technology [7]. In this chapter, we'll discuss some of the interesting HD-QKD protocols and how they operate. We'll also give some intuition behind their security proofs when possible. This is not meant to be an exhaustive survey of all high dimensional quantum cryptographic systems. For further reading, the survey of [7] is useful.

7.1 High-dimensional BB84

We'll start by discussing the HD-BB84 protocol, first introduced in [6]. As in the qubit case, this protocol is fairly straightforward to understand and security, as it turns out, is not too difficult to prove, given modern entropic uncertainty relations. Consider the computational basis, now in dimension d, spanned by $|0\rangle, \cdots, |d-1\rangle$. Let X be some other basis that is mutually unbiased, spanned by $|x_0\rangle, \cdots, |x_{d-1}\rangle$. By mutually unbiased, we mean $|\langle i|x_j\rangle| = \frac{1}{\sqrt{d}}$ for all i, j. (The X and Z bases are mutually unbiased in dimension two, for instance.)

Quantum Communication and Cryptography. https://doi.org/10.1016/B978-0-44-321532-2.00014-0

The high-dimensional BB84 protocol is shown below in Protocol 7.1. Its similarity to standard BB84, discussed in Section 4.2 should give the reader an indication as to its choice in name.

Protocol 7.1 HD-BB84: [6].

Public Parameters:

- d: Dimension of a quantum state
- N: number of rounds
- p_Z: probability of choosing the Z basis. We define $p_X = 1 - p_Z$ to be the probability of choosing the X basis.

Quantum Communication Stage: For $i = 1, 2, \cdots, N$, do:

1. Alice chooses a random raw key *digit* $r_i^A \in \{0, 1, \cdots, d-1\}$ and random basis $b_i^A \in \{0, 1\}$, choosing $b_i^A = 0$ with probability p_Z. If $b_i^A = 0$, she creates and sends the state $|r_i^A\rangle$; otherwise, she sends the state $|x_{r_i^A}\rangle$.
2. Bob, on receipt of a *qudit* will choose a basis $b_i^B \in \{0, 1\}$ (independently of Alice, but choosing the Z basis with probability p_Z) and measures the state he received in basis Z if $b_i^B = 0$ or basis X if $b_i^B = 1$. The result is stored as a digit r_i^B.
3. Goto (1)

Alice now discloses her string of basis choices b^A. Bob sends back $b^A \oplus b^B$. Alice and Bob discard all raw key bits where $b^A \oplus b^B = 1$ (as that indicates Alice and Bob choose different bases). This communication is all done over the classical authenticated channel. Let $N_0 \leq N$ be the size of the final raw key denoted r^A and r^B. It is expected that $N_0 = (p_Z^2 + p_X^2)N$. In the biased version of the protocol (which is the version we'll analyze, later), only Z basis rounds are used for key distillation.

Classical Post Processing Stage: This is similar to the classical post processing stage of standard BB84 discussed in Section 4.2.

Proving the security of HD-BB84 can be done using entropic uncertainty and results in a proof not too dissimilar from that used for standard BB84. Namely, first we will consider an equivalent entanglement based version. Then, we will prove security of the entanglement version using entropic uncertainty.

The entanglement based protocol is a straightforward extension of the E91 protocol. Namely, a source will prepare a high dimensional entangled state of the form:

$$|\phi_0\rangle = \frac{1}{\sqrt{2}} \sum_{a=0}^{d-1} |a, a\rangle_{AB}. \tag{7.1}$$

This is a $2d$-dimensional state and Alice receives the A register while Bob receives the B register. Each party chooses, independently, to measure in the Z basis or the X basis. Alice measuring in the Z or X basis is equivalent to her sending in that basis, in the prepare-and-measure variant.

Why does this work? Well, notice that if Alice and Bob both measure Eq. (7.1) in the Z basis, they will, of course, receive a correlated outcome and, furthermore, the outcome is uniform random (i.e., the probability that Alice observes any $|i\rangle$ is $1/d$). As it turns out, if they both measure in an alternative basis, they will also get a correlated outcome, similar to the two-dimensional Bell case we discussed in Section 4.3.2. Thus, it suffices to analyze this version of the protocol, using similar arguments to that for BB84 and the entanglement based version.

Exercise 7.1. Let $|\phi_0\rangle$ be as defined in Eq. (7.1). Prove that if Alice and Bob measure in an alternative orthonormal basis, they will receive the same outcome. Hint: Measuring in an alternative basis is equivalent to first applying a unitary change of basis, and then measuring in the computational basis.

Now that we have an equivalent entanglement based protocol, we turn our attention to deriving a key-rate expression for it which, in turn, will imply a key-rate expression for the prepare and measure HD-BB84. For this, we turn to entropic uncertainty. Recall the uncertainty relation we introduced in Section 2.6.2.2, simplified for our particular operating scenario. Namely, given a quantum state $|\psi\rangle_{ABE}$ and two mutually unbiased bases, it holds that:

$$H(A_Z|E) + H(A_X|B_X) \geq \log_2 d. \tag{7.2}$$

Assuming that only the Z basis is used for key-generation and, recalling from Section 5.3.1 that the raw key-rate in the asymptotic scenario is bounded by the difference between $H(A_Z|E)$ and the error correction leakage (which, itself, is asymptotically $H(A_Z|B_Z)$), we may therefore conclude that the asymptotic key-rate is:

$$\texttt{rate}^{\infty}_{HD-BB84} = H(A_Z|E) - H(A_Z|B_Z) \geq \log_2 d - H(A_X|B_X) - H(A_Z|B_Z). \tag{7.3}$$

Specifically, this gives us an expression based only on observable statistics, namely Alice and Bob's measurement outcomes; an expression which does not require any knowledge of Eve's attack (Eq. (7.3), essentially, assumes Eve performs the most optimal attack available to her given the observed statistics).

Now that we have an expression for the key rate, how do we evaluate it? Let's assume a depolarizing channel, one of the more common channels to evaluate theoretical QKD performance on. Given a d-dimensional quantum state ρ (which may be mixed or pure), a depolarizing channel, with depolarizing parameter Q, will map ρ to $\mathcal{E}_Q(\rho)$ where:

$$\mathcal{E}_Q(\rho) = \left(1 - \frac{d}{d-1}Q\right)\rho + \frac{Q}{d-1}I, \tag{7.4}$$

where I is the completely mixed state (the identity operator of dimension d).

Before continuing, it is worth looking closer at the action of the map described in Eq. (7.4). Let's consider the qubit case with $d = 2$. Then, if a source device is sending BB84 states, it can be shown that Q represents the probability of an error. For instance, assume a source device sends $|+\rangle$. Then, after the above channel, the state arriving at the receiver will be of the form:

$$\mathcal{E}_Q(|+\rangle\langle+|) = (1-2Q)|+\rangle\langle+| + Q(|+\rangle\langle+| + |-\rangle\langle-|), \tag{7.5}$$

where, above, we used the completeness relation to describe I as the sum of outer products of orthonormal basis vectors. It is evident from the above, that the probability of an error, namely the probability of measuring $|-\rangle$ is exactly Q, while the probability of no error is $1 - 2Q + Q = 1 - Q$. This is true for other bases. In higher dimensions, a depolarizing channel, such as this, will place most of the support of the distribution on the no-error case, and uniformly distribute the remaining to error cases, equally.

This allows us to compute $P_{AB}(A = x \wedge B = y)$ which represents the probability that Alice sends x and Bob observes y. Under our depolarizing channel assumption, this distribution will be the same for all bases.

Let's evaluate P_{AB} using statistics that would occur from the prepare-and-measure variant of the protocol and assuming a depolarizing channel as above. $P_{AB}(A = x \wedge B = y) = Pr(B = y|A = x)Pr(A = x) = \frac{1}{d}Pr(B = y|A = x)$. Now, if Alice had sent $|x\rangle$, the depolarizing channel would alter the state to:

$$\mathcal{E}_Q(|x\rangle\langle x|) = \left(1 - \frac{d}{d-1}Q\right)|x\rangle\langle x| + \frac{Q}{d-1}I. \tag{7.6}$$

Recalling from Section 2.6.1 that I can be written as the sum of all orthonormal basis states, namely $I = \sum_z |z\rangle\langle z|$, we easily see that:

$$P(B = y|A = x) = \begin{cases} 1 - Q & \text{if } x = y \\ Q/(d-1) & \text{otherwise} \end{cases} \tag{7.7}$$

It is not difficult to see that the same distribution will hold for X basis measurements (since $I = \sum_z |x_z\rangle\langle x_z|$ also!). This allows us to readily compute $H(A_Z|B_Z)$ and $H(A_X|B_X)$. One may write a computer program to loop over all possible outcomes (namely, compute $H(A_Z B_Z) = -\sum_{x,y} P_{AB}(A = x, B = y)\log_2 P_{AB}(A = x, B = y)$ and so on). However, it can also be simplified to the following:

$$H(A_Z|B_Z) = Q\log_2(d-1) + h(Q) \tag{7.8}$$

Exercise 7.2. Show that Eq. (7.8) is correct.

One may then write a much simpler and efficient program to evaluate the key rate expression for HD-BB84. Some graphs of the key-rate are shown in Fig. 7.1. We note right away that this protocol can tolerate a very high level of depolarizing noise. This is one of the interesting benefits to theoretical HD-QKD. Note that other noise channels may not exhibit this quality; it is also experimentally difficult, compared to qubits at least, to produce HD states, and so in practice the theoretical benefits may not outweigh the experimental complications, at least not yet.

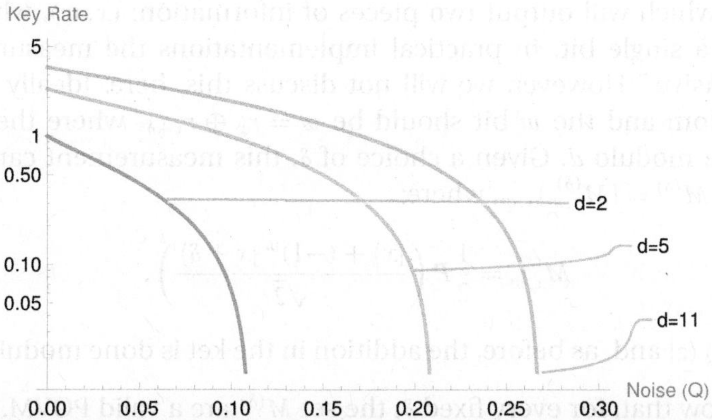

FIGURE 7.1 Showing the key-rate of the HD-BB84 protocol (Eq. (7.3)) for various dimensions, assuming a depolarizing channel. We see that greater efficiency is possible at higher dimensions, at least in theory. Note that, when $d = 2$, we recover the standard BB84 key-rate and 11% noise tolerance.

7.2 Round Robin protocol

One very interesting HD-QKD protocol is the so-called round robin QKD protocol. This is a fascinating protocol, originally introduced in [15], as it demonstrates another unique property of high-dimensional states (HD-BB84 showed how higher noise tolerance is possible). In particular, this protocol allows us to bound Eve's uncertainty based only on the dimension of the system. This is unlike all qubit based protocols, where Eve's information must be bounded by the noise in the channel. Indeed, for this protocol, Eve will always have some uncertainty, so long as the dimension of the system used each round is at least three.

Unlike most prior protocols we looked at, this protocol doesn't think specifically in terms of a fixed set of states and bases (e.g., picking from two bases as in HD-BB84). Instead, the source, Alice, will for every round of the protocol choose a random bit string $r \in \{0, 1\}^d$, choosing each bit uniformly at random. As before, d is the dimension of the quantum system used each round. After choosing r, she prepares and sends the following quantum state:

$$|\psi(r)\rangle = \frac{1}{\sqrt{d}} \sum_{i=0}^{d-1} (-1)^{r_i} |i\rangle, \tag{7.9}$$

where r_i is the i'th bit of r. In practical implementations, this state can be created using time-bin encoding. Namely a single photon existing in a superposition of d different time bins (i.e., $|i\rangle$ represents a photon in the i'th time bin). We'll discuss time bin encoding, briefly, in Chapter 9.

Bob, on receiving this state, will perform a measurement meant to extract the phase difference between two time bins. In particular, he will choose a random δ and perform a measurement which will output two pieces of information: (x, w), where x is an n-bit string, and w is a single bit. In practical implementations the measurement may also output "inconclusive." However, we will not discuss this, here. Ideally the x string will be uniform random and the w bit should be $w = r_x \oplus r_{x+\delta}$, where the addition in the subscript is done modulo d. Given a choice of δ, this measurement can be modeled by POVM operators $M^{(\delta)} = \{M^{(\delta)}_{x,w}\}_{x,w}$, where:

$$M^{(\delta)}_{x,w} = \frac{1}{2} P\left(\frac{|x\rangle + (-1)^w |x+\delta\rangle}{\sqrt{2}}\right), \tag{7.10}$$

where $P(|z\rangle) = |z\rangle\langle z|$ and, as before, the addition in the ket is done modulo d.

Exercise 7.3. Show that, for every fixed δ, the set $M^{(\delta)}$ are a valid POVM. Next, show that, given a state of the form in Eq. (7.9), this POVM will observe a uniform random x and a value w which is equal to $w = r_x \oplus r_{x+\delta}$.

Following his measurement, Bob informs Alice of his choice of δ and the outcome x (but, importantly, *not* the second output of the measurement bit, w). If the measurement is inconclusive (in a practical setup), he informs Alice of this, and both parties discard the round and start over (starting with Alice choosing a new r). The raw key bit will for Alice will be $r_x \oplus r_{x+\delta}$. Alice can compute this once Bob announces x and δ since she knows all of r; Bob will set his raw key bit for this round to be w.

Exercise 7.4. Show that this protocol is completely insecure if $d = 2$. Namely, Eve will be able to learn everything about Alice and Bob's raw key bits without creating any noise.

The protocol is clearly correct, but is it secure? We won't go into explicit details of the proof, which is technical, but instead cover some of the key points from the proof in [15]. The first step of the proof was to consider a "toy" protocol which, though not actually run, exhibits the same amount of information gain for an adversary as the original. Thus, if we can bound the information gain in the toy protocol, we can translate that result directly to the real one. This toy protocol has the nice property that the classical communication from Bob cannot be strongly influenced by Eve and does not leak information. Both are important as we may consider Eve sending her own state (replacing Eq. (7.9)) to Bob and then unduly influencing Bob's messaging.

From this toy protocol, an entropy bound can be derived based only on the dimension of the system. In particular, in [15] it was proven that for $d \geq 3$ (note when $d = 2$ the protocol is insecure, see Exercise 7.4), the key-rate is bounded by:

$$\texttt{rate} \approx 1 - h\left(\frac{1}{d-1}\right) - h(Q_{bit}), \tag{7.11}$$

where Q_{bit} is the error rate in the raw key (needed for error correction). Note that, while the above does require knowledge of Q_{bit}, it does not need this information to bound Eve's information, which is the truly fascinating part. Eve's information is bounded by $1 - h(1/(d-1))$, a function only of the dimension of the system. See Fig. 7.2.

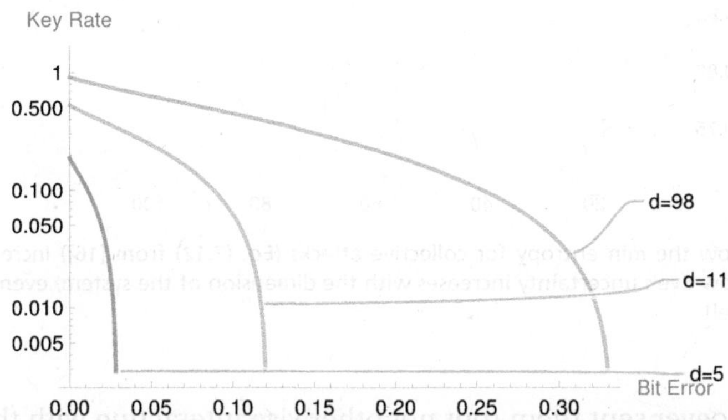

FIGURE 7.2 Showing the key-rate of the Round Robin protocol (Eq. (7.11) from [15]), for different dimensions and raw key error rates (Bit Error). Note that as the dimension increases, the key-rate also increases.

In [16], an alternative proof technique derived a bound on the quantum min entropy of the joint Alice and Eve system under the assumption of collective attacks. This bound was found to be:

$$H_\infty(A|E, \Delta X) = 1 - \log_2\left(1 + \frac{2}{d}\right). \tag{7.12}$$

Above, A is Alice's raw key register while X and Δ are the random variables storing the messages sent by Bob (thus, $A = r_X \oplus r_{X+\Delta}$). Of course, the above is only Eve's information in the collective attack case and does not consider error correction (which would deduct, approximately, an additional $h(Q_{bit})$ bits per round). Note that, when $d = 2$, $H_\infty(A|IJE) = 0$ as expected. See Fig. 7.3.

In summary, the round-robin QKD protocol demonstrates a remarkable property of high-dimensional quantum states, namely that one may bound Eve's information purely based on the dimension of the system and not by the noise level created by her attack. This is unlike qubit based protocols, and also unlike the HD-BB84 protocol discussed above. There, monitoring the X basis noise is required to bound Eve's uncertainty. Of course, in the round-robin protocol one must know the error rate in the final raw key for error correction to operate correctly; however it is not required for a bound on Eve's information.

7.3 High dimensional semi-quantum key distribution

Recall in Section 6.2 we discussed the notion of a semi-quantum key distribution (SQKD) protocol. Here, one party (typically Bob) is said to be *semi-quantum* in nature in that they are restricted to either measuring in the Z basis, sending in the Z basis, or "reflecting"

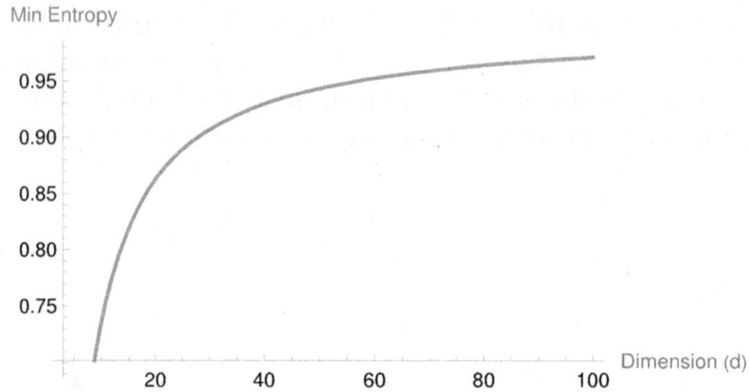

FIGURE 7.3 Showing how the min entropy for collective attacks (Eq. (7.12) from [16]) increases as the dimension increases. This shows how Eve's uncertainty increases with the dimension of the system, even without knowing the error rate of the channel!

qubits back to whoever sent them (but not otherwise interacting with them or extracting any information from them). In general, we combine the measuring and sending into one "Measure/Resend" command whereby Bob will measure in the Z basis and send exactly the same Z basis state he observed. Note that these protocols require a two-way quantum communication channel, as shown in Fig. 6.1.

These notions can be readily extended to HD states where, now, Bob can measure in the high-dimensional Z basis or reflect. Such capabilities lead to a natural extension of the original SQKD protocol from [5] to a HD variant shown in Protocol 7.2.

It is not difficult to see that the protocol, as discussed, is correct. Indeed, in the absence of any noise, if Alice sends a HD Z basis state and Bob chooses to measure/resend, he will observe the same state Alice sent and this can contribute towards both their raw keys. If Bob reflects, then Alice should always observe what she initially sent which can be used to police the channel, similar to the original SQKD protocol from [5] and discussed in Section 6.2. How can we prove security?

Unlike the original SQKD protocol, a direct computation of the von Neumann entropy is challenging since Eve's ancilla dimension will increase quadratically in the first direction (i.e., her first attack) and then quadratically *again* after she attacks the reverse channel. With qubits, when $d = 2$, this is not too bad, but when $d > 2$ this becomes intractable quickly. Instead, an alternative approach may be used [11], which also demonstrates a different proof methodology that can be used for other difficult to prove QKD protocols. This approach is based on the *continuity of von Neumann entropy* [21].

First, we will show how the semi-quantum protocol, which relies on a two-way channel, can actually be turned into a mathematically equivalent one-way *fully quantum* protocol, with Bob sending qudits to Alice, using a construction from [11]. Each round of the protocol has two options: Test, which is used to test the channel and will simulate a "Reflect" event for Bob in the SQKD case; and second Distill which will be used to distill a raw key and will simulate the "measure/resend" case. We will use entropic uncertainty

---

Protocol 7.2 HD-SQKD: (From [11]).

Public Parameters:

- d: Dimension of a quantum state
- N: number of rounds
- p_Z: probability of choosing the Z basis (Alice) or Measure/Resending (Bob).
- $p_X = 1 - p_Z$: probability of choosing the X basis (Alice) or Reflecting (Bob). The X basis is any d-dimensional *mutually unbiased basis* (MUB) with states denoted $\{|x_0\rangle, \cdots, |x_{d-1}\rangle\}$.

Quantum Communication Stage: For $i = 1, 2, \cdots, N$, do:

1. Alice chooses a random raw key *digit* $r_i^A \in \{0, 1, \cdots, d-1\}$ and random basis $b_i^A \in \{0, 1\}$, choosing $b_i^A = 0$ with probability p_Z. If $b_i^A = 0$, she creates and sends the state $|r_i^A\rangle$; otherwise, she sends the state $|x_{r_i^A}\rangle$.
2. Bob, on receipt of a qudit will choose an operation $b_i^B \in \{0, 1\}$ (independently of Alice, but choosing 0 with probability p_Z). If $b_i^B = 0$ he will measure/resend the qudit, subjecting the qudit to a Z basis measurement (of dimension d), and resending the outcome. Otherwise, if $b_i^B = 1$, he reflects the qudit.
3. Alice will measure the qudit in the same basis she originally sent it in.
4. Goto (1)

Alice now discloses her string of basis choices b^A. Bob sends back b^B. Whenever Alice sent in the Z basis and Bob measured/resent, they should share a raw key bit. Other rounds are either discarded or used to estimate the noise in the channel.

Classical Post Processing Stage: This is similar to the classical post processing stage of standard BB84 discussed in Section 4.2. Specifically, error correction and privacy amplification are run, to produce the final secret key.

---

to compute a bound in the Test case and then use continuity of entropy to translate that to an entropy result in the Distill case (which is what we actually want in order to use the Devetak-Winter key rate expression discussed in Section 5.3.1).

The one way protocol is described in Protocol 7.3. Note that it allows for the adversary to influence, slightly, Bob's state preparation – in essence, we give the adversary some additional control over the source device – an important ability needed to prove that security of Protocol 7.3 implies security of the actual semi-quantum one (Protocol 7.2). The necessity of this will be clear momentarily. Note also that the one-way protocol sends $2d$-dimensional states whereas the SQKD protocol sends d-dimensional states. However the necessity of this should be apparent: the one-way protocol is meant to "simulate" the two way protocol, the latter of which sends a d-dimensional state *twice*.

It is not immediately clear why proving security of this protocol (which isn't even semi-quantum) would lead to a proof of security for the SQKD one. However, this highlights one

Protocol 7.3 One-Way Simulated-SQKD QKD protocol: [11].

Public Parameters:

- d: Dimension of a quantum state
- N: number of rounds
- p_Z: probability of choosing the Z basis (Alice) or "simulated" Measure/Resending (Bob).
- $p_X = 1 - p_Z$: probability of choosing the X basis (Alice) or "simulated" Reflecting (Bob).
- $\{p(b)\}_{b=0}^{d-1}$: Probability values set by the adversary but known to all parties. Ideally these should be the uniform distribution.

Quantum Communication Stage: For $i = 1, 2, \cdots, N$, do:

1. Bob chooses, with probability p_z to simulate a measure/resend operation; otherwise he chooses to simulate a reflect operation. In the first case, he prepares the state:

$$|\phi_{MR}\rangle = \sum_{b=0}^{d-1} \sqrt{p(b)}\, |b, b, b\rangle_{A_1 A_2 B} \,. \tag{7.13}$$

   In the second case, he prepares:

$$|\phi_R\rangle = \sum_{b=0}^{d-1} \sqrt{p(b)}\, |b, b\rangle_{A_1 A_2} \otimes |0\rangle_B \,. \tag{7.14}$$

   He then sends the $A_1 A_2$ register to Alice (this is a $2d$-dimensional state).
2. Alice, on receipt of the $A_1 A_2$ register (which potentially passes through adversary Eve in transit), will measure both registers in the Z basis (with probability p_Z) or the X basis.
3. Bob measures his ancilla B. He next reveals his choice to simulate measure/resend or reflect and Alice reveals her measurement basis choice (neither party reveals any measurement outcome, of course). If Bob chose to measure/resend and Alice chose the Z basis, they will use this round for their raw key – in particular, Bob will measure the B register and Alice will use her measurement outcome from the A_1 register. Other events may be used for noise estimation in the natural way.
4. Goto (1)

of the important concepts of cryptography (quantum and classical): if we can define an alternative protocol that leads to the same final density operators as the one we want to show secure, then we're safe analyzing the alternate. Since this is a one-way protocol, its security is "easier" to handle (that's in quotes since there are still some tricks we will need to do).

To prove that security of the fully quantum one-way protocol (Protocol 7.3) implies security of the SQKD one, we will show that, for *every* attack against the semi-quantum protocol, there exists a possible attack against the one-way protocol such that Alice, Bob, and Eve all have the same density operator at the end of the protocol. In particular, this means that $H(A|E)$ will be the same in both cases and any detectable noise will be identical (i.e., Alice and Bob's measurement distributions will be the same in both the one way and semi-quantum protocol). This proof is from [11], and full details can be found there.

Let's consider collective attacks first. Similar constructions can be made for general attacks [10], though the notation is much more complex and does not lead to too many additional insights that can't be gleaned from only the collective attack scenario. As in Section 6.2, we will model a collective attack against the prepare-and-measure SQKD protocol as two unitary attack operators acting on the traveling qudit and Eve's ancilla. The first, U_F will be applied in the *forward* channel and the second U_R will be applied in the *reverse*. We will show how, given any such pair of attack operators, there exists an equivalent attack against the one-way protocol – by equivalent we mean it will produce the same density operator.

The construction of the attack against the one-way protocol follows arguments from [11]. Without loss of generality, we may write U_F's action on basis states $|a\rangle_T$ (where T will denote a transit qudit) in the following manner:

$$U_F\,|a\rangle = \sum_{b=0}^{d-1} |b, e_{a,b}\rangle_{TE}\,. \tag{7.15}$$

Note that we are writing U_F as an isometry – however isometries can be dilated to unitary operators (their action on unneeded input states may be arbitrary), and thus they fall within the realm of possibility for an adversary. The vectors $|e_{a,b}\rangle$ are held private by the adversary, while the T register continues to Bob (recall, this is the two-way protocol we are first considering). We make no assumptions on Eve's ancilla states, other than those imposed by the fact that U_F is an isometry. In particular, this requires for every $a = 0, \cdots, d-1$:

$$1 = \sum_{b=0}^{d-1} \langle e_{a,b}|e_{a,b}\rangle\,. \tag{7.16}$$

Next, Bob will perform some operation on the T system and return it to Eve. We can now begin constructing our attack against the one-way protocol. Recall this attack will consist of two parts: a probability distribution $\{p(b)\}$ which influences Bob's source device in the one-way protocol (where Bob, now, will be fully quantum). The second part is a unitary attack operator U attacking the one way channel (but it will attack a $2d$ dimensional system whereas U_F and U_R act on d-dimensional transit systems).

We let

$$p(b) = \frac{1}{d}\sum_{a=0}^{d-1} \langle e_{a,b}|e_{a,b}\rangle\,. \tag{7.17}$$

This is a valid probability distribution since $p(b) \geq 0$ for all b (this follows from properties of inner product). Furthermore, Eq. (7.16) implies that:

$$\sum_{b=0}^{d-1} p(b) = \frac{1}{d}\sum_b \sum_a \langle e_{a,b}|e_{a,b}\rangle = \frac{1}{d}\sum_a \left(\sum_b \langle e_{a,b}|e_{a,b}\rangle\right) = 1. \tag{7.18}$$

Now, we define a so-called "rewind" operator Rw which is going to simulate U_F's attack against the SQKD protocol, in the one-way protocol. It, in a way, "rewinds" the state Bob actually creates in the one-way protocol, to simulate the state that Alice would have sent in the SQKD protocol if Bob received (in this case sent) what he did. It will act on the A_1A_2 registers and map them into the A_1A_2E register (it will be an isometry, though we must prove that). The action on basis states of the form $|b,b\rangle$ (for $b \in \{0, 1\cdots, d-1\}$) will be:

$$\texttt{Rw}\,|b,b\rangle_{A_1A_2} = \frac{1}{\sqrt{d\cdot p(b)}}\sum_{a=0}^{d-1} |a,b,e_{a,b}\rangle\,. \tag{7.19}$$

We need to show that the above is an isometry – namely that it preserves inner product. Consider $b \neq b'$. Then:

$$0 = \langle b,b|b',b'\rangle = \frac{1}{d\sqrt{p(b)p(b')}}\sum_{a,a'} \langle a,b,e_{a,b}|a',b',e_{a',b'}\rangle = 0. \tag{7.20}$$

(the last equality follows from the fact that $\langle a,b,e_{a,b}|a',b',e_{a',b'}\rangle = \langle a|a'\rangle\langle b|b'\rangle\langle e_{a,b}|e_{a',b'}\rangle$ and that $b \neq b'$). Furthermore, using Eq. (7.17), we have:

$$1 = \langle b,b|b,b\rangle = \frac{1}{d\cdot p(b)}\sum_{a,a'} \langle a,b,e_{a,b}|a',b,e_{a',b'}\rangle = \frac{1}{d\cdot p(b)}\sum_a \langle e_{a,b}|e_{a,b}\rangle = 1, \tag{7.21}$$

This shows that Rw is an isometry and, thus, can be extended to a unitary operator. Finally, let U (which is the actual attack operator we construct) be $U = (I_{A_1} \otimes U_R)\texttt{Rw}$. That is, first apply the Rw operator, then the U_R operator; our claim is that this attack is equivalent to attacking with U_F then U_R in the two-way protocol.

Why does this work? We need to trace the protocol in both the measure/resend case and the reflect case. Let's consider the reflect case (the measure/resend case we leave as an exercise, though it is also worked out in [11]). Consider the SQKD protocol where Alice prepares $\frac{1}{\sqrt{d}}\sum_a |a,a\rangle_{AT}$, sending the T register to Bob. Eve attacks with U_F evolving the state to:

$$|\phi\rangle = \frac{1}{\sqrt{d}}\sum_a |a\rangle_A \otimes \sum_b |b,e_{a,b}\rangle_{TE}\,. \tag{7.22}$$

Next, Bob reflects, sending the state back to Eve who attacks with U_R, evolving the state to $I_A \otimes U_R|\phi\rangle$, whatever that is (turns out we don't need to explicitly write out the action of U_R which makes this problem more tractable).

Now, consider the one-way protocol, along with the attack constructed above. Bob, in the Reflect case, will prepare and send:

$$\sum_b \sqrt{p(b)}\,|b, b\rangle\,, \tag{7.23}$$

where $p(b)$ is defined in Eq. (7.17). Eve attacks with $U = (I_{A_1} \otimes U_R)\texttt{Rw}$. First, the Rw operator is applied, evolving the state to:

$$\sum_b \sqrt{p(b)} \left(\frac{1}{\sqrt{d \cdot p(b)}} \sum_a |a, b, e_{a,b}\rangle \right) = |\phi\rangle\,. \tag{7.24}$$

Since the above is exactly $|\phi\rangle$ in the SQKD case, applying U_R to the A_2 register will evolve it to the exact same state that would have been produced in the SQKD case. Since the state is identical in both cases, (in terms of Alice, Bob, and Eve), it will produce exactly the same measurement outcomes and uncertainty which the SQKD case would have. Of course this is only reflection, but the math for the measure/resend case is similar.

Exercise 7.5. Prove that the constructed attack works in the measure/resend case. In particular, show that it will produce exactly the same state in both SQKD and one-way protocols in the event Bob chooses to measure/resend.

Now that we've shown that for *every* attack against the SQKD protocol, there exists an equivalent attack against the one-way protocol, we need only to prove security of the one-way protocol. Note that there may be more attacks against the one-way protocol – however, based on our analysis above, if we prove security against the one-way protocol, we are definitely including all attacks against the SQKD one.

To prove security of the one-way protocol, we will use entropic uncertainty (discussed in Section 2.6.2.2) and continuity of von Neumann entropy [21]. Let's look at the reflect case. In this instance, Bob prepares a state unentangled with himself and sends it to Alice (after Eve interacts with it). It would be equivalent in this instance had Eve simply created her own state and sent it directly to Alice. Alice then measures her two systems both in either the computational Z basis, or the alternative X basis. This is exactly HD-BB84 and we may use entropic uncertainty to find:

$$H(A_1^Z|E)_\rho \geq \log_2 d - H(A_1^X|A_2^X). \tag{7.25}$$

Here A_1^Z is the random variable induced from measuring the A_1 register in the Z basis (similar for A_1^X and A_2^X). We will use ρ to denote the density operator in the reflect case. In the measure/resend case, we will use σ to denote the density operator.

Of course, to derive the key-rate, we must determine $H(A_1^Z|E)_\sigma$, namely the entropy in the measure/resend case σ. But as it turns out ρ and σ will not differ "too much" in terms of trace distance. Since entropy is continuous [21], the entropy will also not differ too much! Thus, we will bound the entropy of ρ, which is a reflection case but not used for key distillation; by continuity, we can use this to bound the entropy in σ, the measure-resend case which is the density operator actually storing the raw key.

In particular, if we can find a bound δ (which may – and in fact does in this case – depend on the noise in the channel) such that:

$$\frac{1}{2}||\rho - \sigma|| \leq \delta, \tag{7.26}$$

then it will hold that [21]:

$$|H(A_1^Z|E)_\rho - H(A_1^Z|E)_\sigma| \leq \delta \log_2 d + (1+\delta)h\left(\frac{\delta}{1+\delta}\right). \tag{7.27}$$

Note that if δ is small, the difference between the two entropies will also be small. With some work, bounds on δ can be determined. In particular, if one assumes a depolarizing channel, it can be shown [11] that this trace distance can be upper bounded by a function of the noise and the dimension, such that, when the noise is zero, $\delta = 0$. When the noise increases, δ increases smoothly. Deriving this bound is somewhat technical, and requires analyzing the eigenvalues of certain matrices, so we do not go into the details here. However, an "easy" to evaluate expression for the entropy can be determined (see [11] for details) allowing for graphs such as Fig. 7.4 to be created. In particular, this shows HD states can also benefit alternative models of quantum cryptography (i.e., where users are "weakly" quantum or semi-quantum). Note that the key-rate and noise tolerance are not as high as the HD-BB84 protocol we compare with. However, that is not surprising, considering semi-quantum protocols place device restrictions on users. Protocols all have different advantages and disadvantages – indeed, some of the advantages may be from a purely theoretical point of view (e.g., for understanding "how quantum" protocols must be to gain information theoretic security, which is perhaps the most important reason to study semi-quantum cryptographic systems).

7.4 Cryptographic applications of quantum random walks

An interesting theoretical investigation involves the cryptographic uses of quantum random walks. This can be considered another HD-QKD system, since quantum walks involve states living in a high-dimensional state space.

Such walks, which are, unsurprisingly given their name, quantum analogues of classical random walks [12,18], can be used for various quantum cryptographic applications [13,20,19,1,17]. We'll first take a look at the basics of quantum walks, in particular their construction and simulation. There is a very rich theoretical foundation for quantum random walks which lies greatly out of scope of this book, however the interested reader may find their further study enlightening; in this text, we will only cover the very basics of one type of quantum walk. Following this, we'll discuss how they can be used for cryptographic applications.

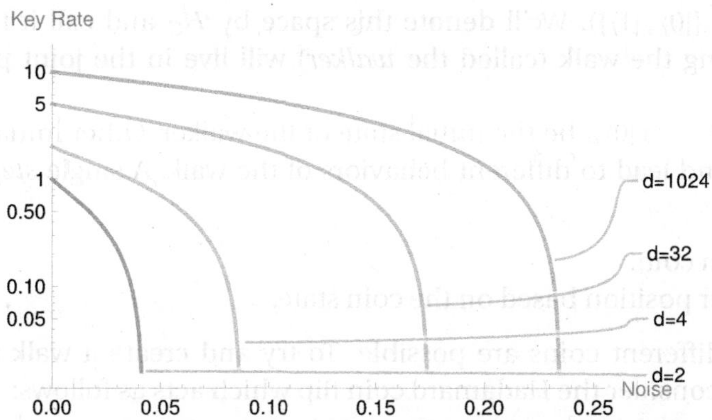

FIGURE 7.4 Showing the key-rate of the HD-SQKD protocol, using results from [11]. Note that, as with the "fully quantum" setting, increasing the dimension can increase the theoretical key-rate and noise tolerance (at least over depolarizing channels). Note that the SQKD protocol is not as efficient as the standard HD-BB84 protocol, discussed in Section 7.1 (see, also, Fig. 7.1). However, this is not surprising, as semi-quantum protocols place several restrictions on the users.

7.4.1 Quantum random walks

Let's consider, first, a basic classical random walk on the number line. Suppose you start at position $x = 0$ and you flip a coin. Heads, you move left to position $x = -1$; tails and you move right to position $x = 1$. You repeat the process again, moving left or right one unit each time, with equal probability. After T steps, where do you land? Well this is a probability distribution, namely $P(x|T)$ and as it turns out this distribution is Gaussian with a mean at the origin and a variance of $\sigma^2 = T$ [12].

Now, let's consider a quantum analogue. We'll consider a "discrete coin" based model (there are also continuous time models [12]). Now, instead of a classical walker moving left or right, we'll have a quantum particle that will move left and right in superposition. The particle will continue to move in superposition, essentially existing at multiple positions simultaneously. At the end of the walk (i.e., after moving for T steps), we'll measure the particle's position – this induces a distribution $P^Q(x|T)$ (note the "Q" superscript to distinguish the Quantum probability from the previously mentioned normal distribution). As it turns out, the quantum induced distribution is far from a normal distribution as it was in the classical case. Instead, the particle tends to find itself at an extreme edge!

Let's see how to model this properly. We need a Hilbert space to model the position of the particle. Because we want to keep all systems finite dimensional in this book, let T be the maximum number of steps we will evolve the particle (e.g., $T = 1000$). In this case, the particle can be in any integer position between -1000 and 1000 (it cannot be in position 1001 for instance since it can only move left or right one unit each step). Thus, we may define a $2T + 1$ dimensional Hilbert space $\mathcal{H}_P$ which we'll call the *Position Space*. It will be spanned by $\{|x\rangle : x = -1000, \cdots, 1000\}$. (In general, it is spanned by $\{|x\rangle : x \in \mathbb{Z}\}$.)

Now, we will need a space to model the "coin" flip which determines whether the particle will move left or right. This will be a two dimensional state spanned by $\{|L\rangle, |R\rangle\}$

(equivalently just $\{|0\rangle, |1\rangle\}$). We'll denote this space by $\mathcal{H}_C$ and call it the *Coin Space.* A particle performing the walk (called the *walker*) will live in the joint position and coin space.

Let $|\psi_0\rangle_{CP} = |L\rangle_C \otimes |0\rangle_P$ be the initial state of the walker. Other initial positions are of course possible and lead to different behaviors of the walk. A single *step* of the walk will consist of:

1. Flip a quantum coin.
2. Shift the walker position based on the coin state.

For the first, different coins are possible. To try and create a walk analogous to the classical one, we consider the Hadamard coin flip which acts as follows:

$$H|L\rangle = \frac{1}{\sqrt{2}}|L\rangle + \frac{1}{\sqrt{2}}|R\rangle$$
$$H|R\rangle = \frac{1}{\sqrt{2}}|L\rangle - \frac{1}{\sqrt{2}}|R\rangle$$

One can think of this as a walker deciding to move left or right with equal probabilities – however, the reader should know by now that these amplitudes can add and cancel each other out, thus it will not result in classical behavior.

Following this coin flip operator, applied only to the coin space, a *Shift* operator is applied which will move the walker's position based on the coin state. In particular, this operator will map $|L\rangle_C \otimes |x\rangle_P$ to $|L\rangle_C \otimes |x-1\rangle_P$; a similar action is found when the coin space is $|R\rangle$ in which case the position increases by one. There are two concerns: what if $x = -T$ (or T) in which case we cannot move further left (or right)? Second, this operator needs to be unitary. In general, the position Hilbert space is infinite dimensional and there will always be a position the walker can move to; however in this case, since we know we'll always stop the walk after T steps, we don't have to worry about what happens and we can define the operator arbitrarily. Let's define it so that $|L, -T\rangle_{CP}$ maps to $|L, T\rangle$ and that it maps $|R, T\rangle$ to $|R, -T\rangle$. I.e., it wraps the line around to itself, creating a cycle (which is actually an important walk model we'll look at later anyway). Given this, we may define the shift operator, denoted S, using operator sum notation, as:

$$S = |L\rangle\langle L|_C \otimes \left(|T\rangle\langle -T| + \sum_{x=-T+1}^{T} |x-1\rangle\langle x| \right) + |R\rangle\langle R|_C \otimes \left(|-T\rangle\langle T| + \sum_{x=-T}^{T-1} |x+1\rangle\langle x| \right) \tag{7.28}$$

The reason for using operator sum notation is to easily prove unitarity. But before doing so, let's show it's correct – namely it correctly models the behavior described above. Given a basis vector $|L, x\rangle$, it is not difficult to see that $S|L, x\rangle = |L, x-1\rangle$ (assuming $x \geq -T+1$, otherwise the output will be $|L, T\rangle$ which is also as discussed). Similarly, one may verify the correctness of the operator if the coin space is in the $|R\rangle$ state.

Exercise 7.6. Show that the shift operator, defined above, is correct when the coin is in the $|R\rangle$ state.

For unitarity, we must show that $\mathcal{S}\mathcal{S}^* = I$. The adjoint $\mathcal{S}^*$ is easily seen to be:

$$\mathcal{S}^* = |L\rangle\langle L|_C \otimes \left(|-T\rangle\langle T| + \sum_{x=-T+1}^{T} |x\rangle\langle x-1|\right) + |R\rangle\langle R|_C \otimes \left(|T\rangle\langle -T| + \sum_{x=-T}^{T-1} |x\rangle\langle x+1|\right) \tag{7.29}$$

Then, it is not difficult to see that:

$$\mathcal{S}^*\mathcal{S} = |L\rangle\langle L| \otimes \left(|-T\rangle\langle -T| + \sum_{x=-T+1}^{T} |x\rangle\langle x|\right) + |R\rangle\langle R| \otimes \left(|T\rangle\langle T| + \sum_{x=-T}^{T-1} |x\rangle\langle x|\right) \tag{7.30}$$

$$= |L\rangle\langle L| \otimes I_P + |R\rangle\langle R| \otimes I_P, \tag{7.31}$$

where I_P is the identity operator on the position space (due to the completeness relation, discussed in Section 2.1.3). Finishing:

$$|L\rangle\langle L| \otimes I_P + |R\rangle\langle R| \otimes I_P = (|L\rangle\langle L| + |R\rangle\langle R|) \otimes I_P = I_C \otimes I_P = I_{CP}, \tag{7.32}$$

where, similar to before, I_C is the identity on the two dimensional coin space and I_{CP} is the $2 \times (2T+1)$ dimensional identity on the joint coin and position space. One readily verifies that $\mathcal{S}\mathcal{S}^* = I$ also. It is therefore a unitary operator and, thus, something that may actually be done to a quantum particle.

The total evolution of the quantum state, then, is to first apply $H \otimes I_P$ (i.e., the coin operator on the coin space, and identity on the position space) followed by $\mathcal{S}$. After T steps, if the initial state of the walker was $|\psi_0\rangle$, then the new state will be:

$$|\psi_T\rangle = (\mathcal{S} \cdot (H \otimes I_P))^T |\psi_0\rangle. \tag{7.33}$$

After evolving for T steps, a measurement is made on the walker's position. How does this distribution compare to the classical case? We work out the relevant distributions for the first five steps in Table 7.1 (for the classical case) and Table 7.2 (for the quantum case). In Exercise 7.7 you are asked to write out the explicit state vector for these four steps to compare and to show *why* the quantum case is behaving differently (in particular, when the walker returns to a position, since it is traveling in superposition, some of the waveform may suffer destructive interference). In Exercise 7.8 you are tasked with writing a simulator to see how the walker evolves over a larger time window, T.

Exercise 7.7. Write out the state vector for the first five steps of the walk. Compare your answer with Table 7.2.

Exercise 7.8. Write a computer simulation to determine the distribution after T steps. Test your simulator with $T = 5$ and compare with Table 7.2. Then try $T = 100$ and compare with results in [12]. Finally, try different initial positions to see how the distribution changes.

Table 7.1 Showing the distribution of a classical random walk on the line for the first five steps. Each row represents a new time step (starting at time zero where the walker is at position 0) while each column represents the probability that the walker arrives at the given position. Note that the distribution is normal. Table recreated from [12].

	−5	−4	−3	−2	−1	0	1	2	3	4	5
0						1					
1					1/2		1/2				
2				1/4		1/2		1/4			
3			1/8		3/8		3/8		1/8		
4		1/16		1/4		3/8		1/4		1/16	
5	1/32		5/32		5/16		5/16		5/32		1/32

Table 7.2 Showing the distribution of the quantum walk for initial state $|L\rangle|0\rangle$. Each row represents a time step while each column represents the probability of observing the walker at that position, if a measurement were to be made at the corresponding time. Note that, initially, the walk follows a normal distribution, however this rapidly changes and the walker becomes skewed. The bias can be altered by changing the initial walk state, $|\psi_0\rangle$. Table recreated from [12].

	−5	−4	−3	−2	−1	0	1	2	3	4	5
0						1					
1					1/2		1/2				
2				1/4		1/2		1/4			
3			1/8		5/8		1/8		1/8		
4		1/16		5/8		1/8		1/8		1/16	
5	1/32		17/32		1/8		1/8		5/32		1/32

7.4.2 Quantum walk based key distribution

Quantum walks can be used for key distribution; here, we'll discuss a protocol and proof introduced in [19]. The idea is to "hide" the key information in a quantum walk's initial state. Exercise 7.8 demonstrates how slight changes to a walker's initial state can greatly affect the final evolved waveform. The second critical element of these systems is that, since the walk operator (Eq. (7.33)) is unitary, it is also reversible, assuming you know the walk parameters.

While there are several potential QW based QKD protocols, the most basic one (which is also provably secure using entropic uncertainty relations), involves Alice and Bob agreeing on walk parameters, where the walk operates over a cycle of size P. The agreed upon parameters include the coin operator, number of iterations T, and the number of nodes

P in the cycle. The walk operator W is then defined to be:

$$W = (S \cdot C \otimes I_P)^T. \tag{7.34}$$

Recall from Section 2.4.1, that unitary operators are, essentially, unitary changes in basis. That is, consider the standard computational basis $\{|0\rangle, |1\rangle, \cdots, |2P-1\rangle\}$ (note that the total dimension of the system is $2P$, since there are P nodes in the cycle, and the coin space is dimension two). Then, we can define the *walk basis* to be spanned by basis states $|w_i\rangle = W|i\rangle$. In this case, we now have, essentially, a BB84 style protocol where, for each round of the quantum communication stage, Alice and Bob perform the following operations:

1. Alice chooses to send either a computational state $|i\rangle$ or a walk state $|w_i\rangle = W|i\rangle$, for a randomly chosen $i \in \{0, \cdots, 2P-1\}$.
2. Bob, on receipt of a quantum state, will choose to measure in either the computational basis, or the walk basis. Let j be the resulting outcome (with an observation of $|w_j\rangle$ being translated to an outcome of j).
3. Alice and Bob disclose their choice of basis. If they do not match, this round is discarded. Otherwise, they keep their outcomes i and j for their raw keys (adding $\log_2(2P)$ bits each round).

The above repeats a sufficient number of times. As usual, parties will disclose full information on a select number of randomly chosen rounds allowing them to estimate the noise in the channel in both the computational and walk basis. Finally, parties run error correction and privacy amplification to distill their final secret key.

Exercise 7.9. Explain how Bob can measure in the walk basis if he knows the walk parameters (i.e., he knows how W is implemented).

How can we prove security? Normally, one would have to determine how much information Eve could extract from a complicated walker state. However, as it turns out, thanks to entropic uncertainty relations (see Section 2.6.2.2), it's actually very straightforward! First, we need to define an equivalent entanglement based version. Alice will prepare a high dimensional Bell state of the form:

$$\frac{1}{\sqrt{2P}} \sum_{a=0}^{2P-1} |a, a\rangle, \tag{7.35}$$

sending the second system to Bob. In the absence of noise, it can be readily verified that if both parties measure in the computational basis they will observe the same outcome; if they both measure in the walk basis, they will also observe the same outcome. Finally, it can only help the adversary if we let her (Eve) create the entanglement and send one system to Alice and one to Bob. Note that the systems sent to parties are both of dimension $2P$ and, thus, the total dimension of the system, not counting Eve, is $(2P)^2$.

Now, let's assume that Alice and Bob only use the computational basis for their raw key, and walk basis states are used for estimating the noise in the channel, similar to the function of the X basis in the biased version BB84. In that case, we need to compute $H(A_Z|E)$, where A_Z is the random variable storing Alice's computational basis measurement outcome. Entropic uncertainty, (see Section 2.6.2.2) however, states that [3]:

$$H(A_Z|E) + H(A_W|B_W) \geq -\log c, \tag{7.36}$$

where A_W and B_W are the random variables storing Alice (respectively Bob's) measurement outcomes in the walk basis, and where c is the maximal overlap:

$$c = \max_{i,j} |\langle i|w_j\rangle|^2. \tag{7.37}$$

The key-rate, then can be derived using the Devetak-Winter keyrate expression (discussed in Section 5.3.1), and can be found to be:

$$H(A_Z|E) - H(A_Z|B_Z) \geq -\log c - H(A_W|B_W) - H(A_Z|B_Z) \tag{7.38}$$

Above, $H(A_W|B_W)$ and $H(A_Z|B_Z)$ can both be determined by Alice and Bob by sampling a suitable subset of their measurement results, to determine the error rate in the Z and W(alk) basis. The value of c is determined by the walk itself and can be evaluated by simulating the walk over all input states, and computing the maximal overlap. Namely, by simulating the state vector $|w_j\rangle = W|j\rangle$ for all j, and then determining the maximal probability of the walk being observed at position $|i\rangle$ if a Z basis measurement were to have been performed. This can all be done by computer simulation – at least for walks that are not "too large." For larger walks, one would require an expression for c based only on the walk parameters (as opposed to a numerical value for c); finding such an expression is an open question. Interestingly, as with other HD-QKD protocols, the noise tolerance of the system tends to increase as the dimension of the quantum walk increased, assuming optimal parameter choices for the walk itself were made. For more details see [19].

Exercise 7.10. Explain how c is related to the probability of observing the walker at a particular position, and coin state, after evolving via operator W at some starting position $|i\rangle$ (in the Z basis).

Exercise 7.11. Write a computer program to simulate the quantum walk described above and use it to calculate c for various walk parameters.

Other cryptographic applications of quantum walks exist. In particular, it was also shown that one can construct *quantum random number generators* (QRNG) from quantum walks. We discuss QRNGs in Section 8.1, however at a high level, now, these are one party protocols whose goal is to distill uniform randomness from a quantum source, that is also secure in the sense that an adversary system has no information on the secret randomness. Various QRNG protocols utilizing quantum walks exist; for more details, see

[14,1]. One can also create more "advanced" cryptographic protocols; for instance, in [13], a scheme supporting a limited form of homomorphic encryption was developed, using quantum walks as the backbone. Quantum walks are fascinating from a theoretical perspective; that they also have cryptographic applications is interesting. The study of such applications may lead to breakthroughs in quantum random walk research and vice versa.

7.5 Group key protocols

The last high dimensional system we'll look at are group key protocols, commonly called *quantum conference key agreement* (QCKA) protocols. These are actually multi-party protocols and the goal here is for all parties to agree on the same secret key (i.e., for $p \geq 3$ parties to all share the same secret key). There are two variants to this: qubit based systems and high-dimensional systems. Technically, qubit based systems require GHZ states (which we consider an HD resource) and so we will consider these first. Later we'll discuss the case when we use higher-dimensional GHZ states.

A GHZ state with p qubits is a state of the form:

$$|g^p(\underbrace{0\cdots0}_{p-1\text{ times}},0)\rangle = \frac{1}{\sqrt{2}}|\underbrace{0\cdots0}_{p\text{ times}}\rangle + \frac{1}{\sqrt{2}}|1\cdots1\rangle. \tag{7.39}$$

We will refer to the first qubit as the A register, and the remaining $p-1$ qubits as the $B = B_1B_2\cdots B_{p-1}$ register(s). In fact, we may define the so-called *GHZ basis* as $|g^p(x,y)\rangle$, where $x \in \{0,1\}^{p-1}$ and $y \in \{0,1\}$. These states are defined to be:

$$|g^p(x,y)\rangle = \frac{1}{\sqrt{2}}|0\rangle_A|x\rangle_B + (-1)^y|1\rangle_A|\bar{x}\rangle_B, \tag{7.40}$$

where $\bar{x}$ is the bit-wise complement of x (i.e., $\bar{x} = x \oplus (1\cdots1)$). Note that, when $p = 2$, these become, simply the standard Bell states and Bell basis.

Exercise 7.12. For any $p \geq 3$, show that the vectors described in Eq. (7.40) form an orthonormal basis of dimension 2^p.

Imagine p parties, denoted Alice, Bob$_1$, Bob$_2$, $\cdots$, Bob$_{p-1}$ are given a GHZ state $|g^p(0,0)\rangle$. If all parties were to measure in the computational basis, they would share a correlated outcome: namely, they would all share a "0" (with 50% probability) or they would all share a "1." This is going to be how they distill a key. However, similar to BB84, how do we check for eavesdroppers? For instance, what if there is a third party Eve, with register E, and the joint state is actually $|g^{p+1}(0,0)\rangle_{ABE}$? Alice and the $p-1$ Bob's will still share a random key bit, however Eve will too!

Recall from BB84, or, more to our point, the entanglement based version discussed in Section 4.3.2. There, parties must measure in the X basis and check the error rate there to determine how much information Eve can have. This worked due to the readily verified fact that:

$$\frac{1}{\sqrt{2}}(|00\rangle + |11\rangle) = \frac{1}{\sqrt{2}}(|{++}\rangle + |{--}\rangle), \tag{7.41}$$

and so any measurement in the Z or X basis should produce a correlated outcome. However, when $p \geq 3$, measuring in the X basis will not produce the same outcomes across all parties. For instance, when $p = 3$ it can be readily shown that:

$$\frac{1}{\sqrt{2}}|000\rangle + \frac{1}{\sqrt{2}}|111\rangle \neq \frac{1}{\sqrt{2}}|{+}{+}{+}\rangle + \frac{1}{\sqrt{2}}|{-}{-}{-}\rangle \tag{7.42}$$

However, even though when measuring in the X basis, the p parties do not all share the same exact outcome, there is a relation. Let $x_j \in \{0, 1\}$ be the X basis measurement outcome of party j (with Alice being party 0 and with an observation of $|+\rangle$ implying $x_j = 0$ and $|-\rangle$ implying $x_j = 1$). Then, if given a p-qubit GHZ state $|g^p(0, 0)\rangle$, it will hold that $\sum_{j=0}^{p-1} x_j \equiv 0(\bmod 2)$. It can also be proven in general, though we leave that as an exercise.

Exercise 7.13. Prove that for any $p \geq 3$ (actually for $p = 2$ also), it will hold that the sum of all X basis measurements of $|g^p(0, 0)\rangle$ will be an even number. More generally, show that if the measurement is made of $|g^p(x, y)\rangle$, for $x \in \{0, 1\}^{p-1}$ and $y \in \{0, 1\}$, then the sum of the X (Hadamard) basis outcomes, modulo two, will be equal to y.

This gives us enough to build the QCKA protocol shown in Protocol 7.4 and originally introduced in [9]. (Interestingly, a third basis can also be used and that was the first developed QCKA protocol in [8], though we will not discuss this variant, here.)

Note the need for pre-shared randomness to allow parties to choose their bases without having any mismatch. This is important as, otherwise, the key-rate would collapse to zero rapidly as the number of users increases (since the probability of all parties measuring in the same basis, needed to get a key-bit, is $p_Z^p \to 0$ as the number of parties $p \to \infty$). The size of this pre-shared secret string is approximately $N \cdot h(1 - p_Z)$ which may be made small if the bias towards Z is high.

How do we prove security? We already have an entanglement based version (we simply assume Eve creates the GHZ states, not Alice) and so there's no need for a reduction. As it turns out entropic uncertainty can work here which was the proof in [9]. We can also use the Bouman Fehr sampling theorem [4], used in Section 5.4.2, to prove security. The latter is the method we'll discuss here, though we will only provide a high-level overview of the proof, leaving some of the technical details as exercises.

7.5.1 Security proof

To prove security, we will need a bound on the quantum min entropy of the system. We'll assume Eve creates the initial GHZ states. To bound the min entropy, we'll use Bouman and Fehr's sampling method, discussed in Section 5.4.2.

To use Bouman and Fehr's work, we need a sampling strategy. Let's say Eve gave the parties N GHZ states (so there are $N \times p$ qubits total). A subset t is chosen, and an X basis

Protocol 7.4 GHZ-based QCKA Protocol: [9].

Public Parameters:

- p: number of parties with a designated leader party "Alice" and follower parties "Bob"
- N: number of rounds
- $m \leq N$: the number of X basis rounds; other rounds will be Z basis rounds

Private Parameters:

- All parties share a small, private, random seed, which they will use to agree on basis choices. This will be refreshed after each session with part of the newly produced secret key. See text for discussion.

Quantum Communication Stage: For $i = 1, 2, \cdots, N$, do:

1. Alice creates N copies of the GHZ state $|g^p(0,0)\rangle$ and sends one qubit to each Bob for each of the N states. Later, we will assume Eve creates this state, entangled with her ancilla, and sends qubits to Alice and each Bob.
2. Based on their private random string, parties choose a subset of the N GHZ states to measure in the X basis. This results in string $q^j \in \{0,1\}^m$ where m is the size of the subset chosen, and j is the party number ($j = 0$ being Alice).
3. The remaining qubits are measured in the Z basis resulting in strings $r^j \in \{0,1\}^n$, where $n = N - m$. This will be their raw key.

Each party $j = 0, \cdots, p-1$ discloses their X-basis measurement string q^j and compute $q = w(q^0 \oplus \cdots \oplus q^{p-1})$, where $w(x)$ is the relative Hamming weight function. Ideally this should be zero (see Exercise 7.13). Higher values of q will be considered noise.
From q, parties determine the amount of information Eve has on their raw key measurements r^j. Error correction and privacy amplification on the raw key are performed. Error correction is done by Alice sending correction information to each Bob.

measurement is performed on all qubits indexed. The result of these measurements are sent to Alice, and the sum, modulo two, is computed by all parties. Ideally, this should be the zero string (see Exercise 7.13). Though, there will be some noise. However, it should hold that the observed noise in the sampled subset should be nearly the same as that in the unsampled portion, *were it to have been measured.* With this in mind, we define our "good words" (see Section 5.4.2 for more details on this sampling method) to be:

$$\mathcal{G}^t = \{(a, b^{(1)}, b^{(2)}, \cdots, b^{(p-1)}) \in \{0,1\}^N \times \cdots \times \{0,1\}^N$$
$$: |w(a_t \oplus b_t^{(1)} \oplus \cdots \oplus b_t^{(p-1)}) - w(a_{-t} \oplus b_{-t}^{(1)} \oplus \cdots \oplus b_{-t}^{(p-1)})| \leq \delta\}. \quad (7.43)$$

Namely, a good word is one where the sum, modulo two, of the observed portions is nearly the same as that in the unobserved portion. What is the error probability (see Eq. (5.44)) of this strategy? We leave the proof of the following lemma as an exercise:

Lemma 7.1. *Consider the above (classical) sampling strategy. Then for any word* $q = (a, b^{(1)}, \cdots, b^{(p-1)}) \in \{0,1\}^N \times \cdots \times \{0,1\}^N$, *it holds that:*

$$Pr(q \notin \mathcal{G}^t) \leq 2\exp\left(\frac{-\delta^2 mN}{N+2}\right) = \epsilon_\delta^{cl}. \tag{7.44}$$

Exercise 7.14. Prove Lemma 7.1. Hint: Try reducing the strategy to one that we analyzed in Section 5.4.2.1.

To promote this to a quantum subspace, we will define the spanning set with respect to the Hadamard basis, namely:

$$\mathcal{G}_Q^t = \text{span}(|a, b^{(1)}, \cdots, b^{(p-1)}\rangle^X \; : \; (a, b^{(1)}, \cdots, b^{(p-1)}) \in \mathcal{G}^t) \otimes \mathcal{H}_E \tag{7.45}$$

Above, we use $|a, b^{(1)}, \cdots, b^{(p-1)}\rangle^X$ (a superscript X) to mean the concatenated string $a||b^{(1)}||\cdots$ written in the X basis, as was done in Section 5.4.2.1.

From this, we can apply Bouman and Fehr's main sampling result (see Theorem 5.1) to show that there exist ideal states of the form $|\phi^t\rangle$, where each $|\phi^t\rangle \in \mathcal{G}^t(X)$ and such that:

$$\frac{1}{2}\left\|\sum_t P_T(t)\,|t\rangle\langle t| \otimes |\psi\rangle\langle\psi| - \sum_t P_T(t)\,|t\rangle\langle t| \otimes |\phi^t\rangle\langle\phi^t|\right\| \leq \sqrt{\epsilon_\delta^{cl}}, \tag{7.46}$$

where ϵ is given in Lemma 7.1, and P_T is the uniform distribution over subsets $t \subset \{1, \cdots, N\}$ of size m.

As in Section 5.4.2.1, we will first determine a bound on the quantum min entropy of the ideal state and then use this to analyze the actual state. For this, let's trace the execution of the protocol on ideal states. Running the sampling stage, and observing a particular q (the XOR sum of all X basis measurements) will cause the ideal state to collapse to something of the form:

$$|\phi^{t,q}\rangle = \sum_{(a,b^{(1)},\cdots,b^{(p-1)})\in J_q} |a, b^{(1)}, \cdots, b^{(p-1)}\rangle^X \, |E^{t,q}_{a,b^{(1)},\cdots,b^{(p-1)}}\rangle, \tag{7.47}$$

where:

$$J_q = \{(a, b^{(1)}, \cdots, b^{(p-1)}) \in \{0,1\}^{n\times p} \; : \; |w(a \oplus b^{(1)} \oplus \cdots \oplus b^{(p-1)}) - q| \leq \delta\}, \tag{7.48}$$

and where $n = N - m$ and m is the size of the subset $|t|$. Note that the Eve vectors, above, are sub-normalized.

Since we want $H_\infty(A|E)$, we trace out all the B systems, yielding:

$$\rho_{AE} = \sum_{b^{(1)},\cdots,b^{(p-1)}\in\{0,1\}^n} Pr(b)P\left(\sum_{a\in J_{q,b}} |a\rangle^X \, |E_a^{t,q,b}\rangle\right) \tag{7.49}$$

where, we use the shortcut b to mean $b^{(1)}, \cdots, b^{(p-1)}$ and where:

$$J_{a,b} = \{a \in \{0,1\}^n \ : \ |w(a \oplus b^{(1)} \oplus \cdots \oplus b^{(p-1)}) - q| \le \delta\}. \tag{7.50}$$

Taking advantage of certain properties of min entropy, in particular the fact that the min entropy of a mixture of the form above can be bounded by the min entropy of the "worst case" outcome, we have:

$$H_\infty(A|E) \ge \min_b H_\infty(A|E)_{\tau^b}, \tag{7.51}$$

where:

$$\tau^b = P\left(\sum_{a \in J_{q,b}} |a\rangle^X |E_a^{t,q,b}\rangle\right). \tag{7.52}$$

Finally, using the fact that the min entropy of a superposition may be bounded using the min entropy of a mixed state, up to a function of the size of the superposition set, (see Eq. (2.127) in Section 2.6.2), we conclude:

$$H_\infty(A|E) \ge n(1 - \max_b \log|J_{q,b}|) \ge n(1 - h(q+\delta)), \tag{7.53}$$

where we used the bound on the volume of a Hamming ball to bound $|J_{q,b}|$ by $2^{nh(q+\delta)}$. This gives us a bound on the min entropy of the ideal state. We may use Lemma 5.2 to finish the proof, similar to how we proved BB84 in Section 5.4.2.1.

Exercise 7.15. Complete the proof of security by deriving a complete key-rate expression for this protocol. Compare to that derived in [9] using standard entropic uncertainty relations.

7.6 Closing remarks

This chapter introduced some of the many high-dimensional quantum cryptographic protocols available today. Many others exist, and we could only cover the tip of the iceberg here. The interested reader should consider the survey [7] for further reading. High-dimensional states offer some interesting benefits over qubit based systems, including the potential for higher noise tolerance, greater efficiency, and interesting applications, making their study highly relevant.

References

[1] Minwoo Bae, Walter O. Krawec, Semi-source independent quantum walk random number generation, in: 2021 IEEE Information Theory Workshop (ITW), IEEE, 2021, pp. 1–6.

[2] Charles H. Bennett, Gilles Brassard, Quantum cryptography: public key distribution and coin tossing, in: Proceedings of IEEE International Conference on Computers, Systems and Signal Processing, vol. 175, New York, 1984.

[3] Mario Berta, Matthias Christandl, Roger Colbeck, Joseph M. Renes, Renato Renner, The uncertainty principle in the presence of quantum memory, Nature Physics 6 (9) (2010) 659–662.
[4] Niek J. Bouman, Serge Fehr, Sampling in a quantum population, and applications, in: Annual Cryptology Conference, Springer, 2010, pp. 724–741.
[5] Michel Boyer, Dan Kenigsberg, Tal Mor, Quantum key distribution with classical Bob, Physical Review Letters 99 (140501) (2007).
[6] Nicolas J. Cerf, Mohamed Bourennane, Anders Karlsson, Nicolas Gisin, Security of quantum key distribution using d-level systems, Physical Review Letters 88 (12) (2002) 127902.
[7] Daniele Cozzolino, Beatrice Da Lio, Davide Bacco, Leif Katsuo Oxenløwe, High-dimensional quantum communication: benefits, progress, and future challenges, Advanced Quantum Technologies 2 (12) (2019) 1900038.
[8] Michael Epping, Hermann Kampermann, Dagmar Bruß, et al., Multi-partite entanglement can speed up quantum key distribution in networks, New Journal of Physics 19 (9) (2017) 093012.
[9] Federico Grasselli, Hermann Kampermann, Dagmar Bruß, Finite-key effects in multipartite quantum key distribution protocols, New Journal of Physics 20 (11) (2018) 113014.
[10] Julia Guskind, Walter O. Krawec, Mediated semi-quantum key distribution with improved efficiency, Quantum Science and Technology 7 (3) (2022) 035019.
[11] Hasan Iqbal, Walter O. Krawec, High-dimensional semiquantum cryptography, IEEE Transactions on Quantum Engineering 1 (2020) 1–17.
[12] Julia Kempe, Quantum random walks: an introductory overview, Contemporary Physics 44 (4) (2003) 307–327.
[13] Peter P. Rohde, Joseph F. Fitzsimons, Alexei Gilchrist, Quantum walks with encrypted data, Physical Review Letters 109 (15) (2012) 150501.
[14] Anupam Sarkar, C.M. Chandrashekar, Multi-bit quantum random number generation from a single qubit quantum walk, Scientific Reports 9 (1) (2019) 12323.
[15] Toshihiko Sasaki, Yoshihisa Yamamoto, Masato Koashi, Practical quantum key distribution protocol without monitoring signal disturbance, Nature 509 (7501) (2014) 475–478.
[16] Boris Skoric, A short note on the security of round-Robin differential phase-shift qkd, IACR Cryptology ePrint Archive, 2017.
[17] S. Srikara, C.M. Chandrashekar, Quantum direct communication protocols using discrete-time quantum walk, Quantum Information Processing 19 (9) (2020) 295.
[18] Salvador Elías Venegas-Andraca, Quantum walks: a comprehensive review, Quantum Information Processing 11 (5) (2012) 1015–1106.
[19] Chrysoula Vlachou, Walter Krawec, Paulo Mateus, Nikola Paunković, André Souto, Quantum key distribution with quantum walks, Quantum Information Processing 17 (11) (2018) 288.
[20] Chrysoula Vlachou, J. Rodrigues, Paulo Mateus, N. Paunković, Andre Souto, Quantum walk public-key cryptographic system, International Journal of Quantum Information 13 (07) (2015) 1550050.
[21] Andreas Winter, Tight uniform continuity bounds for quantum entropies: conditional entropy, relative entropy distance and energy constraints, Communications in Mathematical Physics 347 (1) (2016) 291–313.

8 Alternative security models and cryptographic primitives

So far, we have discussed QKD, including "basic" protocols, and also more advanced ones. However, while the field of quantum cryptography used to be synonymous with QKD, it is, by now, a much wider and more vast area of study, covering many more cryptographic primitives. Here, we will look at some of these primitives, and also investigate what can be done in alternative security models (moving away from the "perfect security" notion, to models where the adversary is bounded in some way, either physically or computationally). As it turns out, once one is willing to place a few assumptions on the adversary, many more new, and exciting, avenues open up (much like in the classical case, as discussed in Chapter 3).

8.1 Quantum random number generation

One of the, perhaps more obvious, applications of quantum information is in *random number generation*. Having access to random numbers is critical in cryptography (along with many other fields and applications). For instance, classical encryption requires randomness to generate secret keys; QKD requires true randomness (not pseudorandomness, as introduced in Section 3.1.2), to make protocol choices (e.g., what state to send, what basis to measure in, and so on). We have waved our hand over this issue up to this point, and simply assumed a source, or pool, of randomness existed. Now we will discuss how quantum measurements and states can be used to actually produce this randomness through a so-called *quantum random number generator* (QRNG).

Early random number sources based on quantum effects include generators based on radioactive decay [31,52] and noise in electronic circuits [29,46]. We refer the reader to an excellent survey in [27] for more details on these systems. We will focus more on optical QRNGs and their theoretical proofs of security.

One of the most basic forms of QRNG involves the preparation of the state $|+\rangle$ and its subsequent measurement in the Z basis. The protocol works over several rounds (as with QKD): on each round, a source device prepares a $|+\rangle$ state and sends it to the measurement device. The measurement device implements a Z basis measurement, reporting the outcome. The final n-bit output is the random string.

However, this basic protocol suffers some drawbacks that need to be addressed before the resulting string can be used as a true random bit-stream. First, the detectors used have numerous device imperfections and introduce noise themselves. For example, detector dead time (which is the time the detector is "blind" after seeing a photon)

Quantum Communication and Cryptography. https://doi.org/10.1016/B978-0-44-321532-2.00015-2

and after-pulsing (when a detector clicks again after seeing a photon, even if none are there) can lead to correlations in subsequent bits. Furthermore, not all detectors are the same and detectors do not always see a photon when it hits it – this all leads to biases in the final output string. In general, some form of post-processing is required to "smooth out" the raw randomness extracted from the measurement device.

Moving beyond such practical concerns as detector inefficiencies, there are also important *cryptographic* concerns to deal with. In particular, how does the user know the measurement is being done correctly? How does the user know the source is preparing $|+\rangle$ and not, say, a classical $\frac{1}{2}|0\rangle\langle 0| + \frac{1}{2}|1\rangle\langle 1|$ using pre-recorded randomness that the adversary shares? In particular, given a QRNG device, how can we *certify* that the output is actually uniform random and secret, in the sense that the random string is independent of any adversary system?

To answer the above, we often consider several different security models, though here, we will only discuss three. The weakest security model (weakest from a user perspective) is the *trusted device* scenario. Here, Alice, the user of the QRNG device, trusts everything about the source and measurements, though the devices may have practical device inefficiencies leading to biased output if using only the raw data.

Next is the *Source Independent* model, where the source device is untrusted (it may be, for instance, a quantum server providing quantum states to measurement devices owned by the users). In this model, the measurement devices are trusted (though, again, may have practical device imperfections). The source being measured may be fully entangled with an adversary system and the goal will be to extract a uniform random string, independent of any adversary (including any adversary controlling the source device).

Third, is the *Device Independent* model, where both the source and the measurement devices are untrusted. This provides the strongest security guarantee, as Alice need not trust anything about the QRNG device (though she must provide input to the device to test it).

In the trusted device scenario, users may trust that their source and measurement devices are honest, though they may be biased and, so, the raw output of the measurement result cannot be used directly. Instead, they must be "smoothed over" using some form of hash function. In general, privacy amplification, namely hashing the raw output through a two-universal hash function as is done in QKD, can lead to a uniform random output. In particular, let's say that the QRNG device outputs a random bit string, the random variable of which we will denote by A. Let's say it's N bits long. Of course, unlike QKD, in this trusted device scenario, there is no adversary system, and so the entire state is simply ρ_A.

Now, with QKD, Alice and Bob must choose a random two-universal hash function, mapping N bits to n bits. If Alice had to do this with QRNG, we would need to account for the additional randomness required to choose this function. After all, one needs randomness to choose the hash function – thus, if a QRNG protocol needed more randomness than it produced, it would be a pretty useless protocol! As it turns out, for

QRNG applications, it is sufficient to choose the hash function once, randomly, and then use it repeatedly [23]. Thus, the randomness needed to choose the hash function is a one-time setup cost of a QRNG protocol, and does not need to be accounted for each time the protocol runs.

Thus, using Eq. (5.9) from Section 5.3, we see that, after the privacy amplification process, it will hold that:

$$\left\| \rho_{F(A)} - \frac{1}{2^n} I \right\| \leq 2^{-\frac{1}{2}(H_\infty(A)_\rho - \ell)}, \tag{8.1}$$

where $\rho_{F(A)}$ is the state of the system after running the raw measurement register, A, through a two-universal hash. The above relies only on the standard, non conditional, entropy. In fact, $H_\infty(A)_\rho$, since ρ_A is a classical state, is simply the standard classical min entropy. It is easy to verify from the definition of min entropy that:

$$H_\infty(A)_\rho = -\log_2 \max_a p(a), \tag{8.2}$$

where $p(a)$ is the probability that the random variable A outputs the n-bitstring a in system ρ_A (before privacy amplification). (In our case, the probability that the trusted device QRNG outputs a after running the protocol, but, of course, before privacy amplification.). Thus, if users can bound $\max_a p(a)$, privacy amplification can be used to "smooth out" the raw distribution, creating a uniform output that can be used in other applications. In general, one can estimate $\max_a p(a)$ through standard classical sampling techniques.

8.1.1 Source device independence

Next on our list of security models is the source independent model. Now, unlike above, we have an adversary system to contend with. Furthermore, unlike QKD, when designing QRNG systems, one must be very careful to account for any and all randomness used by the protocol. Indeed, if the goal of a QRNG protocol is to output randomness, it had better not consume more randomness than it produces. This is unlike QKD protocol design where randomness is essentially considered "for free" and can be used liberally.

The first source independent protocol was introduced in [51] which also introduced the source independent model (though, today, it is often referred to as a *partial source independent* model, due to its assumptions on the dimension of the source state). Nonetheless, it is a good protocol to consider as it shows how entropic uncertainty can be used in the QRNG case.

The protocol is straightforward and is shown, below, in Protocol 8.1. This protocol consists of an untrusted source device sending qubits to a receiver. The receiver consists of an honest measurement device.

Let ρ_{SE} be the state of Alice and Eve (the source) after the protocol is run, where S is the random variable storing the final secret random string s. The security goal of the protocol is that s is uniform random and independent of Eve (even though Eve produced the initial

Protocol 8.1 A partial source independent QRNG protocol from [51].

1. A source device, E, produces a quantum state ρ_{AE}, where the A register consists of N qubits while the E system is arbitrary (and may be entangled with the A system). The A system is sent to Alice (the receiver). If the source is honest and noise free, the state produced should be of the form:

$$|\psi_{AE}\rangle = |+\rangle^{\otimes N} \otimes |0\rangle_E \tag{8.3}$$

namely, N copies of $|+\rangle$ independent of Eve's ancilla. Of course, we make no assumptions on the state prepared other than the dimension of the A system.

2. The receiver will choose a random subset $t \subset \{1, 2, \cdots, N\}$ of size m and measure all qubits, indexed by t, in the Hadamard X basis. The resulting string will be denoted $x \in \{0, 1\}^m$ (where, as usual, a measurement result of $|+\rangle$ will be considered a "0" while an observation of $|-\rangle$ will be a "1"). Note that, ideally, x should be the all zero string; any 1 is considered "noise."
3. The remaining systems will be measured in the computational Z basis, resulting in outcome $z \in \{0, 1\}^n$, where $n = N - m$.
4. Assuming the "noise" is low enough, where noise in this case is defined to be the number of 1's in x, Alice will apply her two-universal hash function to z, outputting the final secret random string $s = f(z)$. As proven in [23], the hash function f may be chosen randomly once and then "hard-coded" into the receiver unit for subsequent rounds, thus we do not need to consume additional randomness for the choice of hash.

state). That, is we want:

$$\left|\left|\rho_{SE} - \frac{1}{2^\ell} I \otimes \rho_E\right|\right| \leq \epsilon, \tag{8.4}$$

for some small ϵ (that the user may choose). Well, to determine a bound on the above trace distance, we may use privacy amplification, same as we use with QKD proofs (see Eq. (5.9)). There are two differences, with this and QKD: One, we don't have to worry about error correction leakage – indeed, Alice never sends any classical information, so there is no information at all leaked over the classical channel, unlike QKD. Second, we must be sure to deduct any of the randomness we used in the protocol from the final random output size, ℓ. In the protocol, above, we need secret randomness to choose the subset t to "test" the noise in the state. Note that this must be chosen independently at random each time the protocol runs – and its choice must be secret. Thus, it must be refreshed after each run.

For the subset choice: we need to choose a subset t of size m, from a total of N possible integers. This requires $c = \log_2 \binom{N}{m}$ bits of initial randomness. Similar to how QKD requires a pre-shared key for the authenticated channel, QRNG requires an initial, random, seed to choose the first subset. After this, the QRNG protocol should output more bits than are needed to replenish these c bits. Namely, the c random bits used to choose a subset must be discarded and subsequently refreshed with new randomness from the resulting QRNG output.

Now, what size should ℓ be set to? Well, we know from Eq. (5.9) that:

$$\left\| \rho_{SE} - \frac{1}{2^\ell} I \otimes \rho_E \right\| \leq 2^{-\frac{1}{2}(H_\infty(A|E)_\rho - \ell)} \tag{8.5}$$

If we set the right-hand side of the above equation equal to the desired security level ϵ, and solve for ℓ, we find that the above security guarantee can be achieved if ℓ is equal to:

$$\ell = H_\infty(A|E) - 2\log\frac{1}{\epsilon} - c \tag{8.6}$$

(Note that subtracting c is not due to Eq. (8.5), but instead due to the need to replenish the randomness needed for choosing the subset t). Thus, as with QKD, security of a QRNG protocol in the source independent model boils down to finding bounds on the quantum min entropy. Once again, entropic uncertainty (see Section 2.6.2.2) can be used here [51]. Bouman and Fehr's sampling approach (discussed in Section 5.4.2) can also be used [55]. Ultimately, one derives a bound of the form:

$$\ell = n(1 - h(Q + \delta)) - c - 2\log\frac{1}{\epsilon}, \tag{8.7}$$

where Q is the relative error rate, namely $w(x)$ (the relative Hamming weight of x), where x is the X basis test string in Protocol 8.1. Above, δ is an error term due to finite sampling effects – its exact form depends on the proof method, though it vanishes asymptotically. Note the similarities to BB84's key-rate equation, with the exception of the error correction deduction (which depends on the noise).

Other source independent QRNG protocols exist. For instance, the protocols in [49,7] utilize quantum random walks (quantum random walks were discussed in Section 7.4.1, but for QKD applications). A protocol in [37] takes source independence to heart and utilizes natural sunlight as a randomness source! While source independent protocols still require trusted measurement devices, they provide a nice compromise, in terms of trust, between the fully device independent model (discussed shortly) and the trusted device model (discussed earlier). They are clearly a stronger security notion than the fully trusted setting; while they are not nearly as strong as the fully device independent model (which, in addition, does not need trusted measurement devices), they are much more practical, allowing for high-speed bit generation rates, measured in Gb/s, using today's technology [6,21].

8.1.2 Device independence and the CHSH inequality

The last security model we'll discuss is the device independent (DI) model, where devices are assumed to be manufactured by an adversary, and cannot be trusted. The notion of such a security model was first introduced in [42], while DI-QRNG protocols were first introduced in [15,16]. Of course, as with all QRNG protocols, one must assume there exists some private, secure, seed randomness – thus, these protocols are sometimes called *private randomness expansion* protocols. (Similar to QKD, which requires an initial secure seed which is expanded by the QKD protocol.)

In the DI setting, we assume the adversary manufactures the source and measurement device. The adversary may even pre-program in some randomness that is fully correlated with the adversary. This box, however, can produce random bits that the user can trust are uniform random and independent of the adversary!

Think, for a second, how powerful this is. Imagine you buy a box built by a company you don't trust. You push several buttons on this box, and it "does something." At the end, the box outputs a stream of bits. How do you know these bits are random and secure? How do you know an adversary didn't just pre-program a bunch of random bits to output, while keeping a copy of these bits herself to use against you later? Clearly classical information cannot make any guarantees here; quantum information and cryptography, however, can. To do so, we need a way to "test" for quantumness.

8.1.2.1 CHSH game

Let's imagine the following scenario, or game (taking a break, for the moment, from QRNG protocols). This is actually a cooperative game with two parties, Alice (A) and Bob (B), and a third party, known as a Referee (R). Alice and Bob are allowed to communicate with one another, initially, to agree on a strategy for the game. They may also, at this point, share information with one another (in the classical case, they may derive some shared randomness which they may use in the game; in the quantum case, they may share entangled particles). After this initial stage, Alice and Bob are no longer allowed to communicate.

Now, in the second stage of the game (when Alice and Bob cannot talk to one another, and must rely on their initial pre-shared communication and strategy agreement), the Referee will pick two random bits x and y, giving x to Alice, and y to Bob. Alice and Bob must now send two bits of their own, denoted a (for Alice's bit) and b (for Bob's bit), back to the Referee. Alice and Bob win the game if the following equation is satisfied:

$$a \oplus b = x \cdot y, \tag{8.8}$$

where $x \cdot y$ is standard multiplication (thus, $x \cdot y = 1$ only if $x = 1$ and $y = 1$; otherwise, the product is zero). If the above equation is not satisfied, then Alice and Bob both lose the game.

The above is known as the CHSH game, after the initial authors Clauser, Horne, Shimony, and Holt [14]. The question, now, is what is the maximal probability of winning? We will first consider the probability of winning if Alice and Bob are restricted to classical resources (e.g., classical communication in the first stage and shared classical randomness). We will then discuss that, given quantum resources, the maximal probability of winning is greater than the classical case. This can be used in a cryptographic application to "check for quantumness." Indeed, one can test that, so long as the CHSH game is won more often than it could be classically, then one is using quantum entanglement, from which secure information may be distilled! (This, of course, is a bit of a high-level explanation, and rigorous proofs are quite technical.)

Let's first consider the classical case. First, we'll consider deterministic strategies, where Alice, given her bit x, will choose some bit deterministically $a = f_A(x)$. Bob, given y, will choose his bit b according to some rule $b = f_B(y)$. It is not difficult to convince yourself that the best chance of winning, in this case, is 75% of the time; see Table 8.1. In general, to win, Alice and Bob must output bits that add to zero (modulo two – that is, they must output equal bits) in three out of four cases; the fourth case they must output different values. There is no way to build a deterministic strategy that will win all of the time – the best you can hope to do, is win in three of the four cases. Try other strategies, and convince yourself of this.

Exercise 8.1. Consider the CHSH game and Table 8.1; consider alternative deterministic strategies and show that you can only win 75% of the time (at best).

Table 8.1 Showing one possible deterministic strategy which wins 75% of the time simply by having Alice and Bob always output 0. In particular, given any x and y, to win, one must output a and b such that $a \oplus b = x \cdot y$. However, Alice (respectively Bob) is only given the value of x (respectively y). Thus, no matter what strategy Alice and Bob use to fill in the "a" and "b" columns, they can only win, at best, 75% of the time.

x	y	x · y	a	b	win?
0	0	0	0	0	Y
0	1	0	0	0	Y
1	0	0	0	0	Y
1	1	1	0	0	N

Of course, the above considered only deterministic strategies; it also did not consider the initial stage, where Alice and Bob are allowed to share information. Well, let's expand this, now, to include randomized strategies, along with pre-shared information. Let's say Alice and Bob share some correlated random variable Λ. This random variable takes some value λ with probability $Pr(\lambda|x, y)$. The value of λ is arbitrary; it can include some shared randomness; it can also include some information on the strategy Alice and Bob will use. Since Alice and Bob are not allowed to communicate once the Referee chooses x and y, we require $Pr(\lambda|x, y) = Pr(\lambda)$ (i.e., Λ must be independent of x and y, since Λ is constructed before the Referee makes its choice).

Now, given a particular λ and an x, y, Alice will play a strategy which outputs a according to: $Pr(a|\lambda, x)$. Bob, meanwhile will choose to output b according to $Pr(b|\lambda, y)$. Note that the value of λ is shared between Alice and Bob; however it is a classical random variable. Later, we will see how things change when Alice and Bob are allowed to share an entangled pair (which is still independent of the choice x and y of course). From this, we can define the conditional probability $Pr(a|x, y) = Pr(a|x) = \sum_\lambda Pr(\lambda)Pr(a|x, \lambda)$. Here, for simplicity, we are assuming a finite Λ – other possibilities are possible, though one must replace the summation over λ with an integral.

Let $\delta(x, y, a, b)$ be a function which outputs 1 if $x \cdot y = a \oplus b$ and otherwise outputs 0. Namely, δ outputs one only if A and B would win given the tuple (x, y, a, b). Given the above, we can write the probability of winning to be:

$$Pr(win) = \frac{1}{4} \sum_{x,y \in \{0,1\}} Pr(win|x, y) = \frac{1}{4} \sum_{x,y} \sum_{a,b \in \{0,1\}} \delta(x, y, a, b) Pr(a, b|x, y) \tag{8.9}$$

Now, $Pr(a, b|x, y) = \sum_\lambda Pr(\lambda)Pr(a, b|x, y, \lambda) = \sum_\lambda Pr(\lambda)Pr(a|x, \lambda) \cdot Pr(b|y, \lambda)$ due to the fact that Alice and Bob cannot communicate after x and y are chosen (and, thus, a cannot be influenced by y for instance). Thus we conclude:

$$\begin{aligned} Pr(win) &= \frac{1}{4} \sum_{x,y,a,b} \delta(x, y, a, b) \sum_\lambda Pr(\lambda)Pr(a|x, \lambda) \cdot Pr(b|y, \lambda) \\ &= \sum_\lambda Pr(\lambda) \underbrace{\left(\frac{1}{4} \sum_{x,y,a,b} \delta(x, y, a, b) Pr(a|x, \lambda) Pr(b|y, \lambda) \right)}_{Pr(win|\lambda)}. \end{aligned} \tag{8.10}$$

The value $Pr(win|\lambda)$ is the probability of winning, assuming a particular λ was chosen by the initial correlated shared randomness Λ. It is not difficult to convince yourself that this value can be no greater than 3/4; indeed, given λ, the strategy can only be a mixture of deterministic strategies, each of which can only win no greater than 75%. Thus, we can conclude that:

$$Pr(win) \le \sum_\lambda Pr(\lambda) \times \frac{3}{4} = \frac{3}{4}. \tag{8.11}$$

Furthermore, of course, there exists a strategy that wins with probability 3/4; it's the simple strategy shown in Table 8.1. Classically, one can do no better. See, for example, [43] for more information on the above derivations.

But what about if we allow for quantum information? Now, instead of communicating in a classical manner (to derive Λ), we allow Alice and Bob to communicate using quantum means. Of course, their communication must still be independent of x and y (which are only chosen *after* all communication, quantum or classical, is done between the two parties). Let's consider a particular quantum strategy:

1. During the communication stage, Alice and Bob share a Bell pair, $|\Phi^+\rangle = \frac{1}{\sqrt{2}}(|00\rangle_{AB} + |11\rangle_{AB})$.
2. When the referee sends to Alice the bit x, she will follow the following strategy to determine her output a:
   - If $x = 0$: Alice will measure her qubit (the A system) in the Z basis and set a to her measurement result
   - If $x = 1$: Alice will measure her qubit in the X basis; if she observes $|+\rangle$ she will set $a = 0$, otherwise she will set $a = 1$.
3. Bob, meanwhile, on receipt of his bit y from the Referee, will do the following:
   - If $y = 0$, Bob will measure in basis $\{|v_0\rangle, |v_1\rangle\}$ where:

$$|v_0\rangle = \cos\frac{\pi}{8}|0\rangle + \sin\frac{\pi}{8}|1\rangle$$
$$|v_1\rangle = -\sin\frac{\pi}{8}|0\rangle + \cos\frac{\pi}{8}|1\rangle$$

   Bob will set b to the observed basis vector index (e.g., $b = 1$ if he observes $|v_1\rangle$).
   - Otherwise, if $y = 1$, he will measure in basis $\{|w_0\rangle, |w_1\rangle\}$, where:

$$|w_0\rangle = \cos\frac{\pi}{8}|0\rangle - \sin\frac{\pi}{8}|1\rangle$$
$$|w_1\rangle = \sin\frac{\pi}{8}|0\rangle + \cos\frac{\pi}{8}|1\rangle$$

Note that Alice and Bob never communicate, once x and y are chosen.

Exercise 8.2. Show that the two bases described above are orthonormal bases. Namely, show that $\langle v_i|v_j\rangle = \delta_{i,j}$ and $\langle w_i|w_j\rangle = \delta_{i,j}$, where $\delta_{i,j}$ is the Kronecker delta function.

What is the winning probability of the above strategy? To determine this, we can calculate the winning probability in all four cases of x and y values, then average them together. We'll do one out, then leave the others as an exercise.

For the $x = 0$, $y = 0$ case, Alice will measure her particle in the Z basis, resulting in outcome $a = 0$ with probability 1/2 or $a = 1$ with probability 1/2. Since this was an entangled pair, Bob's state collapses to $|a\rangle$. Let's consider the $a = 0$ case. In this case, to win, Bob must output $b = 0$. We can calculate the probability of Bob observing $|v_0\rangle$ (which would be required for him to output $b = 0$), as:

$$Pr(b = 0|a = 0, x = 0, y = 0) = |\langle v_0|0\rangle|^2 = \cos^2\frac{\pi}{8} \approx 0.854. \quad (8.12)$$

It can also be shown that $Pr(b = 1|a = 1, x = 0, y = 0) = \cos^2\frac{\pi}{8} \approx 0.854$. Thus, the probability of Alice and Bob winning, in the event the Referee chooses $x = y = 0$, is $\frac{1}{2}(0.854 + 0.854) \approx 85\%$. Exercise 8.3 asks you to compute the winning probability in the other three cases; it can be shown that, in all four cases, the probability of winning is approximately 85%. Thus, in the quantum case, we have:

$$Pr(win|\text{quantum strategies}) \approx 0.85 > 0.75 = Pr(win|\text{classical strategies}). \qquad (8.13)$$

Exercise 8.3. Show that the probability of winning in all four cases of the CHSH game is approximately 85% (we already did the case when $x = y = 0$, above).

Now, we proved, above, that classically, the best you can do is 75%, no matter what strategy you try to use. We also showed that there exists a quantum strategy with a winning probability of 85%. Can we do better with quantum resources? As it turns out, we can't, as proven by Boris Tsirelson in 1980 [13]. It turns out that the best one can hope to win the CHSH game, no matter how complicated the quantum strategy is, is no more than $\cos^2 \frac{\pi}{8} \approx 85\%$. Furthermore, he proved that any quantum strategy which achieves this maximum bound, must be essentially similar to the above discussed strategy.

This shows that entanglement gives us something more interesting to work with than simple classically correlated randomness! Furthermore, this can be used for cryptography: if someone gives you and your friend a quantum state, claiming it is entangled, now you can verify it, by seeing if you can win the CHSH game more often than 75% of the time.

8.1.2.2 Device independent security

Now, returning to the QRNG protocol, how can we use the above to create a device independent QRNG system? As with the source independent scenario, we will assume Alice starts with some initial, secure, random seed; the goal of the protocol will be to create more randomness than is consumed (so that this seed value can be replenished for future runs of the protocol). A typical DI-QRNG protocol will consist of multiple devices that cannot "talk" to one another (they may be separated in space, and isolated so that there is no way for the two devices to communicate). These devices, however, should share an entangled pair (or receive an entangled pair from a source device). All devices are untrusted, however Alice does trust her initial seed value.

As with the source independent model, the protocol will operate in rounds, some of which will be "Test" rounds and others will be "Distillation" rounds. The Test rounds will consume randomness and will be used to see if the devices are winning the CHSH game with probability larger than 75%. If they win maximally (namely, at 85%), it is known that the devices owned by Alice share an entangled pair from which no third party adversary can have any shared entanglement with (i.e., the adversary, even though she created the devices, cannot be entangled with the Bell pair the devices must be using – if she were entangled with the Bell pair, the CHSH game would not be won maximally). Thus, on Distillation rounds, the output of the devices (the a setting, for instance), must be random and uncorrelated with an adversary's system. Of course, that's assuming the devices win with maximal 85% probability. However, so long as the devices are winning with probability greater than can be done classically, there must be some quantum effects which can be used to potentially extract secret randomness from (efficiency will depend on how close to the maximal probability the devices are, amongst other properties).

Proofs of security for device independent systems are non trivial, and the above was only meant to provide a high-level overview of why they are secure. Experimental implementations are also improving, though rates are not yet comparable to weaker security models (such as the source independent model or the trusted device scenario). One must also be careful of how the boxes operate and whether they have memory. For more details and further reading, the reader is referred to [16,47].

8.2 Certified deletion

Before leaving the information theoretic security realm, let's imagine the following scenario: you have a message that you want to save, securely, in "the cloud" (i.e., Internet storage). So, you encrypt your message using some secret key k and a classical encryption scheme (e.g., AES-CBC). Let $c = \mathtt{Enc}(k, m)$ be the resulting ciphertext that you upload. Now, some time later, you want to delete the file, so you kindly request your cloud storage provider to delete file c. You get back a response saying it was taken care of. But was it really? Since classical information can be copied, how do you know c was really deleted and that there are no usable copies out there somewhere? Now, imagine after some time, you accidental release the secret key k. If c really was deleted, it shouldn't matter, since the ciphertext no longer exists to decrypt. However, again, how do you really know that there isn't some copy out there that an adversary has access to and, now that k is leaked, how do you know m cannot be decrypted somehow?

When dealing with classical information, by itself, there is no way to solve the above problem. However, if the ciphertext contains some quantum information, not only can we be certain that no usable copies were made, we can also ask for *proof of deletion*! The first such protocol to provide this capability was proposed in [12]. The protocol is fairly straightforward, at least in hindsight, though the proof is interesting, and makes use of entropic uncertainty, along with some other techniques. It also demonstrates yet another unique application of quantum information and communication to cryptography.

Let m be some n-bit message that one wishes to encrypt. Since this is really a single-party system, the secret key is held only by one party and consists of several parts, minimally $k = (\theta, r^X, u)$, where:

1. Basis choice θ, drawn randomly from the set $\{0, 1\}^N$, where a choice of 0 will mean the computational Z basis and a choice of 1 will mean the Hadamard X basis.
2. $r^X \in \{0, 1\}^x$ chosen randomly, where $x = \#_1(\theta)$ (i.e., the number of ones in θ).
3. $u \in \{0, 1\}^n$, chosen randomly, which will serve as a OTP encryption key.

(Note that the full construction in [12] has more parts to the secret key to support additional features, such as noisy channels.)

To encrypt a message, given the above secret key k, we perform the following process:

1. First, choose $r^Z \in \{0, 1\}^z$, where $z = \#_0(\theta)$. Note that $|r^Z| + |r^X| = N$.
2. Compute $p = f(r^Z)$, for randomly chosen two-universal hash function f. This value p will serve as a *second* OTP key for the message.

3. Output the classical-quantum ciphertext $c = (f, \rho, m \oplus p \oplus u)$, where:

$$|\psi\rangle = H^{\theta} |r\rangle, \tag{8.14}$$

and where r is the $x + z = N$ bit string constructed from r^x and r^z by placing r^X into r, whenever $\theta_i = 1$ and similarly, place r^Z into r, whenever $\theta_i = 0$. Basically, one is creating random BB84 states, based on the choice of θ and the values of r^X and r^Z. Note that r^Z is not saved classically, instead it is "hidden" in the quantum portion of the ciphertext. Thus, the message is protected by two keys: one in the secret key (namely, u), and one hidden in the quantum state (namely p).

The above comment on r^Z not being saved should give some hint as to how deletion will work. Namely, to decrypt, one needs at least some information on r^Z (exactly how much information depends on the hash function f – this is privacy amplification used in QKD!). So if we can somehow "prove" that this state was deleted, we can be sure that the message can never be decrypted, as r^Z will be lost. Before going into details on that, let's just write out the decrypt algorithm:

1. Given a ciphertext $(f, |\psi\rangle, c)$, measure $|\psi\rangle$ using θ as a basis choice for each qubit. (Recall, θ is part of the secret key, which is known to the decryption algorithm.) Let r' be the result of this measurement, translating an observation of $|+\rangle$ to a "0" and a $|-\rangle$ to a "1."
2. Compute r_Z which is the substring of r, indexed by all i such that $\theta_i = 0$. I.e., r_Z is the result of all "Z" basis measurements. In the noiseless case, this should match the r^Z produced earlier, during encryption.
3. Output the message: $c \oplus u \oplus f(r_Z)$.

It is not difficult to see that, in the absence of noise, the above protocol is correct. Namely, if you encrypt a message, then decrypt it with the same secret key, you will output the same message (see the below exercise). It is also secure, since the message is encrypted using one-time-pad (OTP) encryption (see Section 3.1.1).

Exercise 8.4. Show that the above protocol is *correct.* Namely that if one encrypts a message m, and then decrypts the resulting ciphertext using the same secret key, one will recover the original message.

We now discuss the deletion algorithm which is run whenever the user wishes to delete the ciphertext. The algorithm outputs a "tag" t which can be used as proof that the deletion algorithm actually ran correctly and that the message is irrecoverable.

1. To delete: Given a ciphertext $(f, |\psi\rangle, c)$, measure all of $|\psi\rangle$ in the Hadamard basis, resulting in outcome $\tau \in \{0, 1\}^N$ which will be the proof of deletion tag.
2. To verify that the ciphertext was deleted: let t_X be the substring of t indexed by all $\theta_i = 1$. If $t_x = r^X$ (which is part of the secret key), then output "True;" otherwise output "False."

If the above process outputs "True" then Alice will have proof of deletion.

If we take a closer look at this system, we see that it is inspired by QKD (as is much of quantum cryptography) and relies on privacy amplification for security. We won't go through all details of the proof, rigorously. However, we will comment a bit on the high level notions. To show that the ciphertext is securely encrypted is straightforward based on OTP security. The real difficult part of the proof is to show that, given a verified proof of deletion (i.e., the server claims to delete the ciphertext and sends a proof tag τ for which the verify algorithm reports "True"), if the secret key then leaks to the adversary, the adversary cannot learn the message. Such a proof of security is impossible in the classical world since an adversary could always have copied the original ciphertext, so if the secret key leaks, the message can always be decrypted later. This is not the case with the above protocol as the needed r_z token is irrevocably lost in the deletion process. This can be proven with entropic uncertainty: if there is a high degree of correlation in the r_X value (from a measurement), then there must be a lot of uncertainty in the Z basis outcome (needed to decrypt the message).

Security follows some standard QKD methods. First, an equivalent entanglement based version of the protocol is created, where the adversary creates an entangled state $|\psi\rangle_{ABE}$ which will be used as the quantum portion of the ciphertext. Practically, this does not make a lot of sense, however, again, when proving security it is often useful to design "toy" protocols that are not practical, but their security implies security of the "real" system.

In the entanglement-based version, Alice will choose a random basis string θ, as before, however, now, she will measure her A system to attain r_Z and r_X (as opposed to choosing them herself). Eve (who holds both the B and the E portion) wants to learn as much as possible about r_Z. However, she must also provide a proof of deletion. So, Eve must send a classical string τ to Alice (if Eve is honest, this should be an X basis measurement of the B portion). Alice will ensure that r_X and τ are highly correlated. Entropic uncertainty (see Section 2.6.2.2) will then guarantee that Eve's uncertainty on r_Z (which is the result of a measurement made in an alternative basis), will be high. This ensures that $f(r_Z)$ will "look" like a uniform random string (i.e., a OTP key) according to privacy amplification results (see Eq. (5.9)) and, so, the message cannot be recovered.

Obviously the above was a very high level overview of the proof. For further details, the reader is referred to [12]. Before closing this section, we will comment that the protocol, as shown above, is not noise tolerant. However, error correction may be added to the ciphertext to ensure the message can be correctly decrypted. Noise in the deletion token will also lead to Alice becoming, perhaps, "suspicious" that the ciphertext was not deleted; however entropic uncertainty can guarantee a high level of uncertainty for Eve, based on "how noisy" that deletion certificate is. Thus, Alice may accept some noise or errors in the proof of deletion and, so long as the hash function is chosen appropriately (according to how much uncertainty Eve has), security remains. These details are also found in [12].

Certified deletion has since been extended to other cryptographic primitives (sometimes, also, making computational assumptions on the adversary; note that the original protocol, discussed above, made no assumptions on the adversary's capabilities). For instance, much interesting work has been done in public key systems involving certified deletion [28,8]. Certified deletion is a topic of interest and further research, as it truly is a capability impossible to achieve through classical communication alone. It is thus interesting to discover exactly what cryptographic primitives may be constructed which also hold some notion of proof of deletion.

8.3 Bounded storage and noisy storage

We now move beyond security based on the laws of quantum mechanics discussed in Chapter 2 and into a realm where some assumptions are made on the capabilities of the adversary. We consider in this section *physical assumptions*, namely assumptions based on physical device limitations. This is opposed to *computational assumptions*, which place assumptions on an adversary's ability to solve certain algorithmic problems. The latter will be discussed in Section 8.4 and also Chapter 10.

There are two types of physical assumptions we will discus: the first is *noisy storage* (introduced in [53]) and the second is *bounded storage* (introduced in [18]). Both place assumptions on the quantum memory of an adversary. The first assumes an adversary has unlimited quantum memory, but it is noisy and, if not measured withing a certain time period, will degrade to the point where it contains only random noise (and, thus, contains no useful information). The second assumes that the adversary's memory is perfect, but limited in the number of qubits it can store. Both are reasonable assumptions with today's devices, as quantum memory is still rudimentary (as of writing this section) and likely to be highly noisy and very expensive once it does arrive. While these assumptions may not be safe forever, they certainly seem like safe assumptions in the near future. The question, then, is what cryptographic primitives can we solve in this model, that cannot be solved in the "standard model" (i.e., the "unconditional security setting" we considered so far). As it turns out, quite a few interesting things can be done here!

8.3.1 Oblivious transfer (OT)

Let's consider the cryptographic task of *oblivious transfer* (OT). Here, a *sender*, Alice, has two messages, m_0 and m_1. The *receiver*, Bob, has a bit C which is his choice of which message he wants to receiver. Alice and Bob will run the OT protocol, and at the end, Bob will learn m_C (his choice of message) but not m_{1-C}; meanwhile, Alice, will never learn C. OT is an important building block for many other, more complex, cryptographic protocols, and so its study is important. Security is generally considered in two parts: first, one considers a "cheating Bob" who wants to try and learn both messages; second is a "cheating Alice" who wants to learn C. Security should be proven for both. This is known as a "1 out of 2 oblivious transfer" or $1-2$ OT.

OT is impossible in the standard security model, even with quantum resources [38]; instead, to achieve this protocol, we must make some assumptions on the adversary. Classically, we can make computational assumptions to gain security for OT. In the quantum setting, we may achieve secure OT by making physical assumptions, such as noisy storage or bounded storage.

Let's look, first, at the protocol which was actually introduced in [50], then proven secure in the noisy storage model in [53].

1. Alice, the sender, picks a random string $x \in \{0,1\}^n$ and a random basis string $\theta \in \{0,1\}^n$, where 0 will mean the Z basis and 1 will mean the X basis. Let $I_b = \{i : \theta_i = b\}$ (i.e., the indices where basis b was chosen).
2. At some initial time $t = 0$, Alice prepares and sends the quantum state:
$$|\psi_0\rangle = H^\theta |x\rangle \tag{8.15}$$
to Bob. I.e., she sends random BB84 states to Bob. Here, we write H^θ to mean, apply the H (Hadamard) operator to every i'th qubit, where $\theta_i = 1$.
3. Bob will measure all qubits in a randomly chosen basis based on $c \in \{0,1\}$, his desired message choice. In particular, if $c = 0$, he will measure all n qubits in the Z basis; otherwise he will measure all qubits in the X basis. Let $y \in \{0,1\}^n$ be his measurement result (with a $|+\rangle$ being translated to a zero and $|-\rangle$ a one, as usual).
4. Alice waits for a sufficient amount of time (until some time $t = T$), based on the noisy storage assumption. After this amount of time, anything in Bob's quantum memory should be depolarized.
5. Alice picks two random, two-universal hash functions f_0 and f_1 at time $t = T$. At this time, she sends:
   - I_0 and I_1: i.e., she tells Bob which indices she used for Z and X basis qubits
   - f_0 and f_1: the two hash functions chosen
6. Next, she computes x_0 which is the substring of her initial choice x indexed over all i where she chooses the Z basis. Similar for x_1 which is where she chose to send in the X basis. She then computes $s_0 = f_0(x_0)$ and $s_1 = f_1(x_1)$. She sends $m_0 \oplus s_0$ and $m_1 \oplus s_1$ to Bob.
7. Bob computes $s_c = f_c(y_c)$ where y_c is the substring of y (his measurement results) where Alice chooses to send in basis c.

It is not difficult to see that this protocol is correct. Indeed, if both parties are honest, Alice will never learn Bob's choice c and Bob will never learn anything other than m_c (and, of course, he will learn m_c correctly). However, what about adversarial parties?

Exercise 8.5. Trace the execution of the above protocol and show that it is correct, namely that if both parties are honest, Bob will learn m_c and Alice will not learn c.

Typically security proofs of these protocols are broken into two stages: first is to show that it is secure if Alice is adversarial (namely, in that case, she will not learn c, try as she might). Second is to show that it is secure even if Bob is adversarial (in that case, he will never learn both messages). Proving security of the above protocol in the case of adversarial Alice is not too challenging, since Bob never communicates to Alice, there is nothing for her to learn. Thus, the real challenge is in proving security of adversarial Bob.

If we consider the standard security model where only the laws of quantum mechanics apply (which, of course, is the model we have been considering all throughout this book up until this chapter), the above protocol is very clearly *not* secure. Indeed, the attack is trivial: Bob, on receipt of all qubits from Alice, will simply save them in his perfect quantum memory (which does not violate any law of physics). When, later, Alice reveals the basis choice, he can measure the entire state in both bases correctly, thus learning both x_0 and x_1. This will allow him to learn both messages. This attack required quantum memory that was perfect. In the noisy storage model, we assume that if Bob holds on to qubits for "too long" they eventually depolarize and become useless. Thus, Bob must measure at some point and "commit" to some basis choice. When Alice later reveals the basis choices she made, at time T, he either must have measured his qubits, or they are completely depolarized, and so he will learn nothing. That is, he is forced to measure immediately (to learn m_C), or risk learning nothing. Of course, formalizing this requires some technical work; for details, see [53].

Let's look at another OT protocol, this time in the bounded storage model. Technically, this is a *Rabin Oblivious Transfer* protocol (ROT) [48] which transmits a bit message m (chosen by Alice, the sender) or nothing at all. Alice never knows if m was received or not and the receiver, Bob, receives no information about m at least half the time (though he should know if he received the message, or just random data). While this may seem, at first, not very useful, it turns out ROT protocols can be used to construct a $1-2$ OT protocol [17].

The specific quantum ROT protocol we'll look at was introduced and analyzed in [18] and proceeds as follows:

1. The sender picks a random $x \in \{0,1\}^n$ uniformly at random, and a single basis $b \in \{Z, X\}$. Note there are n-bits in x but only one basis is chosen.
2. If $b = Z$, the sender sends $|x\rangle$ to the receiver; otherwise, the sender sends $H^{\otimes n}|x\rangle$, namely $|x\rangle$ in the Hadamard basis.
3. The receiver picks a random basis b' (again, a single random basis), and measures all n qubits in this single, chosen, basis, resulting in outcome $y \in \{0,1\}^n$.
4. The sender chooses a random two-universal hash function f and announces b, f, and $c = m \oplus f(x)$.
5. The receiver outputs $a = 1$ and $m' = c \oplus f(y)$ if $b = b'$ (i.e., if he choose the same basis as the sender, Alice). Otherwise, he outputs $a = 0$ and $m' = 0$.

The flag "a" above is set to one if Bob received the (single bit) message that Alice intended to send. Clearly the protocol is correct – namely if both parties are honest, then Bob will pick the right basis (Alice's basis) only half the time. If he chooses the right basis, he will clearly be able to extract the message m from c (and he will know he received it or not).

For security, it is clear that Alice, the sender, will never know if m was received or not, since no communication goes from the receiver to the sender. Note that, when we say "output," in the protocol above, we mean "output locally" to the party running that sub-routine, not to all parties. Security for dishonest receivers, is where the technical difficulties lie. In [18], a rigorous proof of security is presented, assuming the receiver has bounded quantum storage. Note that the protocol is completely insecure if the receiver has unlimited storage: the receiver may simply store all n qubits, indefinitely, until the basis information is received. At that point, the receiver may simply make the correct message and always learn the message (to be secure, ROT needs to ensure the receiver learns nothing about the message at least half the time).

Security involves, as usual, first reducing the protocol to an equivalent entanglement-based version. The idea, then, is to make sure n, the number of qubits sent, is larger than the memory size of the adversarial receiver. If this is the case, the adversary will be forced to measure at least *some* of the received qubits. This measurement causes uncertainty since the exact basis is not known and, thus, privacy amplification via f, will ensure that $f(y)$ looks random. The only way to receive the message correctly half the time, will be to measure everything right away. Obviously, this is a very high-level idea of the proof – more rigorous details, which involve finding careful bounds on the quantum min entropy, can be found in [18].

This section was only meant to give the reader some idea of the interesting possibilities outside the standard model. The above cryptographic protocols are insecure in the standard "unconditional" security model (where the adversary is bounded only by the laws of quantum mechanics). However, the above primitives are vital building blocks for many more mechanisms. Classically, one needs computational assumptions to make these protocols secure; in the quantum setting, one can use *physical* assumptions. Such assumptions are likely to hold true for a while, as quantum memory is likely to be very noisy, and very limited, for a while. Other cryptographic primitives are also possible to construct within this model of security. See [10] for a survey which discusses these security models (along with others that may be of interest to the reader).

8.4 Quantum pseudo-random states

Pseudorandom quantum states (PRS) were introduced in [32]; they are quantum states that are (1) efficiently producible (i.e., can be produced using a polynomial-sized quantum circuit), and (2) are computationally indistinguishable from a truly random quantum state (drawn according to the Haar measure, a technicality that we won't go into in this book, but think of it as the uniform distribution on quantum states). In some way, they are the quantum analogue of pseudorandom number generators (PRGs) discussed in Section 3.1.2, though with, of course, many differences also. It may be useful to review that chapter, again.

Let G be a polynomial time quantum algorithm which takes, as input, a classical "seed" x of size n-bits. The circuit will then output an N-qubit state $|\psi(x)\rangle$. Let D be some quantum polynomial time *distinguisher* (namely, an algorithm that receives a quantum state and outputs a single classical bit). Then, we say G is a PRS if for every D, there exists a negligible function (see Section 3.1.2) $\eta(n)$ such that:

$$\left|Pr\left(D[G(k)^{\otimes t}]=1\right)-Pr\left(D[|\nu\rangle^{\otimes t}]=1\right)\right|\leq\eta(n), \tag{8.16}$$

where $t=t(n)$ is some polynomial (in n). The probability on the left is over all seed choices k (classical n-bit strings), along with any randomness in D; while the probability on the right is over all N-qubit states chosen uniformly at random (formally, chosen according the Haar distribution – see [32] for more details on the technicalities there) and, again, any randomness in D. Essentially, a PRS is indistinguishable from a truly random quantum state, even if the distinguisher is given (polynomial) many copies of the state.

As it turns out, if quantum-secure one way functions (qOWF) exist, then one can construct a PRS, as we discuss below. A quantum-secure OWF is a OWF (a classical function) which is secure against quantum adversaries (i.e., adversaries with quantum computing capabilities) (see [56]). A OWF is a function that is easy to compute given the input, however, given the output of a OWF, it is hard to find any input which would map to that given output (i.e., the function is "one way" in the sense it is easy to compute but hard to invert).

A functionality similar to a PRS was developed in [4,3] known as *pseudorandom function-like state (PRFS) generators.* These are, in a way, quantum analogues of classical pseudo-random functions (PRFs, introduced in Section 3.1.3) and, as it turns out, are highly useful to cryptographic applications. A PRFS takes as input a secret key k and an input x and outputs a quantum state $|\psi(k,x)\rangle$. Similar to a standard PRF, security will involve a distinguisher that is allowed to make "queries" x and receive either $|\psi(k,x)\rangle$ (the state produced by a polynomial time algorithm, the PRFS generator), or a randomly chosen state $|\nu_x\rangle$. We will forgo writing out the exact definition, as it follows from the classical case, replacing D with a polynomial time *quantum* algorithm, and the response with a quantum state. We do allow D to make superposition queries, though, essentially asking "multiple" questions at once, of the function. Note that in the "randomized case" $|\nu_x\rangle$, if the distinguisher repeats a query, the same state is returned, similar to how a random function is deterministic in its response – the randomness comes from the initial choice.

How can we construct a PRS and a PRFS? Well, let's look at a PRS first. Let F_k be a quantum-secure PRF (a classical PRF that is secure against quantum adversaries; see [56]). Then, a PRS construction, originally suggested in [32] and analyzed in [11,3], is the following:

$$|\psi(k)\rangle=\frac{1}{\sqrt{2^N}}\sum_{x\in\{0,1\}^N}(-1)^{F_k(x)}|x\rangle. \tag{8.17}$$

This is not the only way to construct a PRS, and several other constructions exist by now.

Given a quantum secure PRF , it's also possible to construct a secure PRFS: Namely, given an input x and a secret key k, first compute $y=F_k(x)$. Then output the state:

$$|\psi(k,x)\rangle=\frac{1}{\sqrt{2^N}}\sum_{z\in\{0,1\}^N}(-1)^{F_y(z)}|z\rangle \tag{8.18}$$

The above construction was analyzed in [3]. It is, of course, not the only way to build a PRFS.

While interesting in their own right, PRS and PRFS can be used to build several different cryptographic systems. We'll only discuss one application, below, to give the reader some idea of their utility. We will also discuss a novel quantum cryptographic protocol.

8.4.1 Cryptographic application: quantum money

We'll look at one application of a PRS, namely a *Quantum Money Scheme* using a construction from [32]. Quantum money was actually one of the very first discoveries in quantum cryptography (and led to QKD) [54,9]. Recently, they have received more interest as new constructions and rigorous security proofs are coming into play [1,39,22,2].

A quantum money scheme, of the kind analyzed in [32], involves two primary algorithms (along with a key-generation algorithm which we'll glaze over). The first is the "Bank" algorithm which takes a secret key and produces a *banknote* which is a quantum state $|\$\rangle$. The second algorithm is a "Verify" algorithm, which takes, again, the secret key, along with a candidate banknote state, and either "accepts" or "rejects" the note. Correctness of such a scheme should ensure that the Verify algorithm will only accept a banknote created by the Bank algorithm, so long as both Bank and Verify use the same secret key. (There may be a negligible probability of failure.)

For security, consider the following: Let's say Bank created q legitimate banknotes: $|\phi\rangle = |\$_1\rangle \cdots |\$_q\rangle$ (which we take to be a single joint state). These are given to some adversary, called a *Counterfeiter* C. This adversary (a quantum algorithm) C attempts to output a new quantum state $|\psi'\rangle$. Now, let's define a "Count" algorithm which, given a quantum state consisting of, potentially multiple, candidate banknotes, will output the number of legitimate banknotes in the state (or, rather, the number of times the Verify algorithm returns True). Of course it should be that $\text{Count}(k, |\phi\rangle) = q$. Note that this algorithm needs the secret key as it will be required to run the Verify algorithm. We say that the quantum money scheme is secure if for any counterfeiter algorithm C, it holds that:

$$Pr\left(\text{Count}(k, C(|\$_1\rangle \cdots |\$_q\rangle)) > q\right) \le \text{negl}. \tag{8.19}$$

Note that the counterfeiter algorithm, C, does *not* have access to the secret key. Ultimately, the protocol is secure if the probability of being able to trick the bank into thinking more than q banknotes exist, if given only q banknotes, is small.

Given the above, the scheme using a PRS, is actually fairly straightforward. The Bank algorithm will output a PRS state $|\psi(k)\rangle$. The Verify algorithm will perform a projective measurement using POVM operators $M_0 = |\psi(k)\rangle\langle\psi(k)|$ and $M_1 = I - M_0$. Above, M_0 represents an "Accept" outcome of the measurement. We won't prove security, details may be found in [32]. Security follows from a "no-cloning" theorem of PRS states, proven in that same reference.

Exercise 8.6. Describe a polynomial time algorithm that can perform the above POVM $M_0 = |\psi(k)\rangle\langle\psi(k)|$ and $M_1 = I - M_0$, if you have k.

8.5 Game theoretic security

The last security model we'll consider here is the notion of game theoretic security. Here, we can assume the adversary is not computationally bounded (i.e., we are back to the "standard model" in that sense); however we will assume the adversary is "rational" in some sense. Namely, the adversary has some objective beyond simply destroying everything Alice and Bob are trying to do; and, furthermore, there will be a cost associated with the adversary's actions. This may make sense in the immediate to near future, where, with quantum equipment being very expensive, it is not entirely feasible to try and attack with a strong unitary probe, for example, if that action will provide a small pay off. Game theory has been used to study classical cryptographic systems [33,25,5,35,44,41] and several papers have investigated the use of game theory in the quantum realm (some of which are discussed, below). Thus, it makes sense to see how it can be applied to the quantum setting.

In the quantum setting, game theory has been used to analyze secret sharing schemes [40], bit commitment [57], and quantum state sharing [20]. It has also been used to study QKD security, with various models being proposed [19,30,34,36,26]. We will discuss, here, the model proposed in [36,26] as it is, to our knowledge, the only game theoretic model for QKD security which allows for key-rate computations to be made (thus, allowing for a direct comparison to the performance of QKD systems in the standard model). This model also takes into account "natural noise" which will always occur in any practical QKD system.

Before we can discuss the model, we must introduce some game theory background. This is a very vast field; for more details, the reader is referred to [45,24].

A *Normal Form Game* is a tuple (N, S, u), where:

- N is the set of all players (typically $N = \{1, 2, \cdots, n\}$)
- S is the set of player strategies; namely $S = \{S_1, \cdots, S_n\}$, where each S_i is player i's *strategy space*.
- u is a set of utility functions, namely $u = \{u_1, \cdots, u_n\}$, where each:

$$u_i : S_1 \times \cdots \times S_n \to \mathbb{R}, \tag{8.20}$$

  is player i's utility function, which assigns a utility to that player if all other players choose a particular strategy.

An important concept in game theory is the *Nash Equilibrium* (NE). This is a strategy s^* such that no player can gain more utility by changing their strategy unilaterally, given that all other players' strategies are fixed. That is to say, no rational player should want to deviate from a Nash equilibrium, if that player knows all others will choose the given strategy also. Note that strategies can be *mixed* meaning that players will choose a particular strategy with a certain probability. This is an important necessity for a Nash equilibrium to exist. We will assume that strategies may be mixed.

The model introduced in [36,26] considers the QKD protocol as a normal form game between two players. The first player is the joint "Alice+Bob" player, denoted AB, who will work cooperatively to choose the best strategy for themselves. The second player is, of course, the adversary Eve, denoted E. The AB player's goal is to establish a shared secret key as large as possible with as few rounds of quantum communication as possible. E's goal, however, is to decrease AB's efficiency as much as possible while also remaining "within budget" in her attack strategy. Note that *denial of service* attacks can shrink the efficiency to zero rapidly and are devastating to any point-to-point system. The model did not consider such attacks as they can affect any point to point system equally and which can also affect the standard model just as easily. Thus, such attacks are outside the strategy set for the adversary player in this model.

The following scenario is considered: Parties Alice and Bob broadcast a maximal tolerated noise level Q which they will accept (or, alternatively, this noise tolerance is some publicly known constant). If an error rate is observed above this, parties will simply abort. Alice and Bob then choose to run a QKD protocol, or to do nothing (the latter costs nothing, but also gains them nothing). We will assume there is some natural noise and so error correction will be needed even if Eve does not attack – this gives Eve information "for free." Eve, on the other hand, knowing this value of Q, is allowed to completely ignore the channel (i.e., not attack), or she may choose to perform some kind of attack on a passing qubit (which will cost her some resources). The idea, for Alice and Bob, will be to make it "too costly" to attack; this is done through *decoy rounds* (not to be confused with *decoy states* as in the decoy state BB84 protocol, discussed in Chapter 9). If Alice and Bob can make it so that a rational adversary will not bother to attack, less rounds may be needed overall (even taking into account the additional decoy rounds) as privacy amplification will need to hash the raw key down by less (as Eve's uncertainty is high).

Decoy rounds are indistinguishable from regular rounds and will force Eve to attack a "useless" round, gaining no information while also expending resources. The more decoy rounds, the more costly it will be for Eve to attack. Decoy rounds, however, also cost Alice and Bob as every time they send or measure a qubit, it will cost them. Thus, the question is: can we find an "equilibrium" where Alice and Bob will want to run the protocol and Eve will want to *not* attack.

In a bit more formality, let's consider just the BB84 protocol where, on any particular round, with probability α, Alice and Bob will use that round as a "real" round, performing the BB84 protocol as normal; otherwise, with probability $1 - \alpha$, that round will be a decoy round, where Alice and Bob will send a random state that will not be used for anything (not even for testing the noise in the channel). The strategy space for AB is to run the BB84 protocol with decoy parameter α, or to "do nothing" (i.e., to not run any protocol, at all). Eve, on the other hand, may choose to attack (which will generally be an intercept/resend attack with some probability p) or to "do nothing" (where, by "do nothing" we mean do absolutely nothing across all rounds – same for AB where "do nothing" means don't even run any QKD protocol). We assume AB choose their strategy first, then Eve is allowed to choose her strategy based on what AB chose.

We must now consider the utility functions for the two parties. After choosing a strategy, they will run the QKD protocol for N rounds, after which Alice and Bob will derive a secret key of size $M \le N$ (possible $M = 0$). The utility for party AB will be:

$$u_{AB} = gM - c_{AB}, \tag{8.21}$$

where g is a positive scalar and c_{AB} is a cost for the given strategy. For Eve, the utility will be:

$$u_E = wK - c_E, \tag{8.22}$$

where w is some positive weight, K is Eve's mutual information on Alice and Bob's raw key, and c_E is a (weighted) cost for implementing the strategy Eve chose.

The cost function c_{AB} and c_E can take into account multiple parameters such as the initial cost to set up attack equipment (a one-time cost for Eve); the cost of communicating over the authenticated channel (a cost for Alice and Bob, for each round), and so on. In particular, for AB the cost of running BB84 for N rounds can be parameterized by:

$$c_{AB} = N((3 + h(\alpha))C_R + \gamma_4(C_P + C_M)) + C_{auth}, \tag{8.23}$$

where:

- C_R: The cost to produce a uniform random bit. If the bit is not uniform random, but biased by some α, then the cost is $h(\alpha)C_R$.
- C_P: the cost to prepare a single qubit. $\gamma_4 C_P$ is the cost of operating the hardware necessary to prepare one of 4 possible states (needed for BB84 for instance).
- C_M: Similar to C_P but the cost to measure.
- C_{auth}: The cost for Alice and Bob to use the authenticated channel (a one time cost).

For Eve, we will assume she either does not attack, or she will perform a measure-resend attack, choosing to measure/resend any particular round with probability p (choosing independently and identically each round). In this case, her cost, if she chooses to attack, will be:

$$c_E = N(h(p)C_R + p\gamma_2(C_M + C_P)) + C_S, \tag{8.24}$$

where C_S is the initial cost for Eve to set up her attack equipment. Note that Eve is measuring/resending in a single basis, thus the use of γ_2 above.

Given the above setup, one may analyze the game and find suitable α (depending on the above costs, and the tolerated noise Q), such that the resulting strategy of Alice and Bob running the BB84 protocol and Eve not attacking is, essentially, a NE (one actually considers so-called subgame perfect equilibrium, however that is a detail out of scope of this overview chapter). In particular, it was shown in [26], that, assuming classical communication is free, and that also the "setup cost" for Eve is free, then a "rational" adversary will not attack, if there exists an $\alpha \in (0, 1)$

satisfying:

$$\frac{C_R + (\gamma_4 - \gamma_2)C_P}{\frac{1}{4} + \frac{1}{4}h\left(\frac{2Q}{1+2Q}\right) - \frac{1}{2}h(Q)} \leq \alpha < 5.013\gamma_2(C_M + C_P). \tag{8.25}$$

The above depends on the cost function parameters, and also the maximal channel noise. Interestingly, higher noise tolerances are possible in the game theoretic model, than in the standard model (depending on the above cost functions). More details on all of the above can be found in [26].

So why bother looking at a game theoretic model? On the one hand, it is difficult to put precise values to the utility and cost functions of the above, especially from an adversary's perspective. However, it can produce more efficient systems, and game theory has been used to study classical security problems, as discussed. Ultimately, like any assumption, the "rational adversary" assumption may or may not be practical, depending on the operating scenario. It does, however, open up some interesting research directions that the reader may find relevant.

References

[1] Scott Aaronson, Quantum copy-protection and quantum money, in: 2009 24th Annual IEEE Conference on Computational Complexity, IEEE, 2009, pp. 229–242.

[2] Scott Aaronson, Paul Christiano, Quantum money from hidden subspaces, in: Proceedings of the Forty-Fourth Annual ACM Symposium on Theory of Computing, 2012, pp. 41–60.

[3] Prabhanjan Ananth, Aditya Gulati, Luowen Qian, Henry Yuen, Pseudorandom (function-like) quantum state generators: new definitions and applications, in: Theory of Cryptography Conference, Springer, 2022, pp. 237–265.

[4] Prabhanjan Ananth, Luowen Qian, Henry Yuen, Cryptography from pseudorandom quantum states, in: Annual International Cryptology Conference, Springer, 2022, pp. 208–236.

[5] Gilad Asharov, Yehuda Lindell, Utility dependence in correct and fair rational secret sharing, Journal of Cryptology 24 (1) (2011) 157–202.

[6] Marco Avesani, Davide G. Marangon, Giuseppe Vallone, Paolo Villoresi, Source-device-independent heterodyne-based quantum random number generator at 17 gbps, Nature Communications 9 (1) (2018) 5365.

[7] Minwoo Bae, Walter O. Krawec, Semi-source independent quantum walk random number generation, in: 2021 IEEE Information Theory Workshop (ITW), IEEE, 2021, pp. 1–6.

[8] James Bartusek, Dakshita Khurana, Cryptography with certified deletion, in: Annual International Cryptology Conference, Springer, 2023, pp. 192–223.

[9] Charles H. Bennett, Gilles Brassard, Seth Breidbart, Stephen Wiesner, Quantum cryptography, or unforgeable subway tokens, in: Advances in Cryptology: Proceedings of Crypto 82, Springer, 1983, pp. 267–275.

[10] Mathieu Bozzio, Claude Crépeau, Petros Wallden, Philip Walther, Quantum cryptography beyond key distribution: theory and experiment, Reviews of Modern Physics 97 (4) (2025) 045006.

[11] Zvika Brakerski, Omri Shmueli, (Pseudo) random quantum states with binary phase, in: Theory of Cryptography Conference, Springer, 2019, pp. 229–250.

[12] Anne Broadbent, Rabib Islam, Quantum encryption with certified deletion, in: Theory of Cryptography: 18th International Conference, TCC 2020, Durham, NC, USA, November 16–19, 2020, Proceedings, Part III 18, Springer, 2020, pp. 92–122.

[13] Boris S. Cirel'son, Quantum generalizations of Bell's inequality, Letters in Mathematical Physics 4 (2) (1980) 93–100.

[14] John F. Clauser, Michael A. Horne, Abner Shimony, Richard A. Holt, Proposed experiment to test local hidden-variable theories, Physical Review Letters 23 (15) (1969) 880.
[15] Roger Colbeck, Quantum and relativistic protocols for secure multi-party computation, arXiv preprint, arXiv:0911.3814, 2009.
[16] Roger Colbeck, Adrian Kent, Private randomness expansion with untrusted devices, Journal of Physics A: Mathematical and Theoretical 44 (9) (2011) 095305.
[17] Claude Crépeau, Equivalence between two flavours of oblivious transfers, in: Conference on the Theory and Application of Cryptographic Techniques, Springer, 1987, pp. 350–354.
[18] Ivan B. Damgård, Serge Fehr, Louis Salvail, Christian Schaffner, Cryptography in the bounded-quantum-storage model, SIAM Journal on Computing 37 (6) (2008) 1865–1890.
[19] Bhaskar Das, Utpal Roy, et al., Cooperative quantum key distribution for cooperative service-message passing in vehicular ad hoc networks, International Journal of Computer Applications 102 (16) (2014) 37–42.
[20] Zhao Dou, Gang Xu, Xiu-Bo Chen, Xin Liu, Yi-Xian Yang, A secure rational quantum state sharing protocol, Science China. Information Sciences 61 (2) (2018) 022501.
[21] David Drahi, Nathan Walk, Matty J. Hoban, Aleksey K. Fedorov, Roman Shakhovoy, Akky Feimov, Yury Kurochkin, W. Steven Kolthammer, Joshua Nunn, Jonathan Barrett, et al., Certified quantum random numbers from untrusted light, Physical Review X 10 (4) (2020) 041048.
[22] Edward Farhi, David Gosset, Avinatan Hassidim, Andrew Lutomirski, Peter Shor, Quantum money from knots, in: Proceedings of the 3rd Innovations in Theoretical Computer Science Conference, 2012, pp. 276–289.
[23] Daniela Frauchiger, Renato Renner, Matthias Troyer, True randomness from realistic quantum devices, arXiv preprint, arXiv:1311.4547, 2013.
[24] Drew Fudenberg, Jean Tirole, Game Theory, MIT Press, 2026.
[25] Joseph Halpern, Vanessa Teague, Rational secret sharing and multiparty computation, in: Proceedings of the Thirty-Sixth Annual ACM Symposium on Theory of Computing, 2004, pp. 623–632.
[26] Songyang Han, Walter O. Krawec, Fei Miao, A game-theoretic security framework for quantum cryptography: performance analysis and application, Quantum Information Processing 19 (10) (2020) 349.
[27] Miguel Herrero-Collantes, Juan Carlos Garcia-Escartin, Quantum random number generators, Reviews of Modern Physics 89 (1) (2017) 015004.
[28] Taiga Hiroka, Tomoyuki Morimae, Ryo Nishimaki, Takashi Yamakawa, Quantum encryption with certified deletion, revisited: public key, attribute-based, and classical communication, in: International Conference on the Theory and Application of Cryptology and Information Security, Springer, 2021, pp. 606–636.
[29] W. Timothy Holman, J. Alvin Connelly, Ahmad B. Dowlatabadi, An integrated analog/digital random noise source, IEEE Transactions on Circuits and Systems. I, Fundamental Theory and Applications 44 (6) (2002) 521–528.
[30] Mahboobeh Houshmand, Monire Houshmand, Habib Rajabi Mashhadi, Game theory based view to the quantum key distribution bb84 protocol, in: 2010 Third International Symposium on Intelligent Information Technology and Security Informatics, IEEE, 2010, pp. 332–336.
[31] Masatugu Isida, Hiroji Ikeda, Random number generator, Annals of the Institute of Statistical Mathematics 8 (2) (1956) 119–126.
[32] Zhengfeng Ji, Yi-Kai Liu, Fang Song, Pseudorandom quantum states, in: Annual International Cryptology Conference, Springer, 2018, pp. 126–152.
[33] Jonathan Katz, Bridging game theory and cryptography: recent results and future directions, in: Theory of Cryptography Conference, Springer, 2008, pp. 251–272.
[34] Hargeet Kaur, Atul Kumar, Game-theoretic perspective of ping-pong protocol, Physica A. Statistical Mechanics and Its Applications 490 (2018) 1415–1422.
[35] Gillat Kol, Moni Naor, Cryptography and game theory: designing protocols for exchanging information, in: Theory of Cryptography Conference, Springer, 2008, pp. 320–339.
[36] Walter O. Krawec, Fei Miao, Game theoretic security framework for quantum key distribution, in: International Conference on Decision and Game Theory for Security, Springer, 2018, pp. 38–58.
[37] Yu-Huai Li, Xuan Han, Yuan Cao, Xiao Yuan, Zheng-Ping Li, Jian-Yu Guan, Juan Yin, Qiang Zhang, Xiongfeng Ma, Cheng-Zhi Peng, et al., Quantum random number generation with uncharacterized laser and sunlight, npj Quantum Information 5 (1) (2019) 97.

[38] Hoi-Kwong Lo, Hoi Fung Chau, Is quantum bit commitment really possible?, Physical Review Letters 78 (17) (1997) 3410.
[39] Andrew Lutomirski, Scott Aaronson, Edward Farhi, David Gosset, Avinatan Hassidim, Jonathan Kelner, Peter Shor, Breaking and making quantum money: toward a new quantum cryptographic protocol, arXiv preprint, arXiv:0912.3825, 2009.
[40] Arpita Maitra, Sourya De Joyee, Goutam Paul, Asim K. Pal, Proposal for quantum rational secret sharing, Physical Review A 92 (2) (2015) 022305.
[41] Mohammad Hossein Manshaei, Quanyan Zhu, Tansu Alpcan, Tamer Bacşar, Jean-Pierre Hubaux, Game theory meets network security and privacy, ACM Computing Surveys 45 (3) (2013) 1–39.
[42] Dominic Mayers, Andrew Yao, Quantum cryptography with imperfect apparatus, in: Proceedings 39th Annual Symposium on Foundations of Computer Science (Cat. No. 98CB36280), IEEE, 1998, pp. 503–509.
[43] Logan Meredith, The CHSH game as a Bell test thought experiment, preprint, 2017, pp. 1–9.
[44] Fei Miao, Quanyan Zhu, Miroslav Pajic, George J. Pappas, A hybrid stochastic game for secure control of cyber-physical systems, Automatica 93 (2018) 55–63.
[45] Martin J. Osborne, Ariel Rubinstein, A Course in Game Theory, MIT Press, 1994.
[46] Craig S. Petrie, J. Alvin Connelly, A noise-based ic random number generator for applications in cryptography, IEEE Transactions on Circuits and Systems. I, Fundamental Theory and Applications 47 (5) (2000) 615–621.
[47] Stefano Pironio, Antonio Acín, Serge Massar, A. Boyer de La Giroday, Dzmitry N. Matsukevich, Peter Maunz, Steven Olmschenk, David Hayes, Lefroy Luo, T. Andrew Manning, et al., Random numbers certified by bell's theorem, Nature 464 (7291) (2010) 1021–1024.
[48] Michael O. Rabin, How to exchange secrets with oblivious transfer, Cryptology ePrint Archive, 2005.
[49] Anupam Sarkar, C.M. Chandrashekar, Multi-bit quantum random number generation from a single qubit quantum walk, Scientific Reports 9 (1) (2019) 12323.
[50] Ivan B. Damgård, Serge Fehr, Renato Renner, Louis Salvail, Christian Schaffner, A tight high-order entropic quantum uncertainty relation with applications, in: Annual International Cryptology Conference, 2007, pp. 360–378.
[51] Giuseppe Vallone, Davide G. Marangon, Marco Tomasin, Paolo Villoresi, Quantum randomness certified by the uncertainty principle, Physical Review A 90 (5) (2014) 052327.
[52] C.H. Vincent, The generation of truly random binary numbers, Journal of Physics. E, Scientific Instruments 3 (8) (1970) 594.
[53] Stephanie Wehner, Christian Schaffner, Barbara M. Terhal, Cryptography from noisy storage, Physical Review Letters 100 (22) (2008) 220502.
[54] Stephen Wiesner, Conjugate coding, ACM Sigact News 15 (1) (1983) 78–88.
[55] Keegan Yao, Walter O. Krawec, Jiadong Zhu, Quantum sampling for finite key rates in high dimensional quantum cryptography, IEEE Transactions on Information Theory 68 (5) (2022) 3144–3163.
[56] Mark Zhandry, How to construct quantum random functions, Journal of the ACM 68 (5) (2021) 1–43.
[57] Lu Zhou, Xin Sun, Chunhua Su, Zhe Liu, Kim-Kwang Raymond Choo, Game theoretic security of quantum bit commitment, Information Sciences 479 (2019) 503–514.

9 Practical quantum cryptography

With few exceptions, we have focused primarily on ideal devices. That is, while there may be noise and adversarial influence along the channel connecting the two parties, the devices that any honest user utilizes are trusted and perfect. For instance, when discussing BB84 in Chapter 4, Alice sends qubits to Bob – on each round, Alice can choose to send a qubit and prepare it in a particular state. In practice, however, this may not be what actually happens. For instance, in quantum communication, the qubits we use are really photons, and they are created using a *weak attenuated laser*. This device will not create a single photon (qubit) on demand; instead, when activated, it outputs n photons, with a Poisson distribution:

$$p(n|\mu) = \frac{e^{-\mu}\mu^n}{n!} \tag{9.1}$$

where, above, μ is the intensity of the source (i.e., the mean number of photons per pulse), which may be set by the user. A plot of this distribution, for various μ, is shown in Fig. 9.1. Higher intensity levels mean that with higher probability two or more photons will be created by the source, every time Alice demands a qubit be sent to Bob. This can be detrimental to security as we will discuss momentarily; however it does mean that each round can transmit data to Bob. On the other hand, lower μ (which will be better for security) will mean that, more likely than not, no photons are sent whenever Alice wants to send something to Bob. Of course, if no photon is sent, no data can be transmitted. Thus, a lower μ, though better for security, is worse from an efficiency point of view. One of the goals in practical QKD is to account for device imperfections such as this, ensuring that security goals are met, while also satisfying user efficiency demands.

9.1 Device imperfections: key-rate derivation

Why are multi photon emissions bad for security? Well, let's recall BB84: on each round of the protocol, Alice needs to send a single qubit to Bob of the form $|0\rangle$, $|1\rangle$, $|+\rangle$, or $|-\rangle$. This qubit travels to Bob who makes a measurement. Only after the qubit has traveled does Alice reveal the basis choice. Now, due to the no-cloning theorem, when the qubit travels between parties, Eve cannot make a copy to measure later. If she could make a copy, security would be completely broken as Eve could simply hold on to the copy, send the original to Bob, and then make a measurement, later, once the basis has been announced. At that point, Eve will have no uncertainty on Alice's data and, so, the key-rate will be zero (see Section 5.3.1, specifically, for instance, the key-rate expression in Eq. (5.16)).

Clearly, copies of qubits are bad for security. So, now, let's return to the practical source devices we actually use. If Alice "accidentally" creates two (or more) photons, Alice is,

Quantum Communication and Cryptography. https://doi.org/10.1016/B978-0-44-321532-2.00016-4

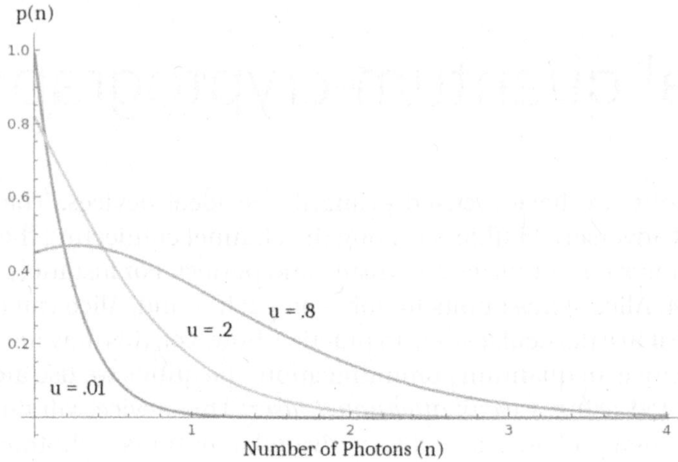

FIGURE 9.1 Showing how the value of μ affects the probability value $p(n|\mu)$ (from Eq. (9.1)).

essentially, creating copies for Eve! Note that, the no-cloning theorem only applies to an unknown state. Since Alice knows the state she is creating, it is not a violation of the theorem for her to make copies (though, in this case, she is making copies unknowingly and unwillingly). Thus, if Alice's device emits two photons, they will both be encoded in her desired state (say the Hadamard "+" state – thus Alice sends to Eve: $|+,+\rangle$, a state no longer living in a two dimensional Hilbert space as we had been assuming so far). Eve, now, receives both qubits (photons) and keeps one of them. The other she transmits to Bob. Since, as discussed in Chapter 5, we may assume all noise is caused by the adversary, this is a "noise free attack" which gives Eve full information.

Clearly, the fact that we are sending multiple photons some of the time is detrimental to security: every round where Alice sends two or more photons, is completely lost as far as security is concerned. On the other hand, every time Alice sends no photons, that round is also lost, as there is no data transmitted. Only on those rounds where a single photon is sent can data be securely transmitted, assuming the noise in such rounds is "low enough." The problem is exacerbated by two facts: (1) Alice never actually knows when she sends zero, one, or two (or more) photons, she only knows the likelihood of such events, according to Eq. (9.1); and (2) Eve is allowed to count the number of photons being sent and can alter her attack strategy based on the number of photons in the channel on any particular round.

Let's consider a particularly strong attack, known as the Photon Number Splitting (PNS) attack [6,7]. Alice, using a weak attenuated laser, sends photons to Bob as dictated by the BB84 protocol. Eve, on each round, will count the number of photons; if that number is one (i.e., the "ideal" case for Alice and Bob), Eve will simply discard the photon (i.e., it will "get lost" as far as Bob is concerned as he will not see anything). If the number of photons is two or more, Eve will forward one of the photons to Bob, while keeping the others.

Now, Alice and Bob understand that they are not using perfect devices and channels. Any realistic quantum channel is going to have some loss to it (we will get back to the expected loss over fiber channels shortly). Even if there is no adversary, some, or many, of the photons sent will be lost. Thus Bob not detecting photons is not too surprising (though, he may be surprised at exactly *how many* of the photons are lost in this particular attack; this is an important detail, but let's not dwell on this for the immediate moment).

However, let's consider the key-rate of the system. Returning to the Devetak-Winter keyrate expression [11,30], discussed in Section 5.3.1, we must compute $H(A|E)$ (Eve's uncertainty) and $H(A|B)$ (Bob's uncertainty). However, these random variables A, B, and quantum ancilla E, are all conditioned on a raw key bit being distilled. Thus, the A register is not Alice's initial raw key selection, but instead her raw key bits, conditioned on Alice and Bob using the round. In the practical BB84 setup, Bob will tell Alice to discard any particular round where he did not see any photons (of course – what else can he do but guess at Alice's bit, which will introduce a 50% error rate in those instances). It is not difficult to see that, conditioned on Alice and Bob using any particular round (i.e., conditioned on Bob seeing something and both parties using the same basis), Eve will know *everything* about that round. This is because such an event can only occur when Alice sends multiple photons. Since Eve keeps the "copies" she will later measure in the correct basis, learning Alice and Bob's raw key bit. Thus, $H(A|E) = 0$. It also holds that this attack creates no noise, thus, also $H(A|B) = 0$. However, this leads to an overall key-rate of zero, implying that the resulting raw keys are completely useless, and Alice and Bob have no other option than to simply abort.

Of course, the above is a very strong attack and, in general, can be detected since the loss will be much higher than Alice and Bob may normally expect. However, Eve can do something more clever: instead of blocking *all* the single-photon rounds, she will only block some of them. She can thus learn a lot of the final raw key, while also creating a loss rate that is "acceptable" to Alice and Bob.

We are left with two questions: one how do we compute the key-rate now? And two, what can be done to improve the situation for Alice and Bob? The first question we will address now; the second we will discuss in Section 9.2.

Let's just consider the asymptotic case for now, assuming collective attacks, in which case we may use the Devetak-Winter keyrate. Let's say Alice and Bob run a QKD protocol, and Eve attacks with some attack strategy; her strategy may depend on the number of photons sent. Now, let's say, if Alice actually sends n photons on any particular round, the final state of Alice, Bob, and Eve can be described (modeled) by the density operator $\rho_{ABE}^{(n)}$. This system models Alice and Bob's system, on a single round, along with Eve's quantum ancilla in this event. In particular, it also models Bob's decision to accept (if he saw something) or reject (if his detectors did not see any photon). Thus this state also has a separate "accept/reject" register. In particular, we may write this state as follows:

$$\rho_{ABE}^{(n)} = p(\text{accept}|n) \, |\text{accept}\rangle \langle\text{accept}| \otimes \rho_{ABE}^{(n,acc)} + p(\text{reject}|n) \, |\text{reject}\rangle \langle\text{reject}| \otimes \rho_{ABE}^{(n,rej)}, \quad (9.2)$$

where $p(\text{accept}|n)$ is the probability that, conditioned on Alice actually sending n photons initially, Alice and Bob choose to "accept" this round, or use it to distill a raw key bit. Otherwise, they will reject (if Bob saw nothing). We save the choice of parties accepting or rejecting the round in this extra classical register as above.

Given the above, the entire state may be written as:

$$\rho_{ABE} = \sum_{n\geq 0} p(n)\rho_{ABE}^{(n)}, \tag{9.3}$$

where $p(n) = p(n|\mu)$ is the probability that Alice sends n photons (e.g., Eq. (9.1)). Using Bayes' theorem, and permuting subspaces, we may write the above state as:

$$p(\text{accept})\,|\text{accept}\rangle\langle\text{accept}| \otimes \left(\sum_{n\geq 0} p(n|\text{accept})\rho_{ABE}^{(n,acc)}\right) \cdot$$
$$+p(\text{reject})\,|\text{reject}\rangle\langle\text{reject}| \otimes \left(\sum_{n\geq 0} p(n|\text{reject})\rho_{ABE}^{(n,rej)}\right) \tag{9.4}$$

Then, finally, conditioning on Alice and Bob accepting, the final state collapses to:

$$\sum_{n\geq 0} p(n|\text{accept})\rho_{ABE}^{(n,acc)} \tag{9.5}$$

Note that, while Alice and Bob know $p(n)$, they do not know, exactly, $p(n|\text{accept})$ as Eve has great influence over this distribution. For instance, she may set $p(n|\text{accept}) = 0$ whenever $n = 1$ (as the above discussed PNS attack will do). This state, above, is the conditional state, modeling Alice and Bob's raw key (along with Eve's quantum side information) in the event Alice and Bob do not discard the system.

We need to compute $H(A|E) - H(A|B)$ of the above. Since the A system is classical, due to properties of von Neumann entropy, we can compute:

$$H(A|E)_\rho \geq \sum_{n\geq 0} p(n|\text{accept})H(A|E)_{\rho^{(n,acc)}} \tag{9.6}$$

Thus, the quantum entropy will be bounded by the average over all n photon events, weighted by the conditional probabilities. For $H(A|B)$, let Q_μ be the average raw key error rate over all n photon events (which Alice and Bob may observe directly). Namely it is the error in their raw key; this is readily seen to be:

$$Q_\mu = \sum_{n\geq 0} p(n|accept)Q_n, \tag{9.7}$$

where Q_n is the error probability conditioned on Alice sending n photons and both parties accepting the round. Note the μ subscript indicates Q_μ's dependence on the power setting.

Let G_μ be the average *gain* of the system, namely:

$$p(\text{accept}) = G_\mu = \sum_{n\geq 0} p(n)p(\text{accept}|n) \tag{9.8}$$

which is the probability that Alice and Bob "accept" any particular round.

Now, it can be shown that $H(A|B) = h(Q_\mu)$ (see Exercise 9.1), namely the binary Shannon entropy of the actual raw key error (which is the error rate, averaged over all n photon events, conditioned on Alice and Bob accepting).

Exercise 9.1. Prove that $H(A|B) = h(Q_\mu)$.

Returning to BB84, we know that there exists an attack which causes Eve to learn everything anytime $n \geq 2$, thus $H(A|E)_{\rho^{(n,acc)}} = 0$, whenever $n \geq 2$. If $n = 0$, Eve can learn nothing (and, thus, $H(A|E)_{\rho^{(0,acc)}} = 1$ since she has maximal uncertainty in this event). Finally, we need $H(A|E)_{\rho^{(1,acc)}}$. However, this is the event where Alice really does send one photon – that is, the event where Alice and Bob run the ideal BB84 system; this was analyzed in Chapter 5 and it was shown that Eve's uncertainty in this case is $1 - h(Q_1)$, where Q_1 is the noise Eve's attack induces in the single photon case. Note that Alice and Bob cannot directly observe Q_1 since they cannot know exactly when Alice sends a single photon. This we will return to shortly.

Thus, Eq. (9.6) becomes:

$$H(A|E) \geq p(0|\text{accept}) + p(1|\text{accept})H(A|E)_{\rho^{(1,acc)}} = p(0|\text{accept}) + p(1|\text{accept})(1 - h(Q_1)), \tag{9.9}$$

since all other terms will be zero.

Now, let m be the size of the raw key and ℓ the size of the secret key. By the Devetak-Winter key-rate expression (see Section 5.3.1), we know that, asymptotically:

$$\frac{\ell}{m} \approx H(A|E) - H(A|B). \tag{9.10}$$

Let N be the total number of signals sent (i.e., the total number of rounds). Then, recalling the definition of G_μ, it is expected that $m = G_\mu N$. Thus:

$$\frac{\ell}{N} \approx G_\mu(H(A|E) - H(A|B)) \geq G_\mu(p(0|\text{accept}) + p(1|\text{accept})(1 - h(Q_1)) - h(Q_\mu)). \tag{9.11}$$

Let

$$Y_n = p(\text{accept}|n). \tag{9.12}$$

Then, by Bayes' theorem, we may write $p(n|\text{accept}) = Y_n p(n)/G_\mu$ (note that $p(n)$ is actually dependent on μ which will be especially important in Section 9.2). Eq. (9.11), then, is simply:

$$\frac{\ell}{N} \approx Y_0 p(0) + Y_1 p(1)(1 - h(Q_1)) - G_\mu h(Q_\mu). \tag{9.13}$$

The above equation and derivation is often known as the GLLP bound, after the original authors who derived it in [14]. The critical element of the above expression is to determine good bounds on Y_0, Y_1, and Q_1. All other variables, above, are known to Alice and Bob (Q_μ and G_μ may be observed, while $p(n)$ is given in Eq. (9.1)). Without additional information, one must take a worst-case bound on these single and vacuum photon events. However, this causes the key-rate to decrease drastically, limiting maximal tolerated distances and noise in practice. It wasn't until later, when the so-called *decoy state protocol* was discovered [17,36,25], that key-rates and, importantly, distances, could be substantially improved.

9.2 Decoy state BB84

The decoy state BB84 protocol [17,36,25] was an important development in mitigating the issues arising from the use of practical devices. Recall that a source device emits n photons with some probability $p(n|\mu)$, where μ is the intensity setting of the laser source (see Eq. (9.1)). The value of μ may be set by the users; importantly for the decoy state protocol, μ may be randomly chosen and changed each round of the protocol. While the adversary may adapt her attack based on the actual number of photons leaving Alice on any particular round, she is not able to determine the actual setting of μ, and thus cannot change her attack based on μ. This is what will allow Alice and Bob to more accurately estimate Y_0, Y_1, and Q_1, needed to evaluate the final key-rate expression, given in Eq. (9.13).

The protocol works, by having Alice choose, randomly on each round, to send a signal using a power level μ chosen from a set of potential power levels $M = \{\mu_1, \mu_2, \cdots\}$. Typically $|M| = 2$ or 3, though we will first consider the "infinite" decoy state protocol which gives optimal asymptotic performance. Interestingly, choosing a more practical setting of two or three provides almost optimal results in many scenarios since $p(n|\mu)$ vanishes, rapidly, for higher-order n.

How does changing the power level help? Well, recall, $p(n) = p(n|\mu)$ is the probability of sending n photons. For every power level that can be chosen, users may, independently, estimate G_{μ_i} and Q_{μ_i}, for each $\mu_i \in M$. Let's see how this can help with determining Y_n. From Eq. (9.8), we know:

$$
\begin{aligned}
G_{\mu_1} &= p(0|\mu_1)Y_0 + p(1|\mu_1)Y_1 + p(2|\mu_1)Y_2 + \cdots \\
G_{\mu_2} &= p(0|\mu_2)Y_0 + p(1|\mu_2)Y_1 + p(2|\mu_2)Y_2 + \cdots \\
G_{\mu_3} &= p(0|\mu_3)Y_0 + p(1|\mu_3)Y_1 + p(2|\mu_3)Y_2 + \cdots \\
&\vdots
\end{aligned} \tag{9.14}
$$

Substituting in the definition of $p(n|\mu)$ (Eq. (9.1)) yields the following system of equations:

$$
G_{\mu_1} = e^{-\mu_1}Y_0 + e^{-\mu_1}\mu_1^1 Y_1 + \frac{e^{-\mu_1}\mu_1^2}{2!}Y_2 + \cdots \tag{9.15}
$$

$$G_{\mu_2} = e^{-\mu_2} Y_0 + e^{-\mu_2} \mu_2^1 Y_1 + \frac{e^{-\mu_2} \mu_2^2}{2!} Y_2 + \cdots$$
$$G_{\mu_3} = e^{-\mu_3} Y_0 + e^{-\mu_3} \mu_3^1 Y_1 + \frac{e^{-\mu_3} \mu_3^2}{2!} Y_2 + \cdots$$
$$\vdots$$

A similar set of equations for Q_n may be expressed using the observed average error Q_{μ_i}. Since the adversary is unable to know the exact power level chosen, only the actual photon number leaving Alice's lab, she cannot change her attack based on the setting of μ. Thus, in the above systems of equations, the Y_n are constant; however each $p(n|\mu_i)$ varies in a way Alice controls. The decoy state method solves these systems of equations for each Y_n. If one chooses an infinite number of decoy states, then one may solve for each Y_n (and also Q_n using Q_{μ_i}) directly.

To summarize, to compute the key-rate of the BB84 protocol under practical devices necessarily involves bounding an expression shown in Eq. (9.13). This requires knowledge of certain values such as Y_0, Y_1 and Q_1. Eve controls these and they are not directly observable. Instead, Alice and Bob can only observe the averages G_μ and Q_μ. Before the decoy state protocol, parties had to make worst case assumptions on Y_0, Y_1, and Q_1 which gave the adversary great advantage. Thanks to the decoy state protocol, one may find either exact (in the infinite decoy state case) or very good (in the finite decoy state case, discussed later) estimates of these three values, leading to more optimistic key-rates.

9.2.1 Simulation

Let's evaluate the resulting key-rate (Eq. (9.13)) to determine how the decoy state BB84 protocol performs. To do this, we need to determine expected values for G_μ and Q_μ (for each power level $\mu \in M$). We will assume an infinite number of decoy states are available; however one power level will be a distinguished "signal" power level. Raw keys will be distilled only when Alice chooses this power level (any other power level will be used for estimating channel statistics only).

For this evaluation, we need to determine how a standard optical QKD implementation might behave. Of course, if the reader has access to actual QKD equipment, one could just run it and observe the various G_μ and Q_μ values! However, when working in theory, we often do not have access to actual equipment and so must test our equations on models that describe expected behavior. For the below derivations, we use results from [26]. We will simulate expected values of G_μ and Q_μ (for any power level μ) by simulating expected values of Y_n and Q_n.

Let's say Alice emits n-photons. The probability that a single photon will be transmitted across the channel (without loss) is denoted t. If we assume fiber channels, over a distance of d km, then:

$$t = 10^{-\alpha d/10}, \tag{9.16}$$

where α is the *loss coefficient* which we may take to be $\alpha = 0.21$ here (this value depends on the setup of the devices). Now, let's say that Bob's detectors have an efficiency of η, meaning that, if a photon hits them, they will detect it with probability η (otherwise the photon is not detected). The total probability, then, of a single photon being received is ηt. In general, one may then assume (for our simulation purposes) that the probability of an n-photon state being detected, is

$$\eta_n = 1 - (1 - t\eta)^n. \tag{9.17}$$

Now, we need G_μ; we will compute the estimated value of this. Again, were one to actually implement a standard optical QKD setup, one could simply observe this quantity. For the expected value, the yield of a zero photon state, namely Y_0 can be expected to be p_d, the *dark-count* rate of the detector (namely, the probability that Bob's measurement device observes a photon, when there actually are none). In practice there may be other factors, such as stray light, for instance however we will only consider the dark count rate, here. Thus, $Y_0 = p_d$ which is typically 10^{-5} or 10^{-6}. The yield of an n-photon state, then, is estimated to be $Y_n = Y_0 + \eta_n(1 - Y_0) \cong Y_0 + \eta_n = p_d + \eta_n$ (the latter follows since p_d and η are both very small typically). This allows us to compute the expected gain as:

$$G_\mu = \sum_n Y_n p(n) = Y_0 + 1 - e^{-\eta_1 \mu} = p_d + 1 - e^{-\eta_1 \mu}. \tag{9.18}$$

These expressions make some assumptions on the independence of photons and background noise, see [26] for details. As for Q_μ, we may set this to be based on the dark-count rate, and the detector misalignment. In particular,

$$G_\mu Q_\mu = \frac{1}{2} Y_0 + e_{det}(1 - \exp(-\eta_1 \mu)) = \frac{1}{2} p_d + e_{det}(1 - \exp(-\eta_1 \mu)). \tag{9.19}$$

Note that the $\frac{1}{2}$ in front of the Y_0 is due to the fact that we assume in the event Alice sends no photons and Bob actually sees something, his result will be random – thus, there is only a 50% chance he will get the right raw key bit in that case (i.e., $Q_0 = 1/2$). The value of e_{det} is the probability that a single photon is observed by the wrong detector.

We also simulate an expected value of $Y_1 Q_1$ to be:

$$Y_1 Q_1 = Q_0 Y_0 + e_{det} \eta_1 = \frac{1}{2} p_d + e_{det} \eta_1. \tag{9.20}$$

See [26] for more details on the above expected values.

This allows us to write the key-rate equation from Eq. (9.13) in the following form:

$$\frac{\ell}{N} \approx \overbrace{p(0)p_d}^{p(0)Y_0} + p(1)\overbrace{(p_d + \eta_1)}^{Y_1}\overbrace{\left(1 - h\left(\frac{\frac{1}{2}p_d + e_{det}\eta_1}{p_d + \eta_1}\right)\right)}^{1-h(Q_1)}$$

$$- \underbrace{(p_d + 1 - \exp(-\eta_1\mu))}_{G_\mu} h \underbrace{\left(\frac{\frac{1}{2}p_d + e_{det}(1 - \exp(-\eta_1\mu))}{p_d + (1 - \exp(-\eta_1\mu))} \right)}_{Q_\mu} \tag{9.21}$$

Above, $p(n) = p(n|\mu)$, for the distinguished signal state power level μ which parties may optimize over. A plot of this is shown in Fig. 9.2.

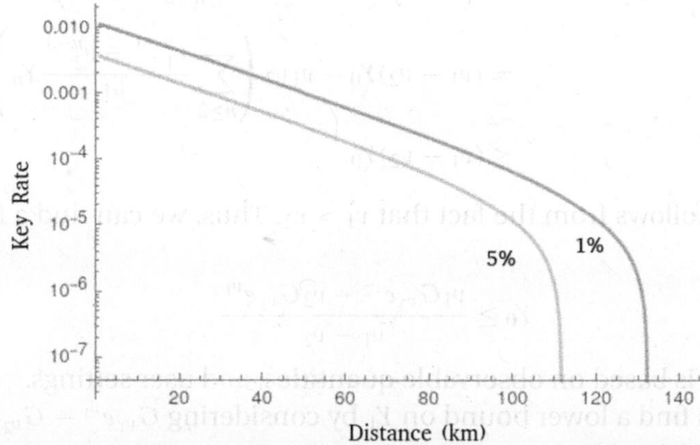

FIGURE 9.2 Evaluating the decoy-state BB84 key rate expression (Eq. (9.21)) in the asymptotic scenario. Here, we evaluate the case where $e_{det} = 1\%$ [Blue (dark gray in print version)] and $e_{det} = 5\%$ [Yellow (light gray in print version)]. For comparison, evaluating Eq. (9.13) without the decoy-state method yields a maximal tolerated distance of under 40 km under these conditions [25].

9.2.2 Three decoy state method

In practice, we cannot use infinite decoy states. Instead, one only has two or three decoy states to choose from; however, since $p(n|\mu)$ vanishes quickly for higher order n, one may determine fairly tight bounds on Y_0, Y_1, and Q_1 even for small $|M|$, given equations of the form shown in Eq. (9.15). Exact details on these derivations can be found in [26], however we summarize some of the key-points here for the $M = 3$ case. Let's assume, as before, that the protocol is such that key bits are only distilled from rounds where Alice chooses a particular power level $\mu \in M$, the signal state power level. Note that, newer key-rate computations actually utilize all power levels for raw key distillation, improving efficiency [24], however the proofs are slightly more involved. In practice, Alice may bias her choice of power level towards μ in order to improve efficiency (similar to how she may bias towards the Z basis).

Let ν_1 and ν_2 be the other two levels in M, such that $0 \le \nu_2 < \nu_1$ and $\nu_1 + \nu_2 < \mu < 1$. These equations must be satisfied in order to derive suitable bounds, as will be clear momentarily.

Recalling that $p(n|\mu) = e^{-\mu}\mu^n/n!$ (and similarly for other power levels), and $G_{\nu_i} = \sum_{n\geq 0} p(n|\nu_i)Y_n$, consider the difference $\nu_1 G_{\nu_2} e^{\nu_2} - \nu_2 G_{\nu_1} e^{\nu_1}$ (see Eq. (9.15)):

$$\nu_1 G_{\nu_2} e^{\nu_2} - \nu_2 G_{\nu_1} e^{\nu_1} = \left(\nu_1 Y_0 + \nu_1 \nu_2 Y_1 + \nu_1 \nu_2 \sum_{n\geq 2} \frac{\nu_2^{n-1}}{n!} Y_n\right) - \left(\nu_2 Y_0 + \nu_2 \nu_1 Y_1 + \nu_2 \nu_1 \sum_{n\geq 2} \frac{\nu_1^{n-1}}{n!} Y_n\right) \tag{9.22}$$

$$= (\nu_1 - \nu_2) Y_0 - \nu_1 \nu_2 \left(\sum_{n\geq 2} \frac{\nu_1^{n-1} - \nu_2^{n-1}}{n!} Y_n\right) \tag{9.23}$$

$$\leq (\nu_1 - \nu_2) Y_0. \tag{9.24}$$

The last inequality follows from the fact that $\nu_1 > \nu_2$. Thus, we can find a lower-bound on Y_0 as:

$$Y_0 \geq \frac{\nu_1 G_{\nu_2} e^{\nu_2} - \nu_2 G_{\nu_1} e^{\nu_1}}{\nu_1 - \nu_2} \tag{9.25}$$

Note that the above is based on observable quantities, and user settings.

Similarly, we can find a lower bound on Y_1 by considering $G_{\nu_1} e^{\nu_1} - G_{\nu_2} e^{\nu_2}$. Then, it can be shown that:

$$G_{\nu_1} e^{\nu_1} - G_{\nu_2} e^{\nu_2} = \left(Y_0 + \nu_1 Y_1 + \sum_{n\geq 2} \frac{\nu_1^n}{n!} Y_n\right) - \left(Y_0 + \nu_2 Y_1 + \sum_{n\geq 2} \frac{\nu_2^n}{n!} Y_n\right) \tag{9.26}$$

$$= Y_1(\nu_1 - \nu_2) + \sum_{n\geq 2} \frac{Y_n}{n!} \left(\nu_1^n - \nu_2^n\right) \tag{9.27}$$

$$= Y_1(\nu_1 - \nu_2) + \sum_{n\geq 2} \frac{Y_n}{n!} \left(\mu^n \left(\frac{\nu_1^n - \nu_2^n}{\mu^n}\right)\right) \tag{9.28}$$

$$\leq Y_1(\nu_1 - \nu_2) + \frac{\nu_1^2 - \nu_2^2}{\mu^2} \sum_{n\geq 2} Y_n \frac{\mu^n}{n!} \tag{9.29}$$

$$= Y_1(\nu_1 - \nu_2) + \frac{\nu_1^2 - \nu_2^2}{\mu^2} (G_\mu e^\mu - Y_0 - Y_1 \mu). \tag{9.30}$$

Exercise 9.2. Using the fact that $x^n - y^n \leq x^2 - y^2$ when $0 < x + y < 1$ and $n \geq 2$, explain how the above bound is derived. You will need the required constraints on the relationship between the three power levels.

Continuing, we determine the following bound:

$$Y_1 \geq \frac{\mu}{\mu\nu_1 - \mu\nu_2 - \nu_1^2 + \nu_2^2} \left(G_{\nu_1} e^{\nu_1} - G_{\nu_2} e^{\nu_2} - \frac{\nu_1^2 - \nu_2^2}{\mu^2} (G_\mu e^\mu - Y_0^L)\right), \tag{9.31}$$

where Y_0^L is the lower-bound for Y_0, from Eq. (9.25). Similar techniques can be used to find *upper* bounds on Q_1. Specifically:

$$Q_1 \le \frac{Q_{\nu_1} G_{\nu_1} e^{\nu_1} - Q_{\nu_2} G_{\nu_2} e^{\nu_2}}{(\nu_1 - \nu_2) Y_1} \tag{9.32}$$

For more details on the above derivations, and for alternative decoy state methods, the reader is referred to [26]. Importantly, three power levels allow us to determine bounds on Y_0, Y_1, and Q_1. Unlike the infinite decoy state protocol, we do not get exact values, but only bounds. However, the following exercise will show you that the performance of this practical three decoy state protocol is not too inferior to the infinite version.

Exercise 9.3. Evaluate the three-decoy state method using simulated expected values G_μ and Q_μ, from the previous section. In particular plot the resulting key-rate when three power levels are used. Try varying the three power levels (make sure they satisfy the given constraints!). Try optimizing them. Compare to the infinite decoy state analysis, discussed in the previous section.

Exercise 9.4. Show how the bound given in Eq. (9.32) is derived. Start with $Q_{\nu_1} G_{\nu_1} e^{\nu_1} - Q_{\nu_2} G_{\nu_2} e^{\nu_2}$.

9.2.3 Alternative protocols

While the decoy state protocol is, today, the standard approach to mitigating practical device imperfections, it is not the only method; earlier work tried to compensate by designing novel quantum protocols. One in particular is the SARG04 protocol [1]. Note that Eq. (9.13) is specific to BB84 style protocols; Eq. (9.6) is general. Given an alternative protocol choice, different expressions may be derived for the various multi-photon entropy terms. One of the critical problems with BB84 in this particular investigation is that two (or more) photons emitted by Alice completely destroy the security of that particular round. This is because, once the basis choice is known, Eve may make the correct measurement on the "copy." However, alternative encoding styles may, and do, mitigate this.

Recall the SARG04 protocol, discussed in Section 4.3.1. This protocol involves Alice choosing a random state (in the Z or X basis) to send, similar to BB84. Now, Bob will make a measurement in a randomly chosen basis, resulting in outcome b. Afterwards, Alice will send a classical message to Bob consisting of a description of the state she sent, along with a description of a randomly chosen state *in the alternative basis.* Note that she sends the description of the Z basis state first, then the X basis state – if she always sent a description of the state she sent first, this would tell Eve everything since, recall, all classical messages are sent over the authenticated, but not secret, classical channel. Bob, now, knows the following: he knows what he observed, b, and he knows two states, one of which is the state Alice actually sent, while the other is a randomly chosen state.

Let's illustrate with an example. Say Alice chose to send $|+\rangle$. Later, she sends the classical message "$0, +$." (She might also have sent "$1, +$.") Now, if Bob measured in the X basis, assuming no noise, he should have observed $|+\rangle$. On the other hand, if he had measured in the Z basis, he might have observed either $|0\rangle$ or $|1\rangle$. If Bob observes $|1\rangle$, this is something that he knows Alice did *not* send (in the first classical message case of "$0, +$"). Since he saw something he *knows* Alice did not send, he therefore *knows* he measured in the opposite basis that Alice sent in! Thus, Alice and Bob will use the basis as the key-encoding: If Alice sends a Z basis state, this will be a key-bit of zero; otherwise it will be a key-bit of one. Bob can learn the key choice by (1) measuring in the wrong basis and (2) seeing something not in the set of states that Alice sent classical descriptions of. Anytime Bob observes a state that Alice described to him in her message, he will consider this "inconclusive" and both parties discard the round. See Section 4.3.1 for more details.

Now, how does this help with the issue of multi-photon sources? Well, at a high level, even if Alice accidentally sends Eve two copies of the state she sent, Eve is not guaranteed to learn the key bit. Indeed, to learn the key-bit with certainty, one must measure in the wrong basis (and the basis is never disclosed this time!) and one must also observe a state not in the disclosed set of two states. Thus, giving Eve a copy "for free" (via a weak attenuated laser source), is not guaranteed to give Eve complete information for that particular round. Namely $H(A|E)_{\rho^{(2,acc)}}$ is not necessarily zero, as it is with BB84 and so even two photon rounds can contribute meaningfully to the final secret key.

This shows the importance of QKD protocol design. Some protocols may work very well in theoretical conditions, others may work well in practical scenarios. As it turns out, SARG04 does often, but not always, work better than *standard* (i.e., the non-decoy state version) BB84 in these practical device conditions [5,37]; however, it typically does not work better (i.e., more efficiently) than the decoy-state BB84 protocol. Today, most practical implementations use the decoy-state BB84 protocol as it is efficient, and allows for very good bounds on Q_1 and the needed conditional probabilities discussed earlier.

9.3 Practical device considerations

We already commented on some of the device imperfections when discussing the decoy-state protocol, above. However there are other imperfections that should be considered. There are also several "encoding" methods that may be used; we'll discuss some of these here, shortly.

As discussed, practical QKD systems utilize photons for qubit encodings. To measure a qubit, in this case, requires the use of photon detectors. In the settings we consider, these are typically *avalanche photodiodes* (APDs). Other technologies are also available, such as superconducting nanowire single-photon detectors (SNSPDs). Note, also, that there are other QKD protocols utilizing so-called *continuous variables* which use different technologies; for more information on these, and other technologies, the reader is referred to [29,3].

APD detectors have several important properties that affect the performance of the system (in particular, the accuracy of the measurements). Some of these have already been discussed, though we summarize them below, while some new ones are also listed (see [29,3] for more details):

- Dark Count Rate: This is the probability that the photon detector "clicks" (or reports the observation of a photon) even when no photon has actually hit it. This is a "false signal."
- Efficiency: This is the probability that the detector actually clicks when a photon hits it. If the efficiency is low, photons may hit the detector, but they will not be detected and the measurement will assume no photon was actually there.
- Dead Time: This is the time required for the detector to reset after a detection. This can affect the overall efficiency of the protocol.
- Afterpulsing: This is the probability that a single photon detection actually causes multiple reports or clicks.
- Timing Jitter: This is the uncertainty of the measurement concerning the actual time at which the photon detector actually detected a photon.

Different technologies for implementing these detectors exist with different properties. For instance, so called InGaAs APDs can operate at telecom wavelengths and do not need to be super-cooled, but can have a very low efficiency (e.g., 20%). On the other end, superconducting nanowire SPDs can operate with very high efficiency of 90% but require cryogenic temperatures. For more details, the reader is referred to the surveys [29,3] and the references within.

9.3.1 Qubit encoding

Now that we've discussed source and measurement devices, we can discuss some of the (many) ways in which qubits can be transmitted, or "encoded" on photons. There are three main methods that we'll discuss: *polarization, phase,* and *time-bin.* Each has their own advantages and disadvantages. Some protocols are designed to work best with a specific encoding method.

Polarization encoding is one of the more basic methods, and we'll spend some time discussing it here. Here, a photon is emitted from a source and its polarization is altered to encode the desired qubit state. A typical encoding scheme has the state $|0\rangle$ being horizontally polarized while a qubit in the state $|1\rangle$ is vertically polarized. Superposition states, such as $|+\rangle$, can then be created using a 45 degree polarization.

Early implementations used multiple laser sources to create various polarized photons, each source connected, optically, to a mechanism capable of creating the desired polarization. For instance, in BB84, where four different qubit states are required, there would be four lasers, one to produce $|0\rangle$, one to produce $|1\rangle$, one for $|+\rangle$, and a fourth for $|-\rangle$. Whenever Alice wanted to send, say, a $|+\rangle$, she would activate that one particular laser device. Many early experimental implementations used this method (see, just for some examples [4,35,23,22]); however such a setup leads to potential side-channel attacks [20] as each laser device has slightly different optical properties. These properties are something that an adversary can potentially utilize to learn which state was sent from Alice.

An alternative, and more secure method of utilizing polarization encoding, is through non-passive optical components. For instance, *electro-optic modulators* which vary the polarization of a photon emitted by a single laser [18,15,2]. This is the preferred way to use polarization encoding. (Though, as with any physical system, classical or quantum, one must still be careful of side-channel attacks!)

To measure a polarized photon, one may use either passive or active methods. In general, a measurement involves having the photon pass through a device called a *polarized beamsplitter (PBS)* which causes the photon to travel in one direction or another based on its orthogonal polarization. On the output of each path of the PBS is a single photon counter (discussed above) which will "click" if a photon hits it. Two detectors are required for each basis measurement, one for each output of the PBS; this implements a basis measurement. Optical components placed before the PBS (such as half-wave plates) may be used to implement alternative basis measurements; see [39,28] for some details on the optical implementation of polarization encoded QKD.

Phase encoding [16,27] is an alternative method to polarization, and is based on the use of interferometers [13]. Here, a source prepares a photon and passes it through the interferometer. After passing through the first beamsplitter, the photon travels, in one of the arms, through a device that can modulate its phase. The signal is recombined (note that one of the arms of the interferometer must be longer than the other to avoid interference at this second beamsplitter), and then later, at the receiver, a similar interferometer, alters the phase again. If the two phase modulators are set so that the difference in their phase shift is zero or 180 degrees, the second (receiver) interferometer's output will be deterministic; otherwise it will be random. Several QKD systems used this method, including the so-called "plug and play" QKD system [33].

Finally, time bin encoding is a process allowing for a photon to be created in multiple time bins, in superposition. This has the added benefit of allowing a single photon to be used to create a high-dimensional qubit state (useful for several QKD protocols as discussed in Chapter 7). For this method, a photon may be emitted in one of *d time bins*. By recording the time of arrival of the photon, one can make the equivalent of a Z basis measurement; namely, creating a photon in time-bin zero will represent a $|0\rangle$, while a photon in time bin one will be a $|1\rangle$. (Higher dimensional states may be created by using additional time bins.)

While this creates qubits (or qudits) prepared in the Z basis, of course for quantum cryptography we need to be able to create states in superposition; otherwise, an adversary could simply measure and record the time of arrival, ultimately making a Z basis measurement and then recreating the photon and sending it to Bob at the correct time-bin for Bob. However, if we can create a photon in the superposition state:

$$\frac{1}{\sqrt{2}}(|\texttt{time-bin-0}\rangle + |\texttt{time-bin-1}\rangle) = \frac{1}{\sqrt{2}}|0\rangle + \frac{1}{\sqrt{2}}|1\rangle, \tag{9.33}$$

then we can implement a $|+\rangle$ (of course, we also need $|-\rangle$ typically too). To create such a state, an unbalanced interferometer may be used. Measurements in this superposition basis can also be achieved through such interferometers; in higher dimensions, one typically requires cascading interferometers [8].

9.4 Quantum networks

While decoy state QKD helps us overcome device imperfections, substantially improving the performance of QKD over a single point to point link, it still does not help us overcome fundamental limitations regarding overall channel loss. Most QKD systems are implemented over fiber channels (free space is, of course, also possible, but requires line of sight). Fiber channels suffer from exponential photon loss as discussed earlier in this section. Due to the need to send single photons (thus, a weak source device), and the fact that loss scales exponentially, it is clear that QKD is highly limited in maximal transmission distances. Since quantum signals cannot be amplified, traditional optical networking devices are not suitable to extend this distance and, thus, new approaches must be investigated.

Quantum networks, like classical networks, consist of multiple *nodes*, interconnected to each other through various links (see Fig. 9.3). In a quantum network, the nodes consist, of end users (e.g., Alice and Bob), along with various other entities, in particular: *optical switches, quantum repeaters,* and *trusted nodes.* These entities allow for complex routing to occur, enabling long distance quantum cryptography, and also allowing multiple source and receiver destinations.

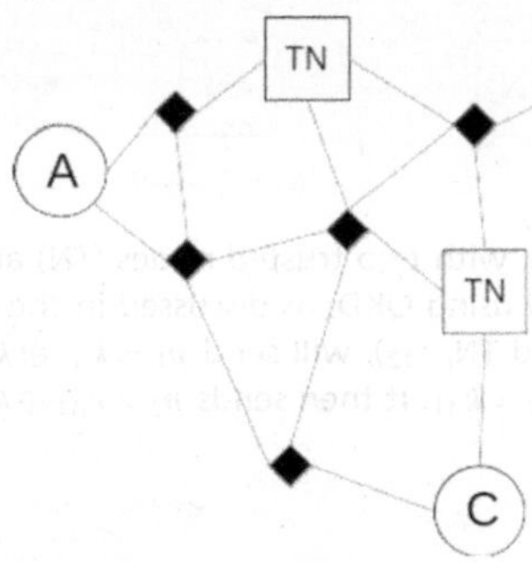

FIGURE 9.3 Showing a basic quantum network, consisting of end-users (A, B, and C); quantum repeaters (the diamonds); and Trusted Nodes (TN).

The very first quantum networks consisted entirely of optical switches [34,12,10]. These are basic devices that route incoming photons from one input port to some other, performing no other operations on the incoming qubits. Essentially, creating a 'telephone' style circuit between different end users. Such switches do not solve any distance limitations of QKD, however, as the photons are essentially routed to a longer circuit; in fact, switches tend to amplify the problems caused by distance, as optical switching devices, themselves, add noise and loss.

9.4.1 Trusted nodes

Today, most QKD networks utilize trusted node (TN) technology [10,28,40,31]. Here, a trusted node is a device that plays the part of an end user, running a full QKD stack with another end-user or trusted node, in order to establish pair-wise secret keys with neighboring nodes and end users. By "QKD stack" we mean the full QKD process consisting, minimally, of the quantum communication stage, sampling, error correction, and privacy amplification. The Trusted Nodes then "encrypts" one secret key using their other secret key, in order to establish pair-wise secret keys between non-adjacent parties.

This process is best illustrated on a chain topology, as shown in Fig. 9.4. Here, A and the first TN (adjacent to A) will establish a shared secret key, k_{A1}, using QKD. Simultaneously, the same TN and the second (right-most) TN will establish a shared secret key k_{12}. Finally, that right-most TN and B will establish a shared secret key k_{2B}. The goal of the network is for Alice and Bob to have a shared secret key. Thus, the first TN will "encrypt" k_{A1} using k_{12} and send the ciphertext to the second TN. That is, the left-most TN sends $k_{A1} \oplus k_{12}$ to the second TN. Since the second TN (nearest Bob) knows k_{12}, this trusted node may "decrypt" the key k_{A1} by XOR'ing the received message with k_{12}. The second TN may then re-encrypt k_{A1} using k_{2B} and send the resulting ciphertext, namely $k_{A1} \oplus k_{2B}$ to Bob. Bob, who knows k_{2B} may now learn k_{A1} thus establishing a key between Alice and Bob. Note that all this classical communication is done over an authenticated channel (e.g., each neighboring node has a quantum link and an authenticated classical channel).

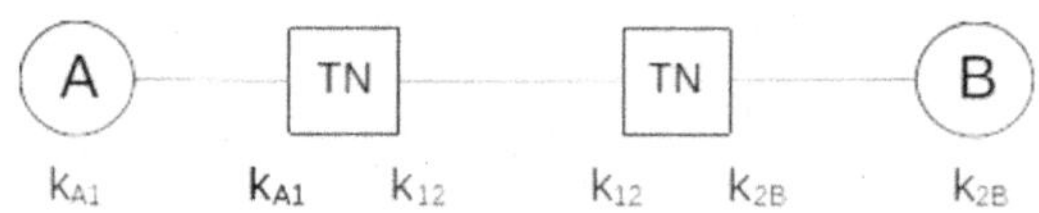

FIGURE 9.4 Showing a basic trusted node chain, with two trusted nodes (TN) and two users, Alice (A) and Bob (B). First, each link will establish a shared secret key using QKD, as discussed in the text. Next, the first TN (who shares a secret key with Alice, k_{A1} and with the second TN, k_{12}), will send $p_1 = k_{A1} \oplus k_{12}$ to the second trusted node. The second TN "decrypts" k_{A1} by computing $p_1 \oplus k_{12} = k_{A1}$. It then sends $p_2 = k_{A1} \oplus k_{2B}$ to Bob. This will allow Bob, who holds k_{2B}, to compute k_{A1} as $k_{A1} = k_{2B} \oplus p_2$.

Exercise 9.5. What are the requirements of the classical communication channels for a trusted node chain, in terms of secrecy and authentication? Do all possible pairs of users need authenticated channels (for instance, in Fig. 9.4, do Alice and the right-most TN need an authenticated channel)? What happens if a channel is not authenticated?

Exercise 9.6. Prove that the trusted node chain protocol is correct. Namely, show that, if there are p TNs in a chain, that Bob will receive the correct key (namely, Alice's key). What happens if, later, an adversary compromises the security of one of the TNs?

Clearly, the above process works – security is guaranteed through the fact that QKD produces uniform random and secret keys and that OTP encryption is perfectly secret (thus, each TN uses one of its keys to encrypt the actual key, k_{A1} to be passed to Bob). Another major advantage of this is that the hardware required to construct a trusted node network is identical to that required for point to point QKD. Namely, trusted node networks are a feasible technology today. This is the reason why current QKD networks utilize this technology. Finally, it is clear that trusted nodes solve the distance problem – when the distance between two parties is too large to efficiently (or at all) establish a QKD key, placing one or more trusted node units between the parties will allow for QKD to operate at speeds roughly matching that of the longest link between two adjacent nodes, since trusted nodes "break" the quantum channel into smaller pieces.

The very obvious downside to trusted node technology is that, as the name implies, the node does indeed need to be *trusted*. This is due to the fact that each trusted node will learn the secret key used by Alice and Bob. If just one trusted node is corrupted on a chain topology, the entire key is compromised. Since this is classical information, if a node is compromised in the future, any information sent using that key in the past is at risk. Clearly this is concerning. A second issue is in practicality: each trusted node must perform a full QKD stack, including the computationally intensive processes of error correction and privacy amplification. In fact, each trusted node must run these processes *twice*! Once with its left neighbor and once with its right. More complex topologies, beyond chains, will lead to even greater computational and communication complexity.

Both problems can be mitigated to some extent, but not fully solved. The security risks can be diluted by using multi-path routing. For example, assume there are two disjoint paths of trusted nodes connecting Alice to Bob. Now, Alice and Bob will use each path to establish two disjoint secret keys k_1 (using the first path) and k_2 (using second path). Trusted nodes on the first path will know k_1 similar to discussed above. Trusted nodes on the second path know k_2. However, trusted nodes on the first path have no knowledge of k_2 and vice versa for the second path and k_1. Thus, if Alice and Bob use as their final secret key $k = k_1 \oplus k_2$, it will be guaranteed that a single compromised trusted node (or even an entire compromised path) cannot compromise the security of the final key. In general, parties may choose to XOR p disjoint paths which will provide a guarantee that the final secret key will remain information theoretic secure as long as no more than $p - 1$ nodes are corrupted.

The second problem, namely the potential efficiency cost due to each node having to execute a full QKD stack, can also be mitigated through the use of so-called *simplified trusted nodes* (STNs). This was a concept introduced in [32] whereby each (simplified) TN will only perform the prepare and measure portion of a QKD protocol, producing a raw key, but no noise sampling, and no error correction or privacy amplification. STNs will broadcast the parity of their produced raw keys (similar to how the TN sends the parity of the produced secret keys). This will allow Alice and Bob to share a raw key from which they will perform error correction and privacy amplification just once, leaving the rest of the network free to continue for other users.

While STNs free up the network nodes from the responsibilities of running potentially expensive classical post-processing operations, there are disadvantages to these systems. Since the raw keys are not error corrected, errors in each link level raw key string will tend to "cascade." In particular, if each link has an expected error rate of Q, then after two hops, the total expected error rate will be $2Q(1-Q)$ (since, to get an error after two hops, one needs an error in one hop but not the other). Furthermore, since the raw key is not a secret key, OTP encryption arguments cannot be used to guarantee secrecy of the underlying message. This means that key-rates tend to suffer if the link level noise is high, as the final key between Alice and Bob will hold more errors than a single link would, while also the broadcasted messages will potentially leak more information to Eve, thus requiring additional privacy amplification (i.e., a smaller final key size).

Security proofs for these networks are still in their relative infancy with only asymptotic proofs of security [32] or finite key-proofs of security but only in the chain topology [21]. Future work exists in further investigating STN performance.

9.4.2 Quantum repeaters

A more complicated solution, from a technological perspective, yet one which, from a security perspective, is far preferred, are quantum repeater networks. A quantum repeater actually performs a Bell swap operation. First, the repeater will attempt to establish shared *link level entanglement*, with its adjacent users. Namely the repeater will share a bell pair $|\Phi^+\rangle_{AR} = \frac{1}{\sqrt{2}}(|00\rangle_{AR} + |11\rangle_{AR})$ with Alice and the Repeater, and a second Bell pair $|\Phi^+\rangle_{RB}$ with the Repeater and Bob. The joint state consists of two separable Bell pairs. At this point, the repeater will perform a Bell measurement on its two qubits (held in the R registers). This will project the AB system to the Bell pair that R observed.

To see why this works, first, note that the state $|\Phi^+\rangle_{AR}|\Phi^+\rangle_{RB}$ can be written, by permuting subspaces so that the two R qubits appear first, followed by the A and B systems, as:

$$|\Phi^+\rangle_{AR} \otimes |\Phi^+\rangle_{RB} \cong \frac{1}{2}\sum_i |\phi_i\rangle_R |\phi_i\rangle_{AB} \tag{9.34}$$

where the above sum is over all four Bell states namely:

$$|\phi_0\rangle = \frac{1}{\sqrt{2}}|00\rangle + \frac{1}{\sqrt{2}}|11\rangle$$
$$|\phi_1\rangle = \frac{1}{\sqrt{2}}|00\rangle - \frac{1}{\sqrt{2}}|11\rangle$$
$$|\phi_2\rangle = \frac{1}{\sqrt{2}}|01\rangle + \frac{1}{\sqrt{2}}|10\rangle$$
$$|\phi_3\rangle = \frac{1}{\sqrt{2}}|01\rangle - \frac{1}{\sqrt{2}}|10\rangle$$

Exercise 9.7. Prove that Eq. (9.34) is correct. Namely that, by permuting subspaces, one may write $|\phi^+\rangle_{AR} |\phi^+\rangle_{RB}$ as depicted in Eq. (9.34).

It is clear from Eq. (9.34) that the repeater system will work; namely, assuming both parties establish shared entanglement with a single repeater node, that repeater can perform a local operation, thus causing entanglement to be formed across the distant parties (who never, directly, communicated with each another). Note that, all four Bell states are equally likely observations from the repeater. Alice and Bob, generally, want a *specific* Bell state shared between them. After the repeater's measurement, technically Alice and Bob share a completely mixed state – totally useless for any operation. However, after measuring, the repeater will send two bits of information to one (or both) of the parties, describing what the observed state was. This message allows a designated "leader" to fix the post-measured state, applying local Pauli operations, to map the conditional state (e.g., $|\phi_i\rangle$) to the protocol-specified state (e.g., $|\phi_0\rangle$). Note that this message from the repeater does not need to be authenticated; instead, an adversary may control and tamper with the classical message. In fact, the repeater may even be completely controlled by the adversary! This will result in potentially the wrong Bell state being distilled, however that can be detected by users, for instance in the QKD sampling stage when using the E91 protocol.

Following the repeater measurement, the Bell state established between Alice and Bob is unentangled with the repeater system. Thus, repeaters give a high level of security: even if the entire network is corrupted, security can still be enforced by checking that the states created between Alice and Bob are proper Bell states, by measuring some in the X and Z basis to test their fidelity. This is unlike the trusted node scenario where a corrupted trusted node learns everything, without being detected.

Quantum repeaters can be chained together, with multiple repeaters performing the necessary Bell measurements and passing along their observations to the leader end-user who must apply local Pauli fixes. They can also be combined in complex network topologies, making for interesting research in quantum network routing. A full-scale network, consisting of only quantum repeaters, is typically called the "Quantum Internet" [19,9,38] and is the ultimate goal in quantum networking; this is because such a network can distribute entanglement over vast distances. Entanglement, can be used for QKD (as seen throughout this book), but it can also be used for a wide variety of other tasks. Trusted node networks, though serving as a useful "first step" towards such a network, really can only serve as QKD systems, since the trusted node breaks entanglement by measuring the received qubits. This makes them practical, today, but perhaps not as useful in the future for general purpose quantum applications.

References

[1] Antonio Acin, Nicolas Gisin, Valerio Scarani, Coherent-pulse implementations of quantum cryptography protocols resistant to photon-number-splitting attacks, Physical Review A 69 (Jan 2004) 012309.

[2] Costantino Agnesi, Marco Avesani, Andrea Stanco, Paolo Villoresi, Giuseppe Vallone, All-fiber self-compensating polarization encoder for quantum key distribution, Optics Letters 44 (10) (2019) 2398–2401.

[3] Omar Amer, Vaibhav Garg, Walter O. Krawec, An introduction to practical quantum key distribution, IEEE Aerospace and Electronic Systems Magazine 36 (3) (2021) 30–55.

[4] Davide Bacco, Matteo Canale, Nicola Laurenti, Giuseppe Vallone, Paolo Villoresi, Experimental quantum key distribution with finite-key security analysis for noisy channels, Nature Communications 4 (1) (2013) 2363.

[5] Cyril Branciard, Nicolas Gisin, Barbara Kraus, Valerio Scarani, Security of two quantum cryptography protocols using the same four qubit states, Physical Review. A, Atomic, Molecular, and Optical Physics 72 (3) (2005) 032301.

[6] Gilles Brassard, Norbert Lütkenhaus, Tal Mor, Barry C. Sanders, Limitations on practical quantum cryptography, Physical Review Letters 85 (6) (2000) 1330.

[7] Gilles Brassard, Norbert Lütkenhaus, Tal Mor, Barry C. Sanders, Security aspects of practical quantum cryptography, in: International Conference on the Theory and Applications of Cryptographic Techniques, Springer, 2000, pp. 289–299.

[8] Thomas Brougham, Stephen M. Barnett, Kevin T. McCusker, Paul G. Kwiat, Daniel J. Gauthier, Security of high-dimensional quantum key distribution protocols using Franson interferometers, Journal of Physics. B, Atomic, Molecular and Optical Physics 46 (10) (2013) 104010.

[9] Marcello Caleffi, Angela Sara Cacciapuoti, Giuseppe Bianchi, Quantum Internet: from communication to distributed computing!, in: Proceedings of the 5th ACM International Conference on Nanoscale Computing and Communication, 2018, pp. 1–4.

[10] Teng-Yun Chen, Jian Wang, Hao Liang, Wei-Yue Liu, Yang Liu, Xiao Jiang, Yuan Wang, Xu Wan, Wen-Qi Cai, Lei Ju, et al., Metropolitan all-pass and inter-city quantum communication network, Optics Express 18 (26) (2010) 27217–27225.

[11] Igor Devetak, Andreas Winter, Distillation of secret key and entanglement from quantum states, Proceedings of the Royal Society A. Mathematical, Physical and Engineering Sciences 461 (2053) (2005) 207–235.

[12] Chip Elliott, Alexander Colvin, David Pearson, Oleksiy Pikalo, John Schlafer, Henry Yeh, Current status of the darpa quantum network, in: Quantum Information and Computation III, vol. 5815, SPIE, 2005, pp. 138–149.

[13] Nicolas Gisin, Grégoire Ribordy, Wolfgang Tittel, Hugo Zbinden, Quantum cryptography, Reviews of Modern Physics 74 (1) (2002) 145.

[14] Daniel Gottesman, Hoi-Kwong Lo, Norbert Lütkenhaus, John Preskill, Security of quantum key distribution with imperfect devices, Quantum Information & Computation 4 (5) (2004) 325–360.

[15] Fadri Grünenfelder, Alberto Boaron, Davide Rusca, Anthony Martin, Hugo Zbinden, Simple and high-speed polarization-based qkd, Applied Physics Letters 112 (5) (2018).

[16] Zheng-Fu Han, Xiao-Fan Mo, You-Zhen Gui, Guang-Can Guo, Stability of phase-modulated quantum key distribution systems, Applied Physics Letters 86 (22) (2005).

[17] Won-Young Hwang, Quantum key distribution with high loss: toward global secure communication, Physical Review Letters 91 (5) (2003) 057901.

[18] M. Jofre, A. Gardelein, G. Anzolin, G. Molina-Terriza, J.P. Torres, M.W. Mitchell, V. Pruneri, 100 mhz amplitude and polarization modulated optical source for free-space quantum key distribution at 850 nm, Journal of Lightwave Technology 28 (17) (2010) 2572–2578.

[19] H. Jeff Kimble, The quantum Internet, Nature 453 (7198) (2008) 1023–1030.

[20] Heasin Ko, Byung-Seok Choi, Joong-Seon Choe, Kap-Joong Kim, Jong-Hoi Kim, Chun Ju Youn, Critical side channel effects in random bit generation with multiple semiconductor lasers in a polarization-based quantum key distribution system, Optics Express 25 (17) (2017) 20045–20055.

[21] Walter O. Krawec, Bing Wang, Ryan Brown, Finite key security of simplified trusted node networks, in: 2024 IEEE International Conference on Quantum Computing and Engineering (QCE), vol. 1, IEEE, 2024, pp. 1777–1787.

[22] Sheng-Kai Liao, Wen-Qi Cai, Wei-Yue Liu, Liang Zhang, Yang Li, Ji-Gang Ren, Juan Yin, Qi Shen, Yuan Cao, Zheng-Ping Li, et al., Satellite-to-ground quantum key distribution, Nature 549 (7670) (2017) 43–47.

[23] Sheng-Kai Liao, Hai-Lin Yong, Chang Liu, Guo-Liang Shentu, Dong-Dong Li, Jin Lin, Hui Dai, Shuang-Qiang Zhao, Bo Li, Jian-Yu Guan, et al., Long-distance free-space quantum key distribution in daylight towards inter-satellite communication, Nature Photonics 11 (8) (2017) 509–513.

[24] Charles Ci Wen Lim, Marcos Curty, Nino Walenta, Feihu Xu, Hugo Zbinden, Concise security bounds for practical decoy-state quantum key distribution, Physical Review A 89 (2) (2014) 022307.

[25] Hoi-Kwong Lo, Xiongfeng Ma, Kai Chen, Decoy state quantum key distribution, Physical Review Letters 94 (23) (2005) 230504.

[26] Xiongfeng Ma, Bing Qi, Yi Zhao, Hoi-Kwong Lo, Practical decoy state for quantum key distribution, Physical Review. A, Atomic, Molecular, and Optical Physics 72 (1) (2005) 012326.

[27] Xiao-Fan Mo, Bing Zhu, Zheng-Fu Han, You-Zhen Gui, Guang-Can Guo, Faraday–Michelson system for quantum cryptography, Optics Letters 30 (19) (2005) 2632–2634.

[28] Momtchil Peev, Christoph Pacher, Romain Alléaume, Claudio Barreiro, Jan Bouda, Winfried Boxleitner, Thierry Debuisschert, Eleni Diamanti, Mehrdad Dianati, James F. Dynes, et al., The secoqc quantum key distribution network in Vienna, New Journal of Physics 11 (7) (2009) 075001.

[29] Stefano Pirandola, Ulrik L. Andersen, Leonardo Banchi, Mario Berta, Darius Bunandar, Roger Colbeck, Dirk Englund, Tobias Gehring, Cosmo Lupo, Carlo Ottaviani, et al., Advances in quantum cryptography, Advances in Optics and Photonics 12 (4) (2020) 1012–1236.

[30] Renato Renner, Security of quantum key distribution, International Journal of Quantum Information 6 (01) (2008) 1–127.

[31] Masahide Sasaki, Mikio Fujiwara, H. Ishizuka, W. Klaus, K. Wakui, M. Takeoka, S. Miki, T. Yamashita, Z. Wang, A. Tanaka, et al., Field test of quantum key distribution in the Tokyo qkd network, Optics Express 19 (11) (2011) 10387–10409.

[32] William Stacey, Razieh Annabestani, Xiongfeng Ma, Norbert Lütkenhaus, Security of quantum key distribution using a simplified trusted relay, Physical Review A 91 (1) (2015) 012338.

[33] Damien Stucki, Nicolas Gisin, Olivier Guinnard, Grégoire Ribordy, Hugo Zbinden, Quantum key distribution over 67 km with a plug&play system, New Journal of Physics 4 (1) (2002) 41.

[34] Paul Toliver, Robert J. Runser, Thomas E. Chapuran, Janet L. Jackel, Thomas C. Banwell, Matthew S. Goodman, Richard J. Hughes, Charles G. Peterson, Derek Derkacs, Jane E. Nordholt, et al., Experimental investigation of quantum key distribution through transparent optical switch elements, IEEE Photonics Technology Letters 15 (11) (2003) 1669–1671.

[35] Gwenaelle Vest, Markus Rau, Lukas Fuchs, Giacomo Corrielli, Henning Weier, Sebastian Nauerth, Andrea Crespi, Roberto Osellame, Harald Weinfurter, Design and evaluation of a handheld quantum key distribution sender module, IEEE Journal of Selected Topics in Quantum Electronics 21 (3) (2014) 131–137.

[36] Xiang-Bin Wang, Beating the photon-number-splitting attack in practical quantum cryptography, Physical Review Letters 94 (23) (2005) 230503.

[37] Zhiyao Wang, Aodhán Corrigan, Norbert Lütkenhaus, Performance of bb84 without decoy states under varying announcement structures, arXiv preprint, arXiv:2603.22448, 2026.

[38] Stephanie Wehner, David Elkouss, Ronald Hanson, Quantum Internet: a vision for the road ahead, Science 362 (6412) (2018) eaam9288.

[39] Henning Weier, Tobias Schmitt-Manderbach, Nadja Regner, Ch Kurtsiefer, Harald Weinfurter, Free space quantum key distribution: towards a real life application, Fortschritte der Physik: Progress of Physics 54 (8–10) (2006) 840–845.

[40] Qiang Zhang, Feihu Xu, Yu-Ao Chen, Cheng-Zhi Peng, Jian-Wei Pan, Large scale quantum key distribution: challenges and solutions, Optics Express 26 (18) (2018) 24260–24273.

10

New and future directions: quantum public key cryptography

Most of the concepts introduced in this book involve information theoretic security. However, there is growing interest in computational security and what quantum information can accomplish in this setting. In this chapter, we'll take a look at quantum public key systems, such as digital signatures and encryption schemes. We'll also discuss some new computational assumptions and what can be constructed from them. This is a highly active research area, as of writing this book, and the material here will rapidly advance. This chapter is not meant to be an exhaustive survey; nor will rigorous proofs be provided. Instead, it is just meant to introduce the reader to alternative topics, and some of the new research being conducted in this area, while giving a high-level idea of how the proofs of security operate for some of the constructions, below.

10.1 New security assumptions and constructions

We already touched on some alternative quantum cryptographic constructions in Chapter 8, when discussing pseudo random state (PRS) generators. However, the study of such alternative constructions and definitions is, as of writing this, a very active, and exciting, area of research. In particular, researchers are interested in finding a *minimal* assumption for quantum cryptography, while also opening the door to new quantum cryptographic primitives. Such work, of course, involves protocols that are not information theoretic secure. However, moving outside of information theoretic security, and into computational security, as in the classical domain (see Chapter 3), allows for many more exciting possibilities, beyond that which can be done in the "perfect security" scenario. In this section, we'll only scrape the surface, however the reader is encouraged to view the references cited to find out more. This is a highly active area, so the frontier is bound to expand by the time this book goes to press!

What is meant by "minimal" assumption? Let's first return to the classical setting, and consider a *one way function* (OWF). Such a function is "easy" to compute (i.e., computing $f(x)$, for input x, can be done in polynomial time), but hard to invert. That is, let's say a random input x is chosen, and you are given the value z where $z = f(x)$. Then, it is hard (i.e., cannot be done in polynomial time) to find some input y, such that $f(y) = z$. Note that y may be the same, or different, from x depending on the function. Such a function is considered the minimal assumption for classical cryptography as (1) it can be used to build a lot of other mechanisms, including pseudo random functions, encryption schemes, and

Quantum Communication and Cryptography. https://doi.org/10.1016/B978-0-44-321532-2.00017-6

digital signatures; and (2) many other cryptographic mechanisms can be used to build a OWF. Thus, it is necessary and sufficient for much of cryptography.

While OWFs cannot build everything (in particular, they cannot build public key encryption schemes [10,8]), they are the cornerstone of modern cryptography. While it is still an open question as to whether or not they exist, it is generally suspected that they do; if they don't, then perhaps we'll all have to switch rapidly to QKD systems and OTP!

In the quantum domain, it is still not clear what the minimal assumption for quantum cryptography is, if it exists at all. Potential candidates are *one way state generators* (OWSG) [13,12] and *Efficient, Far, Indistinguishable* (EFI) pairs [5]. Unlike a Pseudo Random State (PRS) (discussed in Section 8.4), which attempted to generalize classical pseudo random number generators to the quantum domain, both OWSG and EFI are an attempt to generalize OWFs. There are other definitions; we'll only discuss these two as their definitions should be readily understandable from the earlier content in this book.

10.1.1 One-way state generators (OWSG)

A OWSG, as defined in [13,12], consists of three algorithms: `KeyGen`, `StateGen`, and `Verify`. The first, `KeyGen` will output a classical secret key $k \in \{0,1\}^n$. The second, `StateGen` will, on input a secret key, output a (potentially mixed) quantum state $\rho(k)$. Finally, `Verify` will take as input a key k and a quantum state σ, and output `True` or `False`.

Correctness should imply that `Verify`$(k, $`StateGen`$(k)) =$ `True`, with high probability. Security implies that, given multiple copies of the quantum state output, it should be hard to find the given input k that was used to produce the system. In particular, an adversary algorithm is given $\rho(k)^{\otimes t}$, where t is polynomial in the security parameter (see Section 3.1.2), and where k was chosen randomly using `KeyGen`. Then, this adversary, which has polynomial time to run, must output y such that `Verify`$(y, \rho(k)) =$ `True`. A secure OWSG should be such that the probability any adversary can do so is *negligible* (again, see Section 3.1.2, in particular Definition 3.1.2). See Fig. 10.1. Due to a result in [1], security of a OWSG can only be proven when faced with a quantum polynomial time (QPT) adversary: an unbounded adversary can, with high likelihood, determine a valid y. See [12] for more details on this, and the general definition of a OWSG.

Note the similarities between the definitions of a OWSG and a OWF (as also highlighted in Fig. 10.1). Both are given the output of an "easy" to compute function (i.e., a polynomial time function: for the OWSG, it is a polynomial time quantum algorithm, for the OWF it is a polynomial time classical algorithm). Of course, since quantum data cannot be copied, to be fair, the adversary is given polynomial number of copies of the quantum output. Then, the adversary must output some classical information y such that y "maps to the same output". Of course, in the quantum setting, "mapping to the same output" is a bit ambiguous – thus the need for the `Verify` function (classically, this would simply be the "bit-wise equality" test, however for a OWSG, the algorithm must be defined explicitly).

Given a PRS generator (see Section 8.4 for the definition of such a mechanism), then it is possible to construct a OWSG. Indeed, one simply uses the output of the PRS for the OWSG

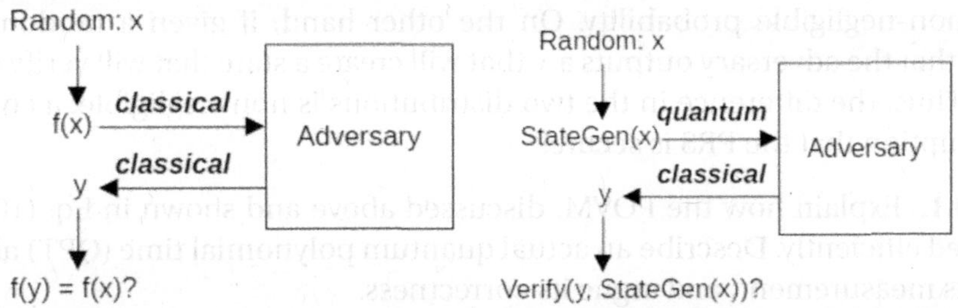

FIGURE 10.1 OWF and OWSG Security Experiments. Left: A function is said to be One-Way if, given $f(x)$, for some randomly chosen input x, no PPT (probabilistic polynomial time) adversary can output a y such that $f(x) = f(y)$, except with negligible probability. Both x and y are classical bit strings. Right: A OWSG is such that, no *quantum* polynomial-time adversary, given a polynomial number of quantum states produced by the `StateGen` algorithm, for a randomly chosen input x, can output a classical y such that the `Verify` algorithm outputs `TRUE` except with negligible probability.

(i.e., the `StateGen` algorithm is simply the circuit G which creates the PRS, using notation from Section 8.4). To run `Verify`, given y, one simply measures using the POVM

$$M_y = \{|\rho(y)\rangle\langle\rho(y)|, I - |\rho(y)\rangle\langle\rho(y)|\} \tag{10.1}$$

(recall, PRS states are pure). Exercise 10.1 will ask you to describe an efficient way to perform the above measurement.

Why is this a secure OWSG? To give a high level overview of the proof, let's assume for contradiction it isn't. Then, we can show that we can actually break the underlying PRS. This style of proof is common in classical cryptography, and also common in quantum cryptography when computational assumptions are required. In particular, if the above is not a secure OWSG, then there must exist some adversary algorithm which can output a value y successfully with some non-negligible probability (see Fig. 10.1). We'll build a new adversary, around this one, to break the PRS.

Security of PRS generators involves the indistinguishability of a state generated via algorithm G, and a state chosen uniformly at random (see Section 8.4). Either way, t copies of a quantum state are given to the adversary. Our goal is to use the adversary which "wins" the OWSG game (Fig. 10.1) to distinguish between a PRS and a random state. This would then be a contradiction since we are assuming the PRS is secure.

First, our new adversary will take its input, and feed it directly to the OWSG adversary (though one copy will be kept safe). The OWSG adversary will, on having time to ponder, output y. Now, if our new adversary was given a PRS, then as far as the OWSG adversary is concerned, it's being given the output of the particular OWSG we are analyzing. Thus, with non-negligible probability, it will output a correct y. On the other hand, if given a random state, then there is no way for the adversary to output a bit string y that just so happens to be "close" to the PRS output for that y. (This last claim is not difficult to show, though requires some technicalities involve the Haar measure – see [5] for details.)

Given this y, the PRS adversary will take one of the saved copies of its input and run the `Verify` function and output the result. If given a PRS, then the adversary will output

True with non-negligible probability. On the other hand, if given a random state, the probability that the adversary outputs a y that will create a state that will verify correctly, is negligible. Thus, the difference in the two distributions is non-negligible: a contradiction in the assumption that the PRS is secure!

Exercise 10.1. Explain how the POVM, discussed above and shown in Eq. (10.1), can be implemented efficiently. Describe an actual quantum polynomial time (QPT) algorithm to perform this measurement, and argue its correctness.

So, it holds that if a PRS exists then so does a OWSG. But what can we build with a OWSG? Well, one application is a digital signature scheme. A digital signature scheme is probably something you are familiar with: one party generates a secret key and a public key pair. (In our case, the public key will be a quantum state.) The secret key holder can "sign" a message – essentially creating a cryptographic token that binds the message to the secret key holder. Later, anyone with a public key may verify the message (namely, be certain that the message came from the secret key holder).

Let's consider a simple *Quantum One Time Digital Signature* scheme whereby the secret key can only be used to sign one message. To define security, we'll, as usual in computational-security based cryptography, define a game or experiment. First, a "challenger," or oracle, creates a secret and public key. The adversary is given a polynomial number of copies of the public key (since they may be quantum states). Next, an adversary sends a message m to the challenger. The challenger signs this message, using the secret key, and sends the signature to the adversary. Now, the adversary's goal is to forge a signature for a new, different, message, even if given information on a correct signature. That is, the adversary must output a signature σ' for a message m' where the adversary can choose any message m' under the single constraint that $m' \neq m$. The adversary "wins" if the signature is correct (i.e., it verifies correctly given a legitimate public key). The probability of winning should be negligible. Note that, for standard digital signatures (not one-time signatures), the adversary is allowed to request the signatures of multiple messages before having to "forge" a new signature (see Section 3.3).

A OWSG can be used to construct the above cryptographic primitive for signing single-bit messages. First, two secret keys are produced k_0 and k_1 using the OWSG function `KeyGen`. Next, public keys are generated, where a public key is $(\rho_0, \rho_1) = (\texttt{StateGen}(k_0), \texttt{StateGen}(k_1))$ (namely, the public key consists of two quantum states). To sign a message m (which is a single bit), the signature is simply k_m. Given a message m and a signature σ, to verify the signature, one simply runs the OWSG verification function $\texttt{Verify}(\sigma, \rho_m)$. Note that the second input is given in the public key.

Why does this work? Well, let's consider an adversary who receives, as input, a copy of the public key (multiple copies can be dealt with, by altering the protocol, as discussed in [12]). The adversary is allowed to ask for the signature of a message m, but must later forge a signature of a different message. Since the messages are only single bits, the adversary may ask for only one signature – either of $m = 0$ or $m = 1$ (if the adversary asks for *both*, the adversary could never win). After this query, the adversary learns k_m. The adversary must now, given this information, output a forgery of the message $1 - m$; namely, the adversary must guess at the value of k_{1-m}. However, since we're using a OWSG, this problem may be reduced to finding an input for which the OWSG function will verify correctly, something which cannot be done, assuming the OWSG is secure.

10.1.2 EFI pairs

EFI, or "Efficient, Far, and Indistinguishable" pairs, introduced in [5], consist of two polynomial time (i.e., Efficient), algorithms whose outputs states are statistically Far, yet computationally Indistinguishable. They are the quantum analogues of EFID pairs, introduced in [7], which are the same, but where there all algorithms and states are classical. In [7], using also results in [9], classical EFID pairs were shown to exist if and only if pseudo random generators exist (which, ultimately, implies EFID pairs exist if and only if many other cryptographic primitives exist, making them a fundamental assumption in cryptography). Due to their importance, classically, it is therefore natural to study the quantum analogue.

Formally, an EFI pair is a pair of *mixed* states ρ_0, ρ_1, such that [5]:

1. They can be *efficiently generated*: namely, there exists a QPT algorithm G that, on input b, outputs mixed state ρ_b.
2. They are *statistically far*: that is, the trace distance between ρ_0 and ρ_1 is at least an inverse polynomial.
3. Finally, they are *computationally indistinguishable*: For any QPT distinguisher D, it holds that D cannot distinguish between ρ_0 or ρ_1, except with negligible probability.

Note that item two implies that, from an information theoretic perspective, the two states are actually very distinguishable. However, it is computationally infeasible to distinguish between them, according to item three.

What cryptographic primitives can be built from an EFI? Several, it turns out, including commitment schemes and oblivious transfer. (We discussed oblivious transfer in Section 8.3.1.) Importantly, the existence of these schemes also imply the existence of an EFI pair (e.g., given a bit commitment scheme, one may build an EFI pair). This provides some evidence that EFI pairs are potentially a foundational assumption in quantum cryptography. We won't go into great detail, in this chapter, however we will discuss one of the above constructions, using results from [5].

A *Bit Commitment* scheme, informally, is a two-party protocol which consists of two stages. In the first stage, one party, say Alice, chooses a bit b and runs some algorithm `Commit`(b) which creates some state (classical or, in our case, quantum). This state is sent to the other party, Bob. Essentially, this state "binds" Alice to her choice b (she cannot later change her mind, without getting caught), and also "hides" her choice from Bob (Bob, who is given the state from Alice, cannot determine whether Alice committed to $b = 0$ or $b = 1$). Later, during the second stage of the protocol, Alice can reveal her choice by sending some information to Bob. This information allows Bob to determine the bit b and also, in combination with the information he received in the first stage, will give him proof that Alice did not change her mind between the two stages.

How can such a scheme be built from an EFI pair? First, note that the output of an EFI pair is a mixed state (see Exercise 10.2). The first step in the protocol is to "purify" the generation algorithm. Specifically, let $G(b)$ be the EFI algorithm that produces ρ_b. One may design a unitary circuit U_b which acts on two registers C and R, such that, on input all zeros (namely, $|0\cdots0\rangle_{CR}$), it holds that:

$$G(b) = \rho_b = tr_R(U_b\,|0\cdots0\rangle\,\langle0\cdots0|_{CR}\,U_b^*). \tag{10.2}$$

Namely, one "purifies" ρ_b by adding an additional ancilla R. Now, in the first stage (the *commitment stage*), Alice will choose her bit b, run U_b on $|0\cdots0\rangle_{CR} = |\psi_b\rangle$. She will then send the C register to Bob (keeping the R register private, for now). Bob receives $tr_R(|\psi_b\rangle\,\langle\psi_b|)$ which is exactly ρ_b. Due to the computational indistinguishability of ρ_0 from ρ_1, Bob, who is computationally bounded, cannot determine which bit Alice committed to. Proof of binding (that Alice cannot change her mind, later), requires a little more technical machinery, though not too much – a proof can be found in [5]. Later, in the second stage of the protocol (the "reveal" stage), Alice will send the ancilla R to Bob, along with the actual bit b that she chose. Bob now holds both C and R ancillas; he then applies the inverse unitary U_b^* and measures in the computational basis. If Alice is honest, his measurement should produce all zeros; otherwise, he aborts as Alice lied. (Note that noise tolerance is not supported.)

Other cryptographic primitives are possible from EFI pairs, including oblivious transfer. Furthermore, it can be shown that bit commitment implies the existence of an EFI pair; namely, if one has a secure bit commitment scheme, then one can build an EFI pair.

So why bother studying these alternative assumptions and their applications? First, it is a very important theoretical question to discover the "minimal" cryptographic assumption, necessary and sufficient for the construction of multiple, useful, quantum cryptographic tasks (e.g., the "quantum version" of the OWF). Second, it is useful to discover alternative assumptions, in case one is ever found incorrect, we have alternative mechanisms to switch to. Finally, it's interesting to discover what new cryptography is possible through the use of minimal computational assumptions. The above was only meant as a brief introduction to this fascinating topic; the referenced papers should provide the interested reader with additional material to study.

Exercise 10.2. Argue that an EFI pair must consist of mixed states. If the states were pure, and if there was an efficient algorithm G_b that created them (in this case, G_b is a unitary circuit), show that they could be easily distinguished if their trace distance is large (i.e., the "Far" part of EFI.).

Exercise 10.3. Argue that if $G(b)$ is polynomial time and produces a mixed state ρ_b, then one can construct a QPT unitary U_b which behaves according to Eq. (10.2).

10.2 Quantum public key encryption

Public key encryption is everywhere in our society – yet the security of all classical systems rely on strong computational assumptions. Can quantum information do better? As it turns out, the answer is yes! In particular, public key encryption systems can be built using only OWFs, or even potentially weaker assumptions than that. Note that, classically, it is *impossible* to build a secure public key encryption system using only a OWF [10,8]. Thus, even in the computational setting, quantum information and communication can provide an advantage over purely classical cryptographic systems.

A quantum public key encryption system consists of four algorithms: SKGen which produces a secret key; PKGen which, on input a secret key, produces a public key ρk (this may be a quantum or a classical-quantum state, depending on the construction); $\mathtt{Enc}(\rho k, m)$ which takes a (quantum or classical-quantum) public key and a message and produces a ciphertext (which may be classical or classical-quantum); finally $\mathtt{Dec}(sk, c)$ decrypts a ciphertext using the produced secret key. Note that encryption may destroy the given public key (i.e., a single public key can only encrypt a single message), or it may allow for public key recycling [11], thereby allowing a single public key to be used multiple times. *Correctness* of a scheme should require the original message is reproduced, after decryption, given that the correct secret/public key pair was used. But what about security?

We'll only consider CPA security, leaving the reader to investigate the literature for CCA secure constructions (recall, from Section 3.1.3, CPA is weaker than CCA). CPA security in the quantum setting was defined in [4] to consist of the following game, or experiment, played between an oracle and an adversary (see, also, Fig. 10.2):

Quantum Public Key Encryption Experiment:

1. First, an oracle creates a secret/public key pair using $k = \mathtt{SKGen}$ and $\rho k^* = \mathtt{PKGen}(k)$. This is called the *challenge public key*. If noise tolerance is supported, the challenge key may undergo some noise map [14]. Furthermore, if the public key is a classical-quantum state, an exact copy of the classical and quantum portion of the challenge public key is sent to the adversary.
2. The adversary is allowed oracle access to $\mathtt{PKGen}(k)$, thereby giving the adversary a polynomial number of copies of the quantum public key.
3. If the encryption protocol supports public key recycling, the adversary is allowed oracle access to $\mathtt{Enc}(\rho k^*, \cdot)$. Otherwise, this is not provided. This is necessary as encryption may alter the challenge public key to an adversary's benefit.
4. The adversary outputs two *challenge messages*, m_0 and m_1 of equal length.
5. The oracle chooses a random bit b, and encrypts message m_b, using the challenge public key. The resulting ciphertext is given to the adversary.
6. The adversary, which continues to have access to any of the above mentioned oracles, must, ultimately, output a guess, b' of the message encrypted.
7. The adversary *wins* the game if $b' = b$.

A quantum public key encryption protocol is said to be CPA-secure if, for every quantum polynomial time (QPT) adversary, there exists a negligible function $f(n)$ (where n is the security parameter), such that the probability of winning the above experiment is no greater than $1/2 + f(n)$. Note that the above definition of security does not take into account the public key distribution problem; this is something we'll comment on in Section 10.2.2.

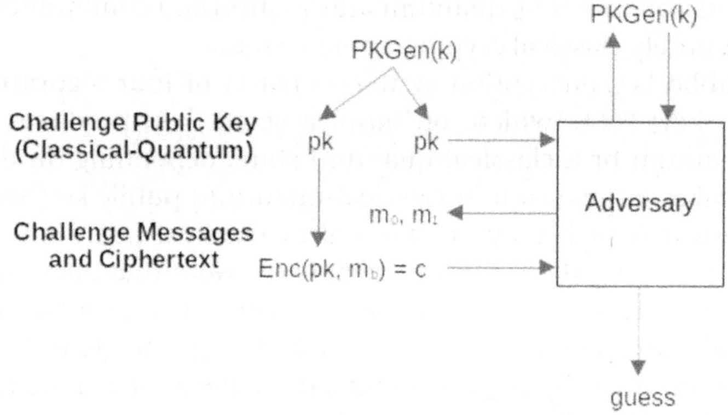

FIGURE 10.2 Showing the standard CPA experiment for quantum public keys, as defined in [4] for quantum public key encryption systems. First, an oracle generates a "challenge quantum public key" (which may be a classical-quantum state). A copy of this is given to the adversary, while a second copy is kept by the oracle, simulating the successful distribution of the quantum public key (a problem to be discussed, later, in Section 10.2.2). The adversary is allowed *oracle access* to PKGen, thus giving the adversary additional copies of the public key (which will also allow the adversary to encrypt arbitrary messages internally – note that if a single public key can encrypt multiple messages, then oracle access to Enc, using the challenge public key, must be provided [4], though this is not drawn here). Later, the adversary must output two challenge messages m_0 and m_1, one of which will be encrypted by the oracle, using the challenge public key. The resulting ciphertext is returned to the adversary who must now guess which of the two messages was encrypted. A system is said to be CPA secure if no quantum polynomial time (QPT) adversary can guess correctly, except with negligible advantage over one-half.

By now, there are many constructions of secure quantum public key encryption systems. Many rely on the existence of a OWF; some rely on the existence of a PRS or OWSG. We'll cover some constructions, below, but the reader should note that this is a highly active research area, and even at the time of writing, there were numerous new constructions arriving in the literature.

10.2.1 Constructions

Let's look at the construction from [6] which assumed the existence of a quantum secure pseudo random function (PRF). PRFs were introduced in Section 3.1.3; a quantum-secure version [15] is similar except that the distinguisher is a quantum polynomial time circuit, and is allowed to make "superposition" queries to its oracle. As shown in [15], such a PRF can be constructed assuming the existence of a quantum secure OWF. Let F_k be a quantum secure PRF, mapping n-bit inputs to 1 bit outputs. The secret key is simply a secret key, k, for the given PRF. The public key algorithm takes k as input and creates the following state:

$$\texttt{PKGen}(k) = \frac{1}{\sqrt{2^n}} \sum_{x\in\{0,1\}^n} (-1)^{F_k(x)} |x\rangle = \rho k. \tag{10.3}$$

Now, to encrypt a message, m, namely $\texttt{Enc}(\rho k, m)$, which for this scheme is a single bit (i.e., $m \in \{0, 1\}$), the following process is performed:

Encryption algorithm, $\texttt{Enc}(\rho, m)$**:**

1. First, a random $r, z \in \{0, 1\}^n$ are chosen (uniformly at random).
2. Pauli Z gates are applied, based on z to map the given public key to the following state:

$$\rho k \mapsto \frac{1}{\sqrt{2^n}} \sum_{x\in\{0,1\}^n} (-1)^{F_k(x)+x\cdot z} |x\rangle = \rho k(z). \tag{10.4}$$

3. The following ciphertext is output:

$$c = (\rho k(z), r, (z\cdot r)\oplus m), \tag{10.5}$$

where $z \cdot r$ is the bit-wise dot product (modulo two), namely $z_1r_1 \oplus z_2r_2 \oplus \cdots \oplus z_nr_n$. Note that to recover the message, z is required; the main security result of [6] is to show that z is "hidden" in the quantum portion of the ciphertext.

Finally, given a ciphertext c (which is a classical-quantum state), and the secret key k, the decryption algorithm $\texttt{Dec}(k, c)$ runs as follows:

Decryption algorithm, $\texttt{Dec}(k, c)$**:**

1. Apply the unitary operator $U_k |x\rangle = (-1)^{F_k(x)} |x\rangle$ to the quantum portion of the public key. This should map $\rho k(z)$ to:

$$\rho k(z) \mapsto \frac{1}{\sqrt{2^n}} \sum_{x\in\{0,1\}^n} (-1)^{x\cdot z} |x\rangle = H^{\otimes n} |z\rangle. \tag{10.6}$$

2. Measure the resulting state in the Hadamard basis to retrieve z. Then, using the classical portion of the ciphertext, extract r, compute $r \cdot z$ and finally extract the message.

Exercise 10.4. Assume that there exists a polynomial quantum circuit U_k such that $U_k |x, y\rangle = |x, F_k(x) \oplus y\rangle$. Show a polynomial circuit for $\texttt{PKGen}$, which creates the public key shown in Eq. (10.3).

Exercise 10.5. Describe the circuit that performs the unitary evolution described by Eq. (10.4).

CPA security of the above system was proven in [6]. Specifically, it was shown that, given the ciphertext shown in Eq. (10.5), recovery of z is only possible with negligible probability (assuming a computationally bounded adversary, of course). Thus the underlying message remains secret. See [6] for more details.

We next look at an alternative construction from [4] which does not require ciphertexts to be quantum. As before, let F_k be a quantum secure PRF but where, now, the input and output are both n-bits long. The secret key generation algorithm will choose two secret keys k_0 and k_1 for the given PRF; the secret key, then, will be the pair $k = (k_0, k_1)$. The public key algorithm will output two quantum states

$$\mathtt{PKGen}(k) = (|\psi_0\rangle, |\psi_1\rangle), \tag{10.7}$$

where:

$$|\psi_b\rangle = \frac{1}{\sqrt{2^n}} \sum_{x\in\{0,1\}^n} |x, F_{k_b}(x)\rangle . \tag{10.8}$$

To encrypt a message m, which, as before, is a single bit, one measures $|\psi_m\rangle$ in the Z basis, yielding a random x and random $F_{k_m}(x)$. The ciphertext, then, is simply $(x, F_{k_m}(x))$. To decrypt, given $c = (c_1, c_2)$, one simply evaluates $F_{k_0}(c_1)$ and $F_{k_1}(c_1)$. Whichever evaluation matches the second portion of the ciphertext, c_2, yields the bit that was originally encrypted.

An alternative method, using a public key similar to Eq. (10.8), above, was also introduced in [4]. Now, the secret key is a single PRF key k, while the public key is also a single quantum state:

$$\frac{1}{\sqrt{2^n}} \sum_{x\in\{0,1\}^n} |x, F_k(x)\rangle . \tag{10.9}$$

To encrypt, given the public key, one measures as before, in the Z basis, yielding x, $F_k(x)$. Now, one uses a CCA-secure symmetric key encryption algorithm (a classical encryption algorithm) to encrypt the message (which may be arbitrarily long, now), using the secret key $F_k(x)$. The ciphertext, then, is $c = (x, \mathtt{Enc}(F_k(x), m))$. Security of this construction follows due to the security properties of the underlying quantum-secure PRF and the classical CCA-secure encryption algorithm. Furthermore, this construction leads to a CCA-secure public key system.

Exercise 10.6. How does the output length of the PRF F_k, used in the above scheme, particularly Eq. (10.8), relate to the probability of decryption failure?

Exercise 10.7. Describe the decryption algorithm for the single-key public key encryption system, described above (with public key defined in Eq. (10.9)).

The above constructions required the use of a quantum secure PRF. Such functions exist, if quantum-secure OWFs exist. However, interestingly, in quantum cryptography, one can build systems with potentially weaker assumptions than this! We'll show yet another construction from [4], which replaces the PRF requirement, with a pseudo-random function like state (PRFS) generator.

Before discussing the encryption algorithm, let us first recall PRFS generators. Introduced in [2], and discussed in Section 8.4, these consist of a state generation algorithm which, given a secret key k, and a classical input x, will output a quantum state $|\psi_{k,x}\rangle$. Think of this as the quantum analogue of a PRF. Security is defined similarly to a PRF: an adversary is allowed to make queries x, will receive either $|\psi_{k,x}\rangle$ or a randomly generated quantum state $|\phi_x\rangle$, and no QPT adversary will be able to distinguish between the two scenarios. Note that, generally the adversary is allowed to make superposition queries, as with quantum-secure PRFs.

Given the above, a public key encryption protocol may be constructed as follows (from [4]): the secret key k, is a secret key for the PRFS generator, while the public key will be a state of the form:

$$\mathsf{PKGen}(k) = \frac{1}{\sqrt{2^d}} \sum_{x\in\{0,1\}^d} |x\rangle_X \, |\psi_{k,x}\rangle_Q^{\otimes t}. \tag{10.10}$$

Above, d is the size (in bits) of the classical input space of the PRFS generator; note that for every input, t copies of the state are made. Due to the security properties of a PRFS generator, even with these t copies, the state will still look like (t copies of) a random state.

Now, given a public key, to encrypt a bit message m, one first measures the X register of the public key state in the computational basis to retrieve a random x^*; the Q portion collapses to $|\psi_{k,x^*}\rangle^{\otimes t}$. Next, a new secret key "z" for the PRFS generator is chosen. Finally, the cipher text will be:

$$c = \begin{cases} (x^*, |\psi_{k,x^*}\rangle^{\otimes t}) & \text{if } m = 0 \\ (x^*, |\psi_{z,x^*}\rangle^{\otimes t}) & \text{if } m = 1 \end{cases} \tag{10.11}$$

Note that, in both cases, the observed x^* is provided; the only difference between the two messages is whether or not the "real" Q portion of the public key is given, or a newly produced one is provided. Regardless, the "new" secret key z is discarded.

To decrypt, given the above ciphertext, along with the actual secret key k, a *SWAP* test (a simple, efficient, algorithm that determines how close two quantum states are if given multiple copies of the state) is performed to test whether or not the quantum portion of the ciphertext is equal to the actual $|\psi_{k,x^*}\rangle$. Note that k is known to the decryption circuit. Security can be proven by taking advantage of the security of the underlying PRFS and through a series of so-called "hybrid arguments." This is a common proof approach in cryptography: one constructs a series of "games" starting with the actual protocol you are trying to prove secure, and leading to one that is known to be secure. Each game is computationally similar to the other (for instance, due to an underlying assumption, such as the secure PRS in this case), so that if the last game is secure, the first one must be also. See [4] for more details on the proof of this protocol.

The above constructions all exhibit certain interesting and unique properties. Some require quantum ciphertexts, some don't. All utilize quantum states as part, or all, of the public key. All of them are proven secure under weak computational assumptions – weaker than is possible classically. Noise tolerance of a quantum public key is an important issue to address. Quantum error correction can help with this, or alternative solutions inspired by QKD may also be fruitful [14]. Again, this is a highly active research area, and new discoveries are made constantly.

10.2.2 Quantum public key distribution

The security definition of a classical public key encryption system does not concern itself with the actual distribution of a public key to an honest user. Indeed, this problem is considered "out of scope" and handled through different mechanisms (e.g., certificate authorities). So far as discussed in this chapter, public key distribution is also not considered in the definition of encryption (or whatever other public key cryptographic primitive one is attempting to create). However, since a public key is now a quantum state, the distribution problem is significantly more challenging to solve, than it is classically.

To see why it is important to ensure the authenticity of a public key before using it, let's go back to classical systems. Alice generates a secret key/public key pair (sk, pk) (classical strings). She sends pk to Bob over an insecure channel. Before the key arrives to Bob, the adversary, Eve, captures pk and produces her own secret key/public key pair $(\widetilde{sk}, \widetilde{pk})$. Recall that public keys "lock" (encrypt) a message and secret keys "unlock" (decrypt) them. Eve forwards her own public key to Bob who subsequently encrypts his message m using $\widetilde{pk}$, unaware that the public key he has is actually the adversary's. When the ciphertext is sent back to Alice, Eve may intercept it, decrypt it using $\widetilde{sk}$, read the message, then even re-encrypt it using Alice's pk so that Alice and Bob remain unaware of this complete break in security. A CPA or CCA secure public key encryption scheme assumes Eve has a copy of the public key, but that Bob also has a legitimate copy.

Classically, ensuring one has a legitimate copy of a public key can be done through certificate authorities. Essentially, a trusted third party will verify the integrity of a given public key and then digitally sign it. The signature can be verified by Bob, later, to ensure he has the right copy of Alice's public key; with "right" meaning an equal copy, down to the bit-level. In the quantum setting, this is not so easily solved – for instance, if a quantum public key is sent over an adversarial network, it may be partially entangled with an adversary; it may be ϵ-close in trace distance to a correct copy; and so on. Furthermore, authentication of an arbitrary quantum state generally requires secrecy of the quantum channel, making the problem even more difficult [3]. For this reason, it almost makes sense to consider the distribution problem at the same time as the main security problem (e.g., encryption), as opposed to treating them separately, as is done classically.

The definition of security for quantum public key encryption, as discussed in Section 10.2, assumes a legitimate copy of the public key is given to Bob (this is the *challenge public key*). A recent work in [11] augments this definition and in doing so, attempts to address the key distribution problem. In their definition of security, denoted `Q-IND-pkT-CPA`, a public key consists of both a classical part and a quantum part. Both are created by the oracle, and given to the adversary; however, only the classical part is kept by the oracle as the challenge public key; the quantum portion is not stored. The adversary may request as many additional classical/quantum public keys as desired (up to a polynomial number, of course). Finally, the adversary outputs two messages, and a *quantum state.* This quantum state may be the real quantum public key, given to the adversary initially; or it can be any other quantum state of equivalent dimension, potentially entangled with an additional ancilla held by the adversary. This simulates the adversary having full control over the quantum portion of a public key, but not the classical portion (which can be distributed using traditional methods, such as certificate authorities). One of the messages is encrypted using a public key consisting of the legitimate classical portion (created initially by the oracle), and the quantum state given by the adversary. The adversary must guess which state was encrypted. Note that encryption, now, is allowed to "abort" if it detects something is wrong with the given quantum state, using data from the classical portion. If encryption aborts, the adversary automatically loses. See Fig. 10.3.

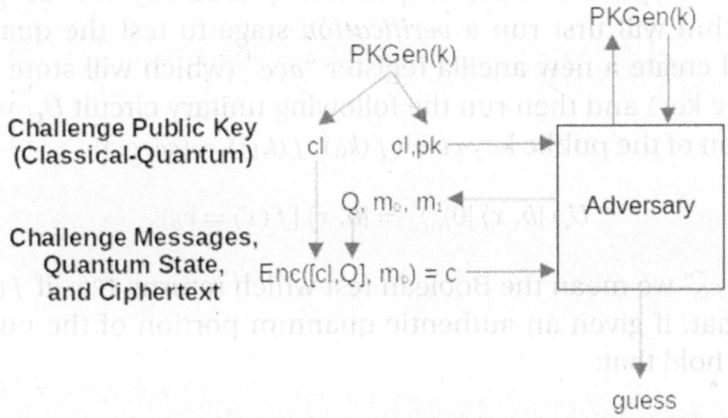

FIGURE 10.3 A diagram of the security experiment for Q-IND-pkT-CPA security, as defined in [11]. Note the similarities, and important differences, from standard CPA security, shown in Fig. 10.2. Now, a public key is a classical-quantum state and only the classical portion is assumed to be safely distributed while the quantum portion is given to the adversary. Later, the quantum portion, used to encrypt the challenge message, is actually produced by the adversary (register Q in this picture). If the adversary is honest, the register Q can simply be the original quantum portion produced by the oracle. Thus this is a stronger definition of security than that shown in Fig. 10.2 and it models the scenario where only the classical portion of the public key must be safely distributed, while the quantum portion can be sent over an insecure quantum channel, controlled by an adversary.

Clearly any scheme proven Q-IND-pkT-CPA secure will also be CPA secure according to the standard definition discussed in Section 10.2. The opposite is clearly not true (see Exercise 10.8). The question is: can such a scheme be built? In [11] a protocol was developed, using OWFs, which utilized the classical portion to store information that can be used to partially authenticate or check, the quantum portion of the public key (which is now provided by the adversary).

Exercise 10.8. Prove, by showing an actual attack, that the public key encryption constructions, discussed earlier in Section 10.2.1, are not Q-IND-pkT-CPA-secure (even though they are CPA-secure according to the standard definition given in Section 10.2).

We will first discuss the protocol of [11], assuming only one copy of the public key is ever made. The scheme can then be extended to support a polynomial number of public keys. Let f be a quantum secure OWF. The secret key generation algorithm of this encryption protocol will choose two random keys k_0 and k_1 which are inputs to f. The secret key is $k = (k_0, k_1)$. To generate a public key, an algorithm is run which produces the following quantum state:

$$\mathtt{PKGen}(k) = \left(\underbrace{(f(k_0), f(k_1))}_{\text{classical portion}}, \underbrace{\frac{1}{\sqrt{2}}(|0, k_0\rangle + |1, k_1\rangle)}_{\text{quantum portion}} \right) \tag{10.12}$$

Note that, given the classical portion $f(k_0)$, $f(k_1)$, the adversary should not be able to find k_i due to the OWF security guarantee.

Now, later, to encrypt, the classical-quantum public key will be provided and the encryption algorithm will first run a *verification* stage to test the quantum portion. In particular, Enc will create a new ancilla register "acc" (which will store a flag whether to accept or reject the key) and then run the following unitary circuit U_{cl} which depends on the classical portion of the public key $cl = (f(k_0), f(k_1)) = (c_0, c_1)$:

$$U_{cl} |b, x\rangle |0\rangle_{acc} = |b, x\rangle |f(x) = c_b\rangle . \tag{10.13}$$

Above, by "$f(x) = c_b$" we mean the Boolean test which returns TRUE if $f(x) = c_b$ and FALSE otherwise. Note that, if given an authentic quantum portion of the public key (namely, Eq. (10.12)), it will hold that:

$$U_{cl} \left(\frac{1}{\sqrt{2}}(|0, k_0\rangle + |1, k_1\rangle) \right) |0\rangle_{acc} = \frac{1}{\sqrt{2}}(|0, k_0\rangle + |1, k_1\rangle) |\mathtt{TRUE}\rangle_{acc} . \tag{10.14}$$

However, if the quantum state were altered, it should hold that the accept register will not be separable *and* in the state $|\mathtt{TRUE}\rangle_{acc}$; this is due to the security of the underlying OWF.

After running the above verification process, a measurement is made of the "acc" register; if this is "FALSE," then the encryption process will abort (and thus the adversary will not win the Q-IND-pkT-CPA game, when considering security). If the result is TRUE, then encryption continues. Let $m \in \{0, 1\}$ be the message to encrypt (a single bit); the encryption circuit will apply the gate Z^m to the first qubit of the quantum state (i.e., the quantum portion of the public key). That is, the encryption circuit will apply the Pauli Z gate only if $m = 1$. Thus, the quantum state becomes:

$$\frac{1}{\sqrt{2}}(|0, k_0\rangle + |1, k_1\rangle) \mapsto \begin{cases} \frac{1}{\sqrt{2}}(|0, k_0\rangle + |1, k_1\rangle) & \text{if } m = 0 \\ \frac{1}{\sqrt{2}}(|0, k_0\rangle - |1, k_1\rangle) & \text{if } m = 1 \end{cases} \tag{10.15}$$

Following this transformation, a measurement in the X basis is performed, resulting in a bit string $d \in \{0, 1\}^{n+1}$, where $n + 1$ is the number of qubits in the quantum portion of the public key, and where, of course, an observation of $|+\rangle$ is translated to a zero, and $|-\rangle$ is translated to a one. It holds (see Exercise 10.10), that $m = d \cdot (0||k_0 \oplus 1||k_1)$, where "||" is bit string concatenation. The final ciphertext will be $c = (d)$. To decrypt, since the secret key k_0 and k_1 are known, it is easy to solve for m.

Of course, it is fairly clear that the above scheme is insecure if the adversary is allowed multiple copies of a public key. Indeed, even with two copies, it is likely that a simple computational Z basis measurement will reveal both k_0 and k_1 (thus revealing the entirety of the secret key and allowing the adversary to correctly forge additional copies itself). To remedy this, additional randomness is used in each public key copy, along with a *digital signature scheme.*

Recall that a digital signature scheme allows a secret key holder (we'll denote the digital signature secret key by sk_{sig}) to sign messages, and a public key (which we'll denote by pk_{sig}) holder to verify them (see Section 3.3). In particular, there is an algorithm $\texttt{Sign}(sk_{sig}, m) = \sigma$ and an algorithm $\texttt{Verify}(pk_{sig}, m, \sigma)$ which returns TRUE or FALSE. An adversary should not be able to forge its own signatures for messages never seen before.

Returning to the quantum public key encryption system, the secret key will be sk_{sig} while a public key will now be of the form:

$$\texttt{PKGen}(k) = \left((r, pk_{sig}), \frac{1}{\sqrt{2}}(|0, \texttt{Sign}(sk_{sig}, \text{"}0||r\text{"})\rangle + |1, \texttt{Sign}(sk_{sig}, \text{"}1||r\text{"})\rangle) \right) \tag{10.16}$$

Note, now, that each public key will have a different r, and so the signatures should also be different, with high probability. Thus, even if an adversary has multiple public keys, measuring them will not provide useful information on the challenge public key portion. Security can be proven, similar to how it is shown for the "one-time" system, discussed earlier. See [11] for details.

Exercise 10.9. Show that the operator U_{cl}, defined in Eq. (10.13), is unitary. Discuss how it could be implemented as a polynomial time circuit, (note that the OWF f can be computed in polynomial time).

Exercise 10.10. Let $|\psi\rangle = \frac{1}{\sqrt{2}} |0, x\rangle + (-1)^z |1, y\rangle$, where $x, y \in \{0, 1\}^n$. Assume an X basis measurement is made of all $n + 1$ qubits. Prove that the resulting string d satisfies the equation:

$$z = d \cdot (0||x \oplus 1||y), \tag{10.17}$$

where $a \cdot b$ is the bit-wise dot product (modulo two).

Exercise 10.11. Assume that Eve produces $|\psi\rangle$, and the encryption circuit runs U_{cl}, followed by a measurement of the "*acc*" register in the TRUE state. Describe the resulting post-measured state. Is it guaranteed to be exactly equal to the quantum portion of Eq. (10.12)? If not, argue that any post-measured state will still hold similar security properties of the actual public key. Though decryption may produce an incorrect result.

10.3 Closing remarks

Quantum public key cryptography using computational assumptions is a relatively new area of research, and is becoming highly active. The use of novel quantum computational assumptions is leading to new breakthroughs, new primitives, and new protocols. This chapter could only cover the very tip of the iceberg and many more fascinating research results are out there, yet to be discovered. It is hoped that this chapter provided some insight into this area, and the references contained here will provide for some additional reading for the interested reader.

References

[1] Scott Aaronson, Shadow tomography of quantum states, in: Proceedings of the 50th Annual ACM SIGACT Symposium on Theory of Computing, 2018, pp. 325–338.

[2] Prabhanjan Ananth, Luowen Qian, Henry Yuen, Cryptography from pseudorandom quantum states, in: Annual International Cryptology Conference, Springer, 2022, pp. 208–236.

[3] Howard Barnum, Claude Crépeau, Daniel Gottesman, Adam Smith, Alain Tapp, Authentication of quantum messages, in: The 43rd Annual IEEE Symposium on Foundations of Computer Science, 2002. Proceedings, IEEE, 2002, pp. 449–458.

[4] Khashayar Barooti, Alex B. Grilo, Loïs Huguenin-Dumittan, Giulio Malavolta, Or Sattath, Quoc-Huy Vu, Michael Walter, Public-key encryption with quantum keys, in: Theory of Cryptography Conference, Springer, 2023, pp. 198–227.

[5] Zvika Brakerski, Ran Canetti, Luowen Qian, On the computational hardness needed for quantum cryptography, arXiv preprint, arXiv:2209.04101, 2022.

[6] Andrea Coladangelo, Quantum trapdoor functions from classical one-way functions, arXiv preprint, arXiv:2302.12821, 2023.

[7] Oded Goldreich, A note on computational indistinguishability, Information Processing Letters 34 (6) (1990) 277–281.

[8] Russell Impagliazzo, A personal view of average-case complexity, in: Proceedings of Structure in Complexity Theory. Tenth Annual IEEE Conference, IEEE, 1995, pp. 134–147.

[9] Russell Impagliazzo, Leonid A. Levin, Michael Luby, Pseudo-random generation from one-way functions, in: Proceedings of the Twenty-First Annual ACM Symposium on Theory of Computing, 1989, pp. 12–24.

[10] Russell Impagliazzo, Steven Rudich, Limits on the provable consequences of one-way permutations, in: Proceedings of the Twenty-First Annual ACM Symposium on Theory of Computing, 1989, pp. 44–61.

[11] Fuyuki Kitagawa, Tomoyuki Morimae, Ryo Nishimaki, Takashi Yamakawa, Quantum public-key encryption with tamper-resilient public keys from one-way functions, in: Annual International Cryptology Conference, Springer, 2024, pp. 93–125.

[12] Tomoyuki Morimae, Takashi Yamakawa, One-wayness in quantum cryptography, arXiv preprint, arXiv:2210.03394, 2022.

[13] Tomoyuki Morimae, Takashi Yamakawa, Quantum commitments and signatures without one-way functions, in: Annual International Cryptology Conference, Springer, 2022, pp. 269–295.

[14] Nishant Rodrigues, Walter O. Krawec, Brad Lackey, Deb Mukhopadhyay, Bing Wang, Quantum public key encryption for nisq devices, in: 2025 IEEE International Conference on Quantum Computing and Engineering (QCE), vol. 1, IEEE, 2025, pp. 100–110.

[15] Mark Zhandry, How to construct quantum random functions, Journal of the ACM 68 (5) (2021) 1–43.

Index

A

Action
  operator, 32, 33
Active attacks, 5, 85
Adjoint operator, 20, 38
Advanced Encryption Standard (AES), 66
Adversarial
  activity, 153
  attacks, 87, 102, 111
  influence, 221
  manufacturer, 10
  model, 112
  network, 254
  noise, 113–115
  probing, 87
  repeater, 211
  states, 145
Adversary
  action, 154
  algorithm, 244, 245
  attack, 49, 112
  capabilities, 2
  computational power, 65
  system, 52, 59, 160, 188, 196, 197
Alternative
  basis, 84, 154, 171, 231
  basis measurements, 234
  protocols, 231
  security models, 3, 9
  security proof techniques, 131
  SQKD protocols, 159
Ancilla, 109, 116
Arbitrary quantum state authentication, 254
Asymmetric key systems, 76
Asymptotic
  analyses, 129
  bound, 151
  key rate, 121, 171
  rate, 120
  scenarios, 56, 120–122, 171, 229
  security proof, 120, 122
  setting, 55, 127
Attack
  models, 62, 112, 117, 118
  types, 115, 117
Authenticated classical channel, 80, 83, 86, 89, 94, 97, 102, 111, 143, 161, 236
Authentication, 8, 79–81
  arbitrary quantum state, 254
  protocols, 1
Avalanche photodiode (APD), 232, 233

B

B92 QKD protocol, 88
Basic
  collective attack security proof, 154
  linear algebra, 13
  post processing techniques, 97
  protocols, 195
  QKD protocols, 88
Basis, 15
Basis encoding, 88, 91, 92
Basis measurement, 24, 27, 28, 35, 85, 93, 131, 149, 155, 160, 190, 195, 206, 207, 234, 257
  inner product rule, 26
BB84
  collective attack security, 154
  protocol, 83, 90, 169
    asymptotic security proof, 120
    description, 84
    security application, 135
  QKD protocol, 56
Bell
  basis, 143, 189
  measurement, 161, 162, 239
  pair, 94, 117, 142, 203, 204, 238

state, 39, 40, 50, 51, 142–144, 161, 162, 165, 187, 189, 238, 239
state superposition, 162
swap operation, 238
Bell-diagonal operator, 122
Binary
search algorithm, 99
Shannon entropy, 44, 225
Shannon entropy function, 56
Bit commitment, 3, 214, 248
scheme, 247, 248
Bounded
adversary, 80, 83, 252
quantum storage, 211
storage, 208, 209
storage model, 208, 210
Braket, 17
Brute-force search attack, 63

C

Carter-Wegman codes, 80
Cascade protocol, 98
Cauchy-Schwarz
induced constraint, 127
inequality, 18, 31, 158
Certified deletion, 2, 205, 207
Challenge
ciphertext, 67, 71
messages, 67, 70, 71, 78, 249, 250, 255
public key, 249, 254, 257
quantum public key, 250
Chosen Ciphertext Attack (CCA)
secure constructions, 249
secure encryption scheme, 80
secure public key encryption scheme, 254
security, 66, 70
Chosen Plaintext Attack (CPA)
secure encryption protocol, 80
secure encryption system, 74
secure scheme, 74
secure system, 71
security, 66, 70, 249, 251, 255
security game, 71, 74
Cipher space, 64, 65
Ciphertext, 62, 68, 70, 71, 77, 249
Classical
advantage distillation, 102
authenticated channel, 80, 83, 84, 89, 105, 112, 170
computation, 1, 9
failure rate, 133
operations on quantum data, 152
post-processing methods, 83
post-processing stage, 87, 89, 97
public key encryption system, 254
sampling strategy, 133
Classical-quantum state (cq-state), 49, 55, 56
Clauser, Horne, Shimony, and Holt (CHSH)
game, 6, 11, 200, 201, 204
classical winning probability, 6
inequality, 199
Coin space, 184, 185
Collective attacks, 116–118, 154, 155, 158, 162, 175, 176, 179, 223
Combining systems, 35
Commitment
schemes, 247
stage, 248
Complement, 133
Completely mixed state, 48, 51, 59, 171
Completeness relation, 20, 28, 48, 185
Complex vector space, 20, 21
Composable security, 109
Computational
assumptions, 1, 2, 61, 65, 66, 70, 207, 208, 211, 243, 249, 253
basis, 3, 4, 8, 22, 23, 25, 92, 93, 152, 154, 187, 248, 253
distinguisher, 73
EAV security, 66
security, 66, 70, 243
Conclusive round, 90
Conditional
entropy, 45, 54, 59, 121, 148
quantum min entropy, 56
Conjugate transpose, 15–17, 20
Constructions, 243, 250
Continuous variables, 232
Correctness, 249
Correctness guarantee, 77

Counterfeiter, 213
Cryptographic
 algorithm, 2
 applications, 213
 concerns, 196
 key, 85
 primitives, 2, 7, 61, 69, 85, 207, 208, 246–248
 protocols, 1, 22, 23, 49, 85
 systems, 2, 61, 62
Cryptography, 1

D
Dark-count rate, 228
Decoy rounds, 215
Decoy state, 93, 215
 alternative protocols, 231
 BB84 protocol, 226
 evaluation, 227
 methods, 227, 231
 protocol, 91, 226
 QKD, 235
Decryption algorithm, 77, 206, 251, 252
Denial of service attack, 215
Density operator, 46–48
 measurement, 47
 unitary evolution, 47
Detectable noise, 91
Deterministic
 algorithm, 73
 function, 72, 74
 output, 23
 polynomial time algorithm, 74
 strategies, 201
Devetak-Winter key rate expression, 121, 177, 188, 223, 225
Device
 imperfections, 195, 196, 221, 231, 235
 independence, 199
Device independent (DI) model, 10, 196, 199
Diagonalizable operator, 21, 47
Digital signature, 79, 81, 243, 244, 246
 schemes, 81, 246, 257
 signing, 81
 verification, 81
Dirac notation, 13, 14, 16, 17, 110
Direct calculation method, 122, 127, 129
Distinguisher, 69, 73
Dot-product, 16

E
E91 protocol, 94
EAV
 secure encryption protocols, 69
 secure encryption scheme, 101
 security, 62, 66, 68, 70, 71
EAV-Privacy experiment, 67
Eavesdropping
 attacks, 62, 66
 model, 67
Efficiency, 169
Efficient Far, Indistinguishable (EFI) pairs, 244, 247, 248
Eigenvalue, 21
Eigenvector, 21
Electro-optic modulator, 234
Encryption, 1, 61, 62
 algorithm, 77, 251–253, 256
 protocol, 62, 64, 65, 249
 schemes, 71, 205, 243
 security definitions, 64
Ensemble of states, 46
Entangled state, 11, 39, 207
Entanglement, 5, 7, 38
Entanglement-based protocols, 94, 162
Entropic uncertainty, 58, 59, 122, 127, 128, 170, 171, 188, 190, 197, 199, 205
 relations, 169, 186, 187, 193
Error correction, 83, 87, 97, 98, 119, 120, 162, 187
 information, 119
 leakage, 120, 148, 198
 protocol, 86, 89, 97–100, 143, 153
 verification, 100
Error rate, 7, 89, 98, 114, 143, 148, 162, 175, 199, 215, 224, 225, 238
Eve's entropy calculation, 122
Extended Euclidean Algorithm, 78

F
Failure rate, 100, 133

False ciphertexts, 70
Finite key
  scenario, 120
  security, 121, 129
  setting, 56, 142
Forward channel, 179
Full information, 145, 160
Fully ϵ_{EC}-correct protocol, 98
Fully quantum one-way protocol, 176, 178

G
Gain, 225
Game theoretic security, 214
General attacks, 117, 121, 122, 129, 179
GHZ basis, 189
GHZ state, 189–191
Good
  states, 134
  words, 133–135
Gram Schmidt process, 47
Greatest common divisor function, 78
Group key protocols, 189
Grover's algorithm, 66
Guess function, 133, 135
Guessing probability, 56

H
Hadamard
  basis, 4, 8, 25, 85, 144, 192, 198, 206, 251
  operator, 32, 33, 39
  plus state, 23
  transform, 164
Hermitian operator, 20, 47
High-dimensional (HD)
  BB84, 169
  quantum states, 169
  semi-quantum key distribution, 175
High-dimensional quantum key distribution (HD-QKD), 169
Hilbert space, 21, 22, 53, 169, 183, 184, 222
Historical cryptography, 61

I
Ideal state, 109, 118, 119, 136, 149, 150, 192
Identity operator, 20, 28, 42, 142, 171, 185
Inconclusive round, 90
Individual attacks, 116, 117
Infinite decoy state protocol, 226
Information extraction, 23
Inner norm preservation, 34
Inner product, 16, 17, 26, 126
  function, 16
  space, 16, 18
Intercept/Resend attacks, 116, 117

J
Joint
  entropy, 45
  measurement, 40

K
Key distribution, 2
  problem, 2, 7, 66, 250, 254
  protocol, 2
Key round, 93
Key space, 62, 64
Key-generation algorithm, 213
Key-rate, 120, 128
Kronecker
  delta function, 18, 34, 203
  product, 36–38

L
Linear
  independence, 34
  operators, 19, 37
Link level entanglement, 238
Loss coefficient, 228

M
Matrix
  functions, 21
  representation, 19, 32
Max-entropy, 60
Measurement
  device outputting, 23
  operators, 28
  quantum state, 23
  qubit, 3

Measuring, 23
Mediated SQKD protocol, 152, 161
Message
authentication, 1
space, 64
Message authentication code (MAC), 79–81, 84
protocols, 80
tag, 79
Minimal assumption, 243, 244
Mismatched
bases, 94
measurements, 87, 94, 156–158, 160
rounds, 160
Mixed states, 46, 54, 58, 110, 111, 145, 150, 247
Modern cryptography, 61, 66, 244

N
Nash Equilibrium (NE), 214, 216
Natural noise, 85, 87, 112, 113, 214
Negligible function, 68, 244
No-cloning theorem, 5, 8, 42, 85, 91, 221, 222
Noise, 8, 85, 112
Noise tolerance, 102, 115, 129, 156, 162, 169, 215, 217, 253
Noiseless coding theorem, 45
Noisy storage, 208
assumption, 209
model, 208–210
Normal
form game, 214, 215
operator, 21, 52
Normalized
states, 17
vector, 18

O
Oblivious transfer (OT), 10, 208, 247
Observable disturbance level, 94
One time pad (OTP), 1, 2, 65, 66
One-way function (OWF), 1, 9, 70, 243, 244
One-way state generator (OWSG), 244
Operator action, 32, 33
Opposite bases, 88, 90
Optical switches, 235
Optimal Asymmetric Encryption Padding (OAEP), 79
Optimal asymptotic performance, 226
Optimization problem, 88, 121, 128
Oracle access, 70, 71, 73, 79, 249, 250
Orthogonal vectors, 17
Orthonormal
bases, 18, 34
states, 17
Outer product representation, 19

P
Partial
measurements, 40, 42
source independence model, 197
trace, 50, 52
trace operation, 51
Pauli operators, 33, 34
Perfect secrecy, 64, 80
Phase, 233
Photon number splitting (PNS) attack, 91, 92, 222, 224
Physical assumptions, 61, 208, 211
Ping Pong protocol, 141, 154
Plain text, 62
Polarization, 233
Polarized beamsplitter (PBS), 234
Polynomial time quantum algorithm, 212
Position space, 183, 185
Positive
operator, 21
semi-definite operators, 21, 56
Positive Operator-Valued Measure (POVM), 29, 30
Post-quantum cryptographic systems, 7
Practical device considerations, 232
Prepare-and-measure protocols, 94
Privacy amplification, 83, 86, 87, 101, 120, 162, 187
protocol, 87, 143, 153
Private key
cryptographic systems, 1
encryption, 1, 64
setting, 62, 76
Private randomness expansion protocols, 200

Probabilistic polynomial time (PPT)
algorithm, 68–70
Probability amplitudes, 22–24
Projective basis measurements, 29
Proof by reduction, 61, 74
Proof of deletion, 205–207
Pseudo Random Function (PRF), 70, 71, 73, 250
Pseudo random function-like state (PRFS), 212
Pseudo Random Generator (PRG), 66, 69
seed, 69
stretch property, 70
Pseudo Random Quantum State (PRS), 211
generators, 243, 245
Public authenticated channel, 135
Public key
cryptographic primitive, 254
cryptography, 76
distribution problem, 76
encryption, 9, 76
encryption protocol, 253
encryption schemes, 244
Pure state, 46, 50, 51, 54
Purification, 60

Q

Quantum
algorithm, 212, 213
ancilla, 116
asymptotic equipartition property, 120
bits, 3, 22
channel, 83
communication, 3
communication stage, 87, 88
computation, 2, 13
computers, 7
cryptographic protocols, 52
cryptographic systems, 5
cryptography, 2, 3
entanglement, 6, 201
entropy, 53
error correction, 254
information theory, 2, 43
max entropy, 131
mechanics laws, 7
memory, 86
noisy, 10
min entropy, 46, 56
money scheme, 213
networks, 235
nodes, 235
One Time Digital Signature scheme, 246
operation, 42
pseudo-random states, 211
public key, 9
distribution, 254
encryption, 249
encryption protocol, 250
random walks, 183
cryptographic applications, 182
repeaters, 235, 238, 239
state, 5, 21, 22, 255
state sharing, 214
system, 46
teleportation, 7, 114
teleportation protocol, 114
walk based key distribution, 186
Quantum conference key agreement (QCKA) protocols, 189
Quantum key distribution (QKD), 3, 7, 61, 66, 83, 221
classical processes mandatory, 97
goal, 83
implementation, 227
protocol, 56, 57, 83, 87, 91, 97, 109, 223, 232
ϵ-secure, 119
asymptotic scenario, 120
basic, 88
stages, 87
variants, 93
security, 46
security proofs, 55
systems, 59, 141, 232
Quantum polynomial time (QPT)
adversary, 244
algorithm, 246, 247
Quantum random number generation (QRNG), 188, 195
Quantum-secure one way functions (qOWF), 212

Qubit, 3, 22
encoding, 233
measuring, 3, 23
multi-qubit state space, 37
Qudit, 169, 170, 177

R
Rabin Oblivious Transfer (ROT) protocol, 210
Random
function, 72, 73
number generation, 10, 195
Randomized ciphertext, 71, 78
Raw key, 87
Real state, 109, 118, 136, 138
Receiver, 208, 210
Relative Hamming weight, 89
Replay attacks, 80, 81
Resource requirements, 83
Reverse
channel, 179
reconciliation, 146, 148, 156
Round Robin protocol, 169, 173
RSA
algorithm, 77
assumption, 77
crypto system, 77
encryption, 61, 78
key generation, 78

S
Sampling strategy, 133, 134, 148, 149
SARG04 protocol, 91
Scalar multiplication, 14, 15
Scheme correctness, 249
Secret key, 2, 7, 62
length bounding, 118
Secret sharing schemes, 214
Security, 63
assumptions, 243
definition, 61, 254
finite key proof, 148
goal, 109
model, 83
parameter, 68
proof, 1
alternative techniques, 131
Seed, 69
Semi-quantum
cryptography, 141, 152
protocols, 162
Semi-quantum key distribution (SQKD) protocols, 153, 159, 175
Sender, 208
Separable states, 39
Shannon
conditional entropy, 121
entropy, 43, 44, 54
axiomatic approach, 44
noiseless coding theorem, 45
security definition, 64
Shift
cipher, 62
operator, 184
Shor's algorithm, 2
Simplified trusted node (STN), 237, 238
Single-qubit source, 83
Smooth min entropy, 57, 119, 131
Source
device independence, 197
independent model, 196
Spanning set, 15
Standard
basis, 16, 17
computational basis, 35, 187
State
collapse, 3, 23, 24
global, 41
discrimination, 30
encoding, 88
evolution, 32
space, 21, 22
vector, 21
Strategy space, 214, 215
Superconducting nanowire single-photon detector (SNSPD), 232
Superposition, 3, 22, 23
SWAP test, 253

Symmetric key
  primitive, 81
  setting, 76
  systems, 77
System gain, 225

T
Target function, 133, 135
Tensor product, 35, 37
  combining systems, 35
  distributive property, 36
Tensor state space, 40
Test round, 93
Three-state BB84 protocol, 93
Time bin, 233, 234
  encoding, 163, 173, 174, 234
Trace distance, 52
Transpose, 15
Triangle inequality, 53
Tripartite
  state, 41
  systems, 59
Trusted device, 196, 197
Trusted Node (TN), 235–237
Twin field QKD (TF-QKD) protocols, 163
Twin field (TF) protocols, 141, 163
Two-universal hash function, 100, 102
Two-way QKD protocols, 141

U
Unit trace, 47, 48
Unitary change in basis, 34
Unitary operator, 20, 32, 34

V
Vector, 14, 15
  space, 15
    dimension, 16
  subspace, 15
Verification
  algorithm, 79
  stage, 256
von Neumann entropy, 46, 53–55, 58
  conditional, 55
  continuity, 176

W
Walk basis, 187
Walker, 184
Weak attenuated laser, 221, 232

X
XOR operation, 6, 65

9780443215322